The Making of a Japanese Periphery, 1750–1920

Kären Wigen

UNIVERSITY OF CALIFORNIA PRESS

Berkeley / Los Angeles / London

University of California Press
Berkeley and Los Angeles, California

University of California Press, Ltd.
London, England

© 1995 by
The Regents of the University of California

Library of Congress Cataloging-in-Publication Data

Wigen, Kären, 1958–
    The making of a Japanese periphery, 1750–1920 / Kären Wigen.
        p.   cm.
    Includes bibliographical references and index.
    ISBN 0-520-08420-9
    1. Ina Valley (Japan)—History.   I. Title.
DS894.I53W56   1994
952'.1—dc20   93-36270
                                                                     CIP

Printed in the United States of America

9 8 7 6 5 4 3 2 1

The Making of a Japanese Periphery, 1750–1920

Twentieth-Century Japan: The Emergence of a World Power

Irwin Scheiner, Editor

For Thomas Smith

Contents

Maps, Figures, and Tables

FIGURES

TABLES

Preface and Acknowledgments

Japanese development poses one of the more insistent puzzles of modern history: how an isolated and decentralized state, far from the European heartland, managed to metamorphose in a few short decades into a formidable global power. If the often-heated debate over that phenomenon shows no sign of abating soon, it is partly because there is no broad consensus on how to frame its central question. Some have sought to explain Japanese ascendancy in narrowly economic terms, pointing to premodern commercial development, early capital accumulation, and subsequent investment and trade patterns as the critical levers facilitating the nation's rise. Others argue persuasively for a more institutional explanation, stressing the role of a vigorous state apparatus in mobilizing and directing resources during the formative decades of the late 1800s and beyond. Still others insist that the critical issue is one of social justice, asking not simply how Japan has succeeded, but what the costs of that success have been, and who has paid them.

All of these lines of inquiry have informed the present study, which confronts the origins of Japanese industrial and imperial might in the period from 1750 to 1920. Rather than privileging one or another agenda, I have attempted here to interweave economic, political, and social concerns in the context of a regional study. Behind this avowedly eclectic strategy is the conviction that an integrative vision is needed if we are to put together a meaningful picture of Japanese development. A central principle of that integration, as I envision it, is a geographical perspective.

Little in the theoretical framework deployed here will be new to geographers. While the particular synthesis is my own, I have foraged widely in the literature of the discipline in search of intellectual tools. More novel is the application of environmental, spatial, and regional analysis to the political economic history of Japan. For the freedom to work in this largely uncharted territory, I thank my graduate advisers at Berkeley: the geographers Michael Watts and Allan Pred and the historian Thomas Smith. Each provided crucial insights, along with sharp criticisms, while this study germinated as a doctoral thesis in geography.

Numerous scholars shaped the environment within which this project was conceived. The faculty and graduate students who came together in the Berkeley geography department during the 1980s created a lively and challenging intellectual community. I am especially grateful to James Parsons, whose love of local complexity proved infectious, and to Mary McDonald and Karl Zimmerer, whose probing questions continue to provoke reflection. G. William Skinner's generosity in including me in the Nōbi Regional Project workshops created an indispensable institutional context for research, and sharpened my thinking about regions.

Field and archival work in Japan from 1987 to 1988 was made possible by a Fulbright-Hays grant from the Department of Education. Additional support was provided by the International Center at Keiō University, and by Hayami Akira, Takahashi Junichirō, Itō Yoshiei, Takagi Isao, and others on the staff of Keiō's Faculty of Economics. Furushima Toshio and a group of his former students offered much-appreciated intellectual guidance during a return trip in 1990, funded by the Northeast Asia Council of the Association for Asian Studies. Nakajima Masamichi in particular gave unstintingly of his time and expertise during that visit.

Extended stays in Iida on both occasions were highly rewarding, and my debts there are many. Kitabayashi Noboru shared the resources of his office at Shinyōsha and made the initial introductions that opened doors in the community. The patient and knowledgeable staffs of the Iida City Library and the Shimoina Kyōikukai made their excellent collections accessible and answered repeated questions. Officials at the Iida City Office arranged interviews with local scholars; among the many historians who took the trouble to share their knowledge of the area, I would especially like to thank Kinoshita Motsumi, Kusakabe Shinichi, Makino Mitsuya, Ōsawa Kazuo, Sakurai Ban, and Tanaka Masataka. Finally, home stays with the Gotō, Kitahara, Kubota, Miyazawa, and Nagai families were the highlight of the year. I cannot adequately describe, much less discharge, the debts I owe these wonderful hosts. Particular thanks to the good-natured women in those families for putting up with the impositions of a long-term foreign guest.

In transforming the ill-formed mass of field notes with which I returned from Japan into a coherent manuscript, I have leaned heavily on the financial and intellectual resources of the American scholarly community. Successive rounds of writing and mapping were supported by grants from the University of California at Berkeley, the Social Science Research Council, the Duke University Research Council, and Duke's Asian/Pacific Studies Institute. To the extent that the book has moved beyond the dissertation, I am indebted to careful readings by John Agnew, Gary Allinson, Gail Lee Bernstein, Andrew Gordon, Saitō Osamu, Irwin Scheiner, and Anne Walthall, all of whom suggested substantive improvements to the text. Any oversights or errors that remain are my own responsibility. Thanks also to Martin Collcutt, Albert Craig, Arif Dirlik, Micaela Janan, Vasant Kaiwar, Rebecca Karl, and Susan Thorne, as well as participants in the Geographic Perspectives in History seminar, whose pointed questions have helped me to rethink parts of the argument. Sheryl Freeman and Patricia Neumann share credit for the cartography. And heartfelt thanks to Sheila Levine, Betsey Scheiner, Peter Dreyer, and the rest of the talented crew at the University of California Press for shepherding the manuscript through the lengthy publication process.

Finally, I am happy to acknowledge the people who have kept life joyful while this project has evolved. The Duke community has been a wonderful place to be during the last phases of writing, and the extended Wigen and Lewis clans have shared in this project from the beginning, in ways too various to mention. Without them, none of this would have been possible. To these friends and family, and especially to Martin, my fondest appreciation.

Abbreviations

NCS *Nagano-ken chōson shi, nanshin hen.* Edited by Nagano-ken. Nagano: Nagano-ken, 1936.

NSTI *Nagano-ken Shimoina-gun tōkei ippan, dai jūshichikai.* Edited by Shimoina-gun. Iida: Shimoina-gun, 1923.

CHAPTER ONE

Introduction

The years between 1750 and 1920 witnessed a profound transformation of Japan's political and economic terrain. At the beginning of this period, the Japanese archipelago was a patchwork of protectionist baronies, militarily at peace but economically at war. United under the suzerainty of the Tokugawa shogun, and subject to the latter's strict controls on foreign trade, the islands nonetheless remained divided into more than 250 separate domains, which were stubbornly mercantilist in their dealings with one another and only fitfully drawn together into a national community.[1] With the arrival of European and American traders in Japanese waters in the mid nineteenth century, however, domainal rivalries were quickly superseded by national imperatives. The new regime that seized power in the Meiji Restoration of 1868 set about converting an alliance of competing fiefdoms into a unified industrial and imperialist state, capable of holding its own in an intensely competitive global economy. The extent to which they succeeded remains remarkable. By 1920, a decade into the reign of the Taishō emperor, Japan was well on its way to becoming a major industrial and imperial power.

The present study tracks the course of this transformation through the changing landscapes of the southern Japanese Alps. Its focus is the lower Ina valley, or Shimoina, a landlocked inland basin in southern Shinano province (map 1). Its intent is to show how changes within this

1. The intensity of domain nationalism during the Tokugawa period is explored in the context of Tosa *han* by Luke Roberts (1991), who identifies "national prosperity thought" as an ideology of the merchant class as well as of domain officials.

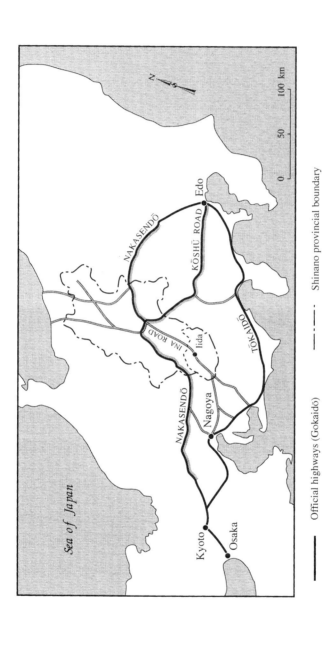

Map 1. Principal Transport Routes in Central Japan, Mid to Late Tokugawa.

——— Official highways (Gokaidō)

⫶⫶⫶ Major private packhorse routes

▊▊▊ Stretches of the Gokaidō
used by private packhorse teams

—·—·— Shinano provincial boundary

········· Shimoina region boundary

valley and its mountainous hinterland were both artifact and agent of Japan's nineteenth-century political and economic revolutions. In particular, I argue that the economy of this region, initially shaped in the crucible of domainal protectionism, was taken apart in the later 1800s and put back together in novel ways for the benefit of the Japanese state as a whole. That process entailed a fundamental rewiring of the circuitry of production and exchange, inverting not only the valley's internal organization but its position in a larger sociospatial order as well. In essential ways, Japan's modern transformation required that Shimoina—and numerous regions like it—be remade as peripheries of a Tokyo-centered national economy.

Yet the making of a periphery in the southern Japanese Alps cannot be considered merely within a national context, for it was also part of a larger and opposite metamorphosis: the remaking of Japan as a new and privileged core in East Asia. In tangible ways, the farmers and silk reelers of central Honshu both gained from and contributed to their nation's rise within an emerging North Pacific economy; indeed, it is precisely in this intersection between the local and the global that their story takes on a broader significance. It is one of my central methodological contentions that to capture this multilevel regional construction requires pursuing an explicitly geographical approach to history. What I have attempted here is to begin mapping out Japan's political and capitalist transformations as a reconstitution of places, and of the relations between places, on a range of scales—castle town and countryside, coast and interior, metropole and provinces, the state and the world. Perhaps the most ambitious project of the book is to reconceive Japan's modern metamorphosis in spatial terms.

Pursuing that project positions this study at a double juncture. Methodologically, it occupies an interface between history and geography; temporally, it spans a divide between the early modern and the modern.[2] Following a brief introduction to the region and an outline of its development, the remainder of this introductory chapter is devoted to exploring these dual axes, which together establish the intellectual coordinates of the study.

2. I use the generic term *early modern* advisedly in this context. Historians often refer to the years from 1600 to 1868 by period labels unique to Japan, designating the seventeenth through nineteenth centuries as the Tokugawa era (after the shogunal dynasty) or the Edo period (referring to the site of government). While these labels are indispensable, I choose the implicitly comparative designation here to suggest what I see as strong commonalities between Tokugawa Japan and contemporary societies in Europe. This comparative perspective—articulated by prewar Marxists as well as by modernization

The Region

In cartographic representations of central Japan, the area known today as Shimoina does not readily spring to the eye. Physically, it constitutes a truncated watershed, straddling only the midsection of the Tenryū River. The heart of present-day Shimoina is the southern half of the Ina graben, or trench, an elongated intermontane basin in central Honshu (map 2). The basin reaches some seventy-two kilometers from north to south but measures only twelve kilometers across at its widest point. Its steplike series of terrace surfaces, which rise gradually toward the north, offer the largest concentration of suitable farmland in the county. While the bulk of Shimoina's 160-odd Tokugawa villages were concentrated in and around the graben, however, this physiographic "core" is in fact considerably off-center within the county, since the Ina basin is bisected by Shimoina's northern border. Culturally, the area is no more clearly defined. Rather than comprising its own distinctive sphere, or even sharing a recognizable identity with a larger culture region, Shimoina lies in a transitional gray area between the two major Japanese culture zones. Whether based on dialect or folk customs, the boundary between eastern and western Japan is consistently drawn near or through this district.[3]

Political history offers no clearer basis for marking off Shimoina as a region. During the Edo, or Tokugawa, period (1600–1868), Shinano province—the sprawling but defunct administrative unit of which Shimoina formed the southern tip—was a political shatter-zone of sorts, subdivided into numerous small fiefdoms. Until well after its reincarnation as Nagano Prefecture in the late 1800s, Shinano province did not mark off an integrated social or political space. Much of its land mass was occupied by the mountain ranges known collectively as the Japanese Alps; between their peaks, which range from 1,500 to over 3,000 meters, the population of Shinano has traditionally been scattered among the handful of intermontane basins shown in map 2. Partly as a result, both economic and political life in the province has historically revolved around smaller population centers within these intermontane basins,

---

theorists in the 1960s, and shared to varying degrees by many contemporary American scholars of Japanese political, economic, and social history—is crucial to the development of the present argument.

3. Amino 1986:123ff.; M. Miyamoto 1966:30; Nagano-ken 1988:47.

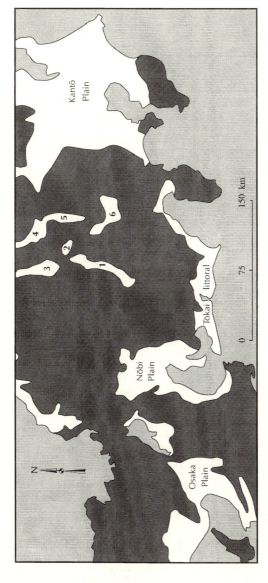

Map 2. Concentrations of Arable Land in Central Japan. Shows the six intermountain basins of Tōsan: (1) Ina, (2) Suwa, (3) Matsumoto, (4) Nagano, (5) Ueda, and (6) Kōfu. (Adapted from Trewartha 1965:23, 411.)

each of which was oriented away from the highland core and toward the nearest coastal lowland.[4] In fact, most of Shinano's constituent basins were in turn administratively subdivided, creating a pluralistic and highly complex political landscape.

The Ina valley itself was never united under a single ruler. On the contrary, the local political terrain was fragmented into a half-dozen small parcels. Largest of these was the domain of Iida, whose 20,000 *koku* in holdings—a paltry fiefdom by national standards—were administered from the castle town of the same name.[5] The remainder of the valley floor was divided among four minor vassals, one nonresident daimyo, and a score of temples and shrines, while the surrounding mountains were for the most part administered by the shogun's government, or Bakufu. Reflecting this history of parcelization, the area's official toponym is relatively recent; Shimoina as a political entity did not come into being until the late 1800s. Throughout the Tokugawa period, the 1,900 square kilometers enclosed within its present boundaries constituted merely the southern portion of Ina district (*gun*), the southernmost and largest of the ten districts constituting Shinano province. Not until 1879 was the area administratively divided in two, yielding an Upper (Kami-) and Lower (Shimo-) Ina.

For a brief but crucial period thereafter, from the first county elections in 1890 until the county system was abolished in 1921, Shimoina was a partially self-governing political entity. An elected assembly with taxation and oversight privileges directed local affairs from offices on the grounds of the former Iida castle. Census statistics were collected and published at the county level as early as 1883, investing Shimoina with considerable importance as a geographical unit for subsequent historiography. At the same time, the region's past also began to be investigated within the new governmental framework. Local historical research was conducted primarily by a county educational association (Shimoina Kyōikukai), a group of school teachers who came together in 1887. Through this group's efforts, which produced works on the geology, biology, prehistory, and folklore of the area as well as its historical devel-

4. Tsukamoto 1979.

5. A *koku* is a measure of volume, equaling approximately 45 U.S. gallons (5.1 bushels) or 180 liters, and translating into roughly 94 kilograms of rice. Conventional wisdom held that one *koku* of grain was enough to feed one person for a year. Although smaller fiefs were allotted to lesser retainers, a daimyo by definition had to hold title to lands with a putative yield of at least 10,000 *koku*; the largest of the 260-odd Edo-period domains ranged upward of fifty times that size. In this context, the 20,000-*koku* Iida domain can be seen for the minor fief that it was.

opment, the county emerged as the preeminent frame of reference for local scholarship.[6]

This institutional legacy lends Shimoina undeniable attractions as a unit for historical analysis. But the geographical determinations for an inquiry into regional economic transformation must be subjected to higher standards than historiographic convenience. The first challenge for such a study is to discern a region with a degree of economic integrity, manifest in a coherent territorial complex of production and exchange. In fact, I argue that, although Shimoina lacked a clear physical, cultural, or even (until recently) political identity, this small upland area had by the middle of the Tokugawa era attained an unambiguous economic identity. Precisely because of its interstitial location, the Ina valley was able to become a channel between the northeast and the southwest, the mountains and the sea, with Iida developing as an entrepôt of sorts in an overland trade network. The resulting ties with nearby population centers in turn facilitated access to the three largest metropolises of Tokugawa Japan: to the west, the old imperial capital of Kyoto and its nearby port of Osaka (financial and trade center of the country), and to the east, Edo, seat of the Tokugawa shogunate and a second home for all of its major retainers.[7] The marked concentrations of wealth in these cities created a market for specialty items from throughout Japan, including mountainous Shinano. By 1750, the area around Iida was beginning to be knit together through the production of luxury goods for exchange within national commercial circuits. It was thus in a dual capacity, as both trade corridor and craft producer, that Shimoina was able to develop a regional economic identity in the Tokugawa period.

Regional Formation and Transformation

At the level of local narrative, the history of southern Shinano from Tokugawa to Taishō might be read as the rise and fall of this incipient small-scale region. Its economy in 1750 was primarily agricultural, with a pronounced bias toward upland crops (the ratio of dry fields

6. Shimoina Kyōikukai 1987.

7. All retainers who held title to lands with a putative yield of 10,000 *koku* or more were obliged to spend every other year in Edo, and to leave their wives and families there as permanent hostages. For more information on this so-called alternate residence (*sankin kōtai*) system, see Tsukahira 1966. On relations among the three dominant cities, see Rozman 1973 and Moriya 1990.

to paddy being estimated at three to one for the Ina valley as a whole). Yet Shimoina also supported a significant industrial enclave, centering on the castle town of Iida, where local merchants oversaw the production of paper, textiles, and lacquerware for export to national markets. The foundations of that commercial enclave lay in the valley's prior development as a cargo throughway. In a sense, the beginnings of the early modern economic region identified here date to the seventeenth century, when pack trains from Shimoina first began to capture the lucrative through-trade between the Pacific coast and interior Shinano.

Shimoina's success in the packhorse trade was both cause and consequence of intense interregional struggle. The neighboring Kiso valley had been designated as the route of the Nakasendō, the government's official turnpike through the mountains (see map 1), and like their counterparts elsewhere, the Nakasendō post stations counted on private cargo to subsidize the semiannual treks of the daimyo and their retinues to and from Edo. But the light hand of the central authorities along the parallel Ina road paradoxically boosted that route's attractions for commercial shippers. Able to escape the burden of subsidizing official travelers, peasant packhorse drivers in Shimoina managed to compete successfully with the post stations along the Nakasendō, creating an alternative shipping service that earned strong support from area merchants.

Trade in turn created important opportunities for commercial development. Although bulk goods could not easily be shipped in and out of a landlocked basin (mandating continued self-sufficiency in food and fuel), local overlords were nonetheless able to take advantage of the valley's regular connections with the major cities of central Japan to establish a number of craft industries, geared to the production of lightweight luxury goods. By 1750, most residents of the area's chief castle town and nearby communities spent at least part of their time producing silk or cotton textiles, lacquered bowls, or paper wares for the national market, or processing agricultural specialty goods such as tobacco and dried persimmons. Over succeeding decades, by gradually extending the resource-procurement and putting-out networks that supported these crafts, Shimoina's merchants, transporters, artisans, and peasant laborers together shaped Iida and its expanding hinterland into an integrated region.

The resulting economic configuration may usefully be conceptualized as protoindustrial, if by protoindustry is meant the widespread development of rural outwork to produce commodities for distant markets. To be sure, this definition is somewhat looser than that common in the European literature, where one prominent work on the subject stipulates that "a large part of the population" must have lived "entirely or to a considerable extent from industrial mass production for inter-regional and international markets."[8] By this standard, Shimoina might better be labeled "subprotoindustrial": that is, characterized by exten-sive, decentralized commodity production, but having "a balanced econ-omy, agriculturally dominated and agriculturally self-supporting . . . , with no built-in tendency to specialization in textiles or other manufac-tures for distant markets and no runaway land sub-division, population explosion or landless rural proletariat."[9] Yet, as I shall argue, settlements within Shimoina's commercial enclave experienced social and demo-graphic developments that in important ways echoed the patterns found in the larger and more specialized early-industrializing regions of Eu-rope. Although Saitō Osamu is right to caution against applying the protoindustrial thesis in a simplistic way to the Japanese past, holding to a European-derived standard would seem equally slavish. Despite the re-stricted scope of Tokugawa trade, the agrarianist ideology of the ruling class, and the "by-employment" character of the country's rural indus-tries, the Tokugawa economy may fruitfully be explored under the pro-toindustrial rubric.[10]

The landscape of protoindustrialization was marked by both contin-uing dynamism and periodic conflict. The commercial economy of the Ina valley evolved steadily through the eighteenth century and the first half of the nineteenth, quickening the region's integration in national commerce and deepening its internal social and spatial differentiation. In the process, the area's economic geography emerged as both source and subject of repeated social conflict. On the one hand, class relations in the commercial core became increasingly antagonistic, aggravated in part by local merchants' price-fixing attempts, but reflecting as well the rise of effective artisanal organizations. At the same time, a second axis of con-flict pitted the domain's licensed merchants against rural competitors. For as the protoindustrial economy expanded, production and exchange circuits increasingly transgressed political boundaries, provoking re-peated attempts to capture the more lucrative processes. Castle-town merchants pressed monopoly claims against their rural counterparts, and domains likewise sought to tax intermediate processing that occurred be-

8. Kriedte, Medick, and Schlumbohm 1981:24.

9. Walton 1989:50; from a description of west Lancashire in the late eighteenth century.

10. Studies undertaken in a similar spirit include D. Howell 1989, 1992; Pratt 1991a, 1991b; Saitō 1983, 1985; Toby 1991; Yasuba and Saitō 1983.

yond their boundaries. Yet such attempts ultimately proved unsuccessful. While control over Shimoina's resources was retained within the general core area, capital increasingly gravitated toward a rural landlord class with close ties to the production process. At the same time, taxes levied on extramural economic activity were ruled illegal, and protoindustrial networks proved surprisingly mobile. In the long run, even Iida, largest of the half-dozen minor territories in the valley, proved unable to keep control of the crafts that owed their beginnings to domainal patronage.

Social tensions in Shimoina escalated through the crisis years of the mid nineteenth century, but the protoindustrial region held together until the opening of Japan's ports. After 1859, however, powerful market forces emanating from Yokohama sent the regional economy into convulsions. In the chaotic years after American gunboats irrupted into the carefully bounded world of Tokugawa trade, opportunistic brokers scrambled through the countryside to gather up silkworm eggs and floss for shipment to the treaty ports. The resulting inflation, aggravated by an influx of debased currencies, played havoc with local commerce, touching off protests throughout Shinano and the nation.

The immediate crisis was resolved in the early 1870s, when the newly established Meiji regime intervened to restore order to both the political and commercial landscapes. But long-term changes had been set in motion that eventually undermined three pillars of the area's traditional political economy. One of these was sovereignty. As power coalesced in the new national capital of Tokyo after 1868, local control over development decisions was sharply curtailed. A second support of the protoindustrial economy, the valley's role in interregional trade, was dealt a decisive blow when superior transport technology was allocated to competing routes. Finally, the increasing mobility of people, money, and things began to loosen the linkages between regional development and regional resources. Strong demand for two local products, timber and silk, began to draw outside capital as well as immigrant labor into the local commercial circuitry.

As this trio of changes worked their way through the local landscape, the Shimoina region was radically redefined in both economic and social terms. On the one hand, as the networks of production and exchange that had given the area's economy its coherence during the Tokugawa period were gradually turned to new purposes, the nucleated economic region centered on Iida split apart and metamorphosed into something distinctly different. By the early twentieth century, the Ina valley and its hinterland had been bifurcated into largely disjunct commodity zones.

Female labor drove a greatly aggrandized sericultural complex in the core, while male lumberers and charcoalers supported an extractive forest industry in the hinterland. In contrast to the craft industries of the earlier era, both silk and timber were fully capitalist enterprises, oriented to outside markets and financed in part by extralocal capital. As a result, an increasing share of the profits accruing from local production—profits that had previously been kept largely within the valley—ebbed out of the region. In the process, Shimoina's identity as an integral, semiautarkic productive unit was ruptured.

Meanwhile, a different but complementary set of forces took hold in the political and social spheres. Paradoxically, even as Shimoina's economic integrity was undermined, the region's social unity was heightened. The political framework for regional identity was supplied when counties were designated as meaningful units in the new administrative hierarchy. For the first time, Shimoina was invested with a significant administrative role; in time, it was also granted electoral representation in Nagano Prefecture, the larger geopolitical unit of which it was part. But this only set the stage for a burst of regionalist rhetoric from below. As early as the 1880s, boosters and elected leaders tried to rally the county's citizens together to secure development funds from prefectural and national coffers. When those efforts failed, local leaders tried to reconfigure the political map, arguing that Nagano Prefecture should either move its capital or divide in two. The invocation of Shimoina affiliation that accompanied these efforts clearly promoted conflict with neighboring territories, a potential threat to the fledgling state. Yet repeated calls for regional unity also served to mask—and in some measure to quell—the potentially more subversive threat of class antagonisms. In the long run, regional rivalries served not to undermine the Meiji social order but to reinforce it.

If territorial competition within the new state remained manageable, one crucial reason was the effective centralization of control over development funds. Local boosters, maneuvering for projects they deemed crucial to the region's interests, were constrained to operate in a sharply hierarchical political universe. To be sure, centralization operated on many levels, some of which augmented local powers. The administrative reach of Shimoina's largest town, for instance, expanded greatly in geographical terms. Home to the new county offices (*gun'yakusho*), Iida now presided over an area several times the size of the domain formerly ruled by the region's largest Tokugawa daimyo. Yet while their territorial reach had grown, local officials' powers were in other ways curtailed.

The Meiji government had effectively converted a once-proud castle town into a low-level arm of the central administration, charged primarily with enforcing the aims of the state. By 1920, not only had the Ina valley been reduced from a corridor to a cul-de-sac, and from a diverse protoindustrial center to a precarious industrial supply zone, its political hub had at the same time been demoted from domainal capital to county seat. Where in 1750 a relatively autonomous economy and polity had existed, a periphery was now in place.

Subordination to a rising industrial state was not without its benefits. On the contrary, as long as silk demand in the United States remained high, Shimoina's new relations with both the national government and the global market brought unprecedented prosperity to the region. For local observers in the 1910s, the Ina valley's incorporation into central Japan's "silk kingdom" was nothing if not a cause for celebration. Sericultural profits boosted farm incomes and government revenues across the county, while new jobs in the forests and filatures enticed thousands to migrate into the county. But the sericultural boom eventually extracted a high price. With more than half of all local resources committed to silk production by the 1920s, the subsequent collapse of American demand for such luxuries proved devastating. During the depression decade, desperate county residents tried to return to subsistence cropping and charcoaling; some even attempted to revive the old crafts. But the newly swollen population of the region simply could not subsist on the local resource base, and many eventually had to leave. Shimoina's mountain villages in particular experienced an acute population decline, with one local settlement giving up more than 1,100 residents to Manchuria during this period.[11] Seven decades later, the population of the county as a whole had barely recovered the level attained in 1920.

Such are the outlines of a narrative in which the history of southern Shinano from Tokugawa to Taishō may be read as the rise and fall of a small-scale politico-economic region. Yet grasping the full significance of this local transformation requires broadening the scope of inquiry beyond the valley's boundaries. Shimoina's modern metamorphosis cannot be understood apart from the national and international context in which the region was articulated.

At the most mundane level, one must reckon with the revolution in transport. The sudden penetration of the Japanese countryside by ve-

11. Kobayashi Kōji 1977.

hicular traffic in the 1870s and 1880s, and the rapid advance of a national rail net in succeeding decades, radically increased the scope of Shimoina's exchange with the rest of Japan, as well as with the world beyond. At the same time, the pattern of exchange that resulted was shaped as much by Japan's aggressive relations with its Asian neighbors as by its newfound trade with North America. By the turn of the twentieth century, the growing population in the Ina valley depended heavily on cheap fertilizer, and later grain, from the Asian mainland: imports that could be extracted at low prices only because of a growing Japanese military presence in Korea and Manchuria. These inputs, effectively an imperial subsidy, were critical in allowing the Ina valley to convert half of its cropland to silk-mulberry orchards without suffering high food prices, which might otherwise have driven its industrial wages out of competitive range. Nor was Shimoina's link with the emerging empire merely passive. Through the foreign exchange earned by its cocoons and floss, Shinano in turn helped finance Japanese military expansion, making the economic realignment of the Japanese Alps an agent as well as a beneficiary of the shifting locus of global power. Only in this dual perspective can Shimoina's transformation be seen for what it was. The southern Japanese Alps were indeed peripheralized: incorporated into larger political and economic circuits in ways that undermined both local autonomy and local accumulation. Yet the process of local peripheralization was deliberately managed to allow Japan as a whole to emerge as a new and privileged core within Asia. It is within this contradictory movement that the meaning of Shimoina's restructuring must be located.

The Conceptual Framework

Analyzing these complicated interactions, both within and across Shimoina's boundaries, requires attending to the spatial dynamics of economic change. From a disciplinary perspective, this entails crossing an artificial but long-standing boundary between history and geography. Precedents for interdisciplinary scholarship in this area are certainly not lacking; historical geographers have constituted a distinct subfield within geography for decades.[12] Yet the geographical impress on social history has not been as profound as it might have been. Apart from

12. Those interested in sampling contemporary research in the field of historical geography are referred to the anthologies edited by Baker (1992), Baker and Gregory

those participating in this specialized subfield, contacts between professional historians and geographers have until recently been minimal. During most of this century, practitioners of these sister disciplines have pursued their respective concerns in virtual seclusion from one another.

Fortunately, momentum appears to be building in both professions for closing the gap. Geographers are increasingly asserting the centrality of historical process. Writing at the end of the 1980s, Paul Knox and John Agnew could categorically state that "the overall trend in academic economic geography has been away from a cross-sectional or time-slice modelling of sectoral (agricultural, industrial, etc.) and regional economic landscapes towards an historical geography of economic development."[13] And even as geographers wrestle with time, historians are evincing a new interest in space.[14] The result is an increasingly voluble dialogue across the disciplinary divide.

The emergence of such a dialogue should provoke less surprise than its long absence. Both history and geography have traditionally been synthesizing disciplines, sharing a concern for social development in the broadest terms. In the words of Donald Meinig, what differentiates the two fields is simply "the proportionate emphasis each gives" to "such common terms as space and time, places and events—pairs that are fundamentally inseparable."[15] Nonetheless, the legacy of a long estrangement is a continuing perplexity in some quarters over just what a geographic perspective in history might entail. Since such a perspective has been conspicuously absent in Anglo-American studies of Japan, a word about the fundamentals of the conceptual lexicon employed here is in order.

In analyzing the regional metamorphosis of Shimoina, this study integrates three contrasting, but complementary, modes of geographical thought. These might be conceived of as a collection of lenses serving to focus attention on distinct, but interrelated, geographical phenom-

(1984), Langton and Morris (1986), Mitchell and Groves (1987), and Genovese and Hochberg (1989). The *Journal of Historical Geography* is also a useful point of departure.

13. Knox and Agnew 1989:63. The literature to which the authors refer is reviewed in N. Smith 1989 and Corbridge 1989.

14. Among recent signs of a rapprochement, one might note the 1989 anthology *Geographic Perspectives in History*, edited by Eugene Genovese and Leonard Hochberg; the December 1990 issue of the *American Historical Review*, featuring three articles said to exemplify "a new interest in combining historical and geographical analysis"; and the creation in 1991 of a Historical Geography Network within the Social Science History Association. For further discussion of this trend, and of a similar "spatial turn" throughout the social sciences, see Wigen 1990:3–6.

15. Meinig 1978:1186.

ena. The first such optic, and the most familiar to historians, is environmental analysis, or the study of human-land interactions over time. On the one hand, this means asking a series of questions about the ways in which particular enterprises—whether in the realm of agriculture or of transport, industry, or politics—might have been encouraged or constrained by the area's resources and physical setting. On the other hand, it also means thinking about the ecological consequences of those developments that did occur: the impact of power configurations and routine economic activities on the fields, forests, and rivers from which the people of the region drew their livelihood.[16]

The second geographical technique employed here might simply be termed spatial analysis: the eliciting and (where possible) mapping of spatial patterns in the organization of social life. Contrary to the widespread notion that geographic inquiry necessarily concerns the physical environment, the subject matter amenable to this kind of treatment is broad; in the present study, questions of spatial order are posed in numerous contexts (from production and trade to gender and class), and on a wide range of scales (from the international to the intraregional and even intravillage levels). In particular, I am interested in plotting the interaction between the state and economic enterprise, which overlap and interpenetrate in complex ways, yet appear to follow fundamentally different principles of spatial organization. Whereas political power tends to be exercised within clearly defined blocks of territory, economic networks are more fluid and "linear," uniting producers, merchants, and markets in complex circuits of production and exchange that might cross a number of political territories. The conflict between these two different sociospatial orders is an essential element of Shimoina's inversion narrative.[17]

16. In the early twentieth century, taking the environment into account in geographical writing typically meant assessing the impact of climate and terrain on social development—most often within the paradigm of environmental determinism (Huntington 1907, Semple 1911). Today, it more often entails assessing the impact of past human societies on their environments. The symposium proceedings published in 1956 as *Man's Role in Changing the Face of the Earth* (2 vols., edited by William L. Thomas, Jr.) were an early landmark in the ascendancy of the latter view. Prominent recent contributors to the quickly growing field of ecological history include William Cronon (1986, 1991) and Albert Crosby (1986); important anthologies include those edited by Richard Tucker and John Richards (1983) and Donald Worster (1988).

17. The notion that commercial organization is essentially linear, in contrast to the areal organization of politico-military administration, is eloquently articulated by Edward Fox (1971). The present formulation has also been influenced by a theoretical statement worked out by Michael Mann (1986: introduction; see also Chris Wickham's 1988 review) concerning the sociospatial organization of what he considers to be the four basic

Finally, the third and most complex geographic perspective deployed here is regional analysis. Empirical regions are the fulcrums where geography and history, environmental resources and competing power networks, and household reproduction and global change come together. Yet determining what constitutes an appropriate piece of terrain for analysis is never easy. As Doreen Massey insists, "'regions' are not necessarily pre-given to the study of intra-national spatial differentiation . . . [but] must be constituted as an effect of analysis."[18] In the present context, the essential task is to identify a functional (or nodal) region: a contiguous area united by a complementarity of economic and political resources, and integrated through the routine interactions among its constituent parts.

The designation of a given area as an economic region should not be taken to imply autarky. While internal economic ties are typically denser and more varied than those across its boundaries, such an area's economic life is rarely sealed off from that of surrounding terrain. The nodal region is rather defined on the basis of its structure, the essential elements being a node or central place, a circulation network connecting the node to a larger area, and a declining intensity of interaction away from the center and in interstitial areas off the main lines of circulation.[19] There is no single scale to which such regions conform; nodal regions can often be delineated on multiple scales in the same landscape, with smaller ones (centering on minor settlements) often "nesting" inside successively larger ones (ultimately organized around major metropolitan areas).[20] Finally, it is essential to distinguish this kind of region from its simpler counterpart, the homogeneous (or formal) region, defined

---

types of social power (economic, political, military, and ideological). Although the present study necessarily touches on ideological and military power, a systematic analysis of their geography is beyond my scope and purpose.

18. Massey 1978:110.

19. See, e.g., Whittlesey 1954. The classic model of how a nodal economic region comes into being derives from J. H. von Thünen (1966 [1826]), who posited that agricultural activities of varying commercial density (such as market gardening, dairying, staple cropping, and ranching) would form a series of rings around the urban core that constituted their primary market. Von Thünen's model has been roundly criticized by James Vance (1970:140ff.) and others for its blindness to historical process and political contingency, yet it has been seminal for generations of geographers, and continues to inform a wide range of work on regional economic development. For very different applications, cf. Skinner 1977, de Vries 1984, and Cronon 1991.

20. I would nonetheless concur with Carole Crumley and William Marquardt (1987:614) in cautioning "against viewing any society as having a single dominant or nested hierarchy, rather than cross-currents of interests and allegiances at a number of scales." For a more extended discussion, see Wigen 1992:15–16.

on the basis of one or more shared characteristics (such as interintelligible dialect, similar environment, shared ancestors, common religion, or the like).

The focus of the present study is a functional rather than a formal region. Shimoina was organized around a central castle town (Iida), and unified not by the sameness but by the complementarity of its several environments. Such functional complementarity is never neutral, however; it rather sustains, and is sustained by, an unequal distribution of power, which gives rise to identifiable cores and peripheries. The choice of this terminology, in preference to an earlier rubric of "advanced" and "backward" areas, is deliberate. Differences across a region are neither incidental nor temporary, and the production and reproduction of those differences—as well as their transformation during periods of restructuring—is an inherently political process. As Anthony Giddens observes, regionalism is ultimately grounded in power; the city that forms the center of a nodal region serves as a "storage container" of administrative resources around which states are built.[21]

Regional analysis thus covers two distinct intellectual processes. First, there is an essentially synchronic look at the region as an integrated economic unit. This requires reconstructing how the region is configured—the territorial anatomy of production, circulation, and consumption within its borders[22]—as well how its articulation into the larger economy takes place: "through what sectors, with what capital and labor requirements, to serve what market, and with what possibility of widening and sustaining development."[23] Equally important, however, is a diachronic analysis of how regions change: "how they are constructed and reconstructed, and how they are just as readily torn apart in the flux of industrialization."[24] Such a perspective, which entails a commitment to understanding "regionalization" (or "the changing nature and influence

21. Giddens 1984:183. This use of the terms *core* and *periphery* is both more pointedly political and less rigorously quantitative than that common in regional systems theory. For further discussion, see chapter 4.

22. The concept of a "territorial anatomy of production" is introduced in Storper and Scott 1986:14. The more inclusive phraseology used here—accommodating circulation and consumption as well as production—springs from my own conviction that the economy is a coherent whole that includes all three activities, and that it is pointless to privilege any one part of the economic circuitry over any other. Similar concerns animate the work of Carol Smith (1976a, 1976b) and Fernand Braudel (1990:460).

23. Pudup 1987:242ff.

24. P. Howell 1991:168. This is the approach of "reconstructed regional geography," whose philosophical underpinnings may be traced to Anthony Giddens (1984) and Allan Pred (e.g., 1986). For reviews, see Gilbert 1988, Johnston et al. 1990, Pudup 1988, Sayer 1989, and Schoenberger 1989.

of regional processes and patterns"), highlights the historical contingency of these sociospatial formations, particularly in times of dramatic restructuring such as the Industrial Revolution.[25] Research in this mode accordingly opens possibilities for situating power, and power struggles, in a central position when theorizing about a region, illuminating the contentious social processes that have created, sustained, and transformed a distinctive local landscape. It also throws light on the landscape as a locus of identity, ideology, and interest, suggesting a need to track the shifting alliances that form and reform around a region as its geography is repeatedly contested.

Together, then, the interlocking phenomena of environmental process, spatial pattern, and regional struggle constitute the historical geography whose changing contours I seek to trace through 170 years of Shimoina's development. As I have suggested, analyzing their intersection requires spanning a methodological cleavage between history and geography. At the same time, reconstructing Shimoina's incorporation into the modern Japanese state entails negotiating another disciplinary divide: the deep cleft within Japanese historiography between the early modern and the modern. It is to this second juncture that the discussion now turns.

The Temporal Framework

The years 1750 and 1920 mark unusual end points for an inquiry into Japanese history. The former date falls midway through the early modern era (1600–1868), also known as the Edo, or Tokugawa, period; the latter, half a century into Japan's modern era (1868–), when the privilege of presiding over the newly centralized nation had passed from the Meiji emperor (r. 1868–1912) to his successor, known posthumously as Taishō (r. 1912–25). To position a regional study between

25. For an appreciation of regionalism in the modern economic history of Britain, see Langton 1984 and Hudson 1989a. Important empirical studies that tackle the multivalent issues of regional transformation during the industrial revolution in Europe include Gregory 1982, Gullickson 1986, Langton and Morris 1986, and the essays in *Regions and Industries: A Perspective on the Industrial Revolution in Britain,* edited by Pat Hudson (Cambridge: Cambridge University Press, 1989). The quotations in the text are taken from Philip Howell's perceptive 1991 review of the latter volume. Studies of related interest on the United States include Cronon 1991, Meyer 1989, Wright 1986, and Hahn and Prude 1985.

Tokugawa and Taishō may not be unprecedented, but it is sufficiently unorthodox to merit explanation.

The motives for this idiosyncratic periodization are both strategic and logistical. Not least is the simple conviction that these years embrace some of the most important puzzles in the Japanese past. However one judges their achievement, the question of what allowed the Japanese to reach industrial and imperial power in a few decades continues to spark lively interest. It has also sparked a controversy over the depth of the mid-nineteenth-century discontinuity, with scholars debating whether primary credit for Japan's modern achievements rests with its Tokugawa heritage or rather with the bold institutional innovations of the Meiji period.

Distinguishing inheritance from invention clearly calls for a vision that spans the Tokugawa/Meiji divide. Yet with a few notable exceptions, it is only as a collectivity that Western historians of Japan have reached for such a vision.[26] Even in longitudinal studies of economic development, the 1860s have figured more often as a terminus than as a centerpoint of inquiry.[27] The present study joins a growing body of work that slices the timeline in different ways, transgressing established temporal boundaries in order to look directly at the period of most pronounced social and economic change.[28] A regional focus makes this approach both feasible and necessary. As profound as their ramifications were, the new policies of the Meiji era neither dissolved nor created regions overnight. Just as it had taken decades to knit the Ina valley economy together under the Tokugawa regime, so it took many decades for industrialization and imperialism to rework the area's infrastructures of production and exchange. An ambitious temporal sweep would seem to offer the best hope of capturing that protracted transformation, which is of central interest here.

In addition to these strategic considerations, there are good logistical reasons as well for anchoring this study in the mid eighteenth and

26. Important collections of essays that cross the Tokugawa/Meiji watershed include Najita and Koschmann 1982, Jansen and Rozman 1986, and Jansen 1989.

27. Even those driven by the search for continuities have traditionally confined their primary research to one side or the other of this watershed (see, e.g., T. Smith 1959, Lockwood 1965, and Hanley and Yamamura 1977). Obvious exceptions include studies of the Reformation itself (Craig 1961, Jansen 1961, and Totman 1980), as well as essays on the midcentury crisis as a liminal moment in the creation of a new national consciousness (Harootunian 1988), but these represent fundamentally different genres from the long-term socioeconomic research that is of interest here.

28. Comparable monographs that bracket the Restoration years include Chambliss 1965, D. Howell 1994, Pratt 1991a, and Waters 1983. Not coincidentally, each of these works is essentially a regional study.

the early twentieth centuries. Both periods afford excellent sources for reconstructing the Ina valley economy with a high degree of spatial specificity. A window on the mid Tokugawa world is provided by a pair of trade surveys undertaken at the behest of the shogunate in 1763 to arbitrate a dispute between public and private overland freighters in the valley. Together, these documents record all commercial packhorse traffic entering and leaving the valley's central castle town during a twelve-month period. Of particular interest is that, in addition to noting the volume and value of cargo in transit, the surveys also indicate its origins and destinations. The resulting data present an unusually comprehensive snapshot not only of local production but of the area's interregional trade patterns as well—patterns that stubbornly elude reconstruction from the scattered sources more typical of this period.

For the early twentieth century, by contrast, the important primary sources are of a more familiar and ubiquitous type—namely, village and county gazetteers and local economic censuses. Perpetuating a tradition that has left thousands of such local records throughout East Asia, the government of Shimoina *gun*—the Meiji administrative district created around Iida—published these surveys at irregular intervals over several decades. Each presents statistical data broken down to the village level, enumerating the county's resources and compiling scores of commercial indices that were of interest to the local government.

The trade surveys of the 1760s and the censuses of the early 1900s highlight different components of the Ina valley economy. The former set out to document interregional exchange; the latter speak primarily to intraregional production. But juxtaposing these materials unambiguously shows that, between 1750 and 1920, two very different economic geographies had crystallized in the Ina valley. In the intervening years, interlocking changes in the local, national, and international order had "inverted" the region in fundamental ways. It is in order to track this metamorphosis that I have chosen to put the tumultuous nineteenth century at the center rather than the periphery of vision.

Within this conceptual and temporal framework I stake the larger claims of the present study. Its principal arguments are three. First, a case is made here simply for bringing more self-consciously geographical analysis to the study of Japanese development. The political and economic revolutions of the nineteenth century were clearly spatial revolutions as well; centralizing the government and industrializing the countryside entailed wrenching established relations between places into new configurations. The Japanese experience fully corroborates the "extraordinary volatility of the landscapes of the early Industrial Revolution" observed by Derek Gregory.[29] If we would understand what Japan's modern transformation meant on the ground, I submit, we must begin to elicit its geography.

A second and more contentious argument is that, both before and during its passage to modernity, the spatial patterns of the Japanese economy were the products of complex social negotiations. Physical environment alone did not dictate the routing of trade or the location of regional industries. Nor were these developments the simple outcome of economic rationality or administrative choice. Rather, the regional structures created in the Ina valley during the Tokugawa period, as much as those that displaced them in the Meiji and Taishō eras, were the products of intense local struggle. Geographies were made, not given, and the process of their production was shot through with politics.

Lastly, the transformation of Shimoina confirms that spatial patterns are more than a passive projection of social process. The inherited matrix of production and exchange in the Ina valley—the spatial configuration of the area's political and economic landscape—both constrained and channeled subsequent investments in important ways. In particular, the extensive commercial relationships engendered during a century of protoindustrial development served as vehicles through which national resources could be quickly mobilized in new directions in the later nineteenth century. While the region was clearly peripheralized in the process, it is also clear that the rapidity of its subordination to the new nation-state actively contributed to Japan's industrial and imperial successes. Replicated in different forms throughout the archipelago, the construction of peripheries was an essential part of the process by which the Japanese were able to invert the political economic order of Asia as a whole. Indeed, Japan's rise to power in the western Pacific was contingent in no small part on the state's ability effectively to redirect existing regional circuits, and refashion domestic landscapes like those of the southern Alps, in the interest of larger national imperatives. It is ultimately this dynamic, I believe, that gives the making of this Japanese periphery a claim on our attention.

29. Gregory 1988:51.

PART ONE

The Region Constructed, 1750–1860

CHAPTER TWO

Ina in the Tokugawa Space-Economy

The Making of a Trade Corridor

For a brief period in the mid 1700s, an obscure castle town in the southern Japanese Alps came under the scrutiny of the highest courts in Japan. At the time, most provincial lords would have found such attention as uncomfortable as it was unwonted. Since the Pax Tokugawa of a century and a half earlier, the archipelago had been loosely united under a stable but circumspect central regime, leaving provincial affairs almost entirely to the discretion of local rulers.[1] In such a setting, coming within the purview of the nation's central authorities invariably signaled trouble. Yet for local officials in the Ina valley, the arrival of the shogun's representatives in 1763 was a moment less of trial than of triumph. The conflict at hand was one in which the local daimyo—or, more precisely, his merchant-magistrates—stood to gain more than they might lose from shogunal interdiction.

1. The Tokugawa political system united some 260 formerly autonomous domains (*han*) into a federation overseen from Edo by the Tokugawa shoguns, a hereditary dynasty of suzerains. Although the Tokugawa directly ruled only a quarter of the national terrain, the officials who staffed their administration (the shogunate, or Bakufu) asserted broad powers to intervene in the affairs of the daimyo, fief holders who governed the remainder of the country. The daimyo retained fiscal autonomy, yet were regulated by the shogunate in such diverse matters as succession, marriage, trade, armaments, and foreign policy. For more on the workings of this peculiar hybrid government, known in Japanese as the *baku-han* system, see Totman 1967.

25

What had brought the cumbersome apparatus of Tokugawa judicature to the valley was a smoldering conflict over transportation rights. Since the mid 1600s, post-station operators in southern Shinano province had been protesting to the local lords that unlicensed peasant competitors along the Ina road were engaging in commercial drayage. If true, this constituted an open defiance of domain law, which promised a handful of designated post stations a legal monopoly over all third-party cargo in exchange for corvée. Deprived of paying customers, the post-station managers could not readily meet the continued demand for those services, and as their finances steadily eroded over the course of the eighteenth century, many had become desperate. By the 1750s, tensions had escalated, in some instances to the point of violence; angry relay operators at more than one station had assaulted their illegal competitors and made off with cargo. With the conflict crossing local rulers' jurisdictions, the Bakufu was finally forced to intervene.

In preparation for rendering a decision in the case—one officials hoped would put an end to over a century of conflict in the valley—a year-long survey of the illegal trade was ordered in 1763. From the provincial city of Iida, Shimoina's sole castle town and the most important distribution point in the trade, officials began documenting private cargo traffic to ascertain its extent. The results abundantly substantiated the accusations of the relay operators: a flagrant illicit trade was indeed found to be flourishing along the Ina road. During the year they were put under surveillance, unlicensed packhorse operators were found to have carried more than 70,000 loads in and out of the castle town—an average of nearly 200 loads per day.

Yet the case was not easily put to rest. The Bakufu's investigators were also presented with documents testifying that this same illicit trade had become integral to a thriving commercial economy in the corridor through which it ran. Merchants in the castle town were able to make a persuasive case that, should the suit end in banning all private drayage through the area, shipping costs would rise, commerce would decline, and dire consequences would ensue, not only for Iida but for the larger region it served. Whatever the self-interest behind these alarmist predictions, the shogunate ultimately saw fit to acknowledge them. In 1764, in the face of clear and incontrovertible evidence of sustained illegal shipping, the Ina road's private packhorse network was licensed and sanctioned. The problem was solved, not by punishing the lawbreakers, but by changing the law.

The present chapter looks at this unusual legal decision from two perspectives. The core of the chapter is essentially historical, chronicling the making of a trade corridor in the Ina valley. Bracketing that story, however, are geographical discussions, positioning the region within the wider Tokugawa space-economy. The link between these temporal and spatial analyses lies not only in the legal decision on which they both pivot, but in the intersection of politics and environment that produced the decision in the first place. The Ina valley's location and shape may have created a pathway for trade. But it was the area's fragmented political terrain that allowed private carriers to gain a foothold—and it was the maneuvers of a merchant-peasant alliance that secured recognition for the resulting network in 1764.

To lay the groundwork for this argument, the chapter begins with an overview of Shimoina's linkages to surrounding metropolitan markets. The Ina valley formed a natural corridor between the Tōkai coast and interior Shinano, through which it was connected in turn to Kyoto, Osaka, and Edo. Yet Ina shippers faced keen competition in the through-trade from their counterparts in the parallel Kiso valley. A look at Shimoina's location on the map of early modern Japan suggests that what was at stake in the 1764 decision was not simply who would carry cargo on the Ina road, but how much of the southern Alps' trade would traverse that route in the first place.

Following a brief sketch of this wider geographical context, the focus narrows to the valley itself. The first question addressed here concerns the Ina valley's need to rely extensively on animal transport. In the preindustrial economy, even the smallest of boats were more efficient than any land-based alternative, and Tokugawa authorities supported Herculean efforts to clear the country's major river channels. Yet the Tenryū River was never fully developed for cargo transport. It is my contention that more than physical geography stood in the way of navigation on the Tenryū. By the mid Edo period, if not earlier, local impediments to riparian transport were socially as well as naturally constituted.

Forced to rely on overland transport, the valley's residents were left with two major drayage options: the official relay system, or unofficial, peasant-run pack trains. An extended comparison of the two highlights the advantages of the peasant network over its official competitors, explaining its rise to favor among the area's merchant class. This raises the issue of the role of cargo brokers, who provided both a marketing nexus and a powerful ally in the legal struggles that culminated in the packhorse operators' legitimation. Here above all, the documents create a vivid sense of the extent to which the commercial landscape of Tokugawa Japan was contested terrain. If the circulation network whose spatial contours were laid bare in the Bakufu's survey was in part the product of topography, it was also a product of organized struggle.

The final section of the chapter reverts once again to a synchronic mode of exposition, analyzing the geography of trade as revealed by the 1763 survey. The data submitted for this court case allow us to revisit the question of Shimoina's position, plotting its coordinates in the national economic landscape with considerable precision. Doing so, moreover, highlights the intimate connections between the realm of circulation, the primary focus here, and that of production, to be taken up in the following chapter.

The story of Shimoina's pack trains thus sets the stage for the regional inversion in two senses. Historically, the creation of a regular yet constrained channel of exchange with the wider Japanese economy facilitated Shimoina's protoindustrialization, encouraging the commercial developments that would both integrate and differentiate the region during the later Tokugawa period. Theoretically, too, the transport struggles traced here foreshadow the central themes of the study. From well before its inversion, the economic infrastructure of this region proves to have been not given but made, and the process of its production was indeed politically charged.

Orientations: Shimoina in Japan

The castle town of Iida, focal point of the lower Ina valley, was located roughly halfway between the largest population centers of Tokugawa Japan (see map 1).[2] Some two hundred kilometers to the west, in the Kinai plain, lay the twin cities of the ancient Japanese heartland: Kyoto, the imperial capital, and nearby Osaka, the country's commercial center. To the east, overlooking the even more extensive Kantō plain, lay the city of Edo (present-day Tokyo), the seat of effective political power. The ancient road threading the Ina valley connected Iida with both of these major hubs of the Tokugawa economy. En route to

2. Unattributed facts in the following discussion are from Heibonsha 1979 and companion volumes in the same gazetteer series.

these metropolises, it passed through two intermediary regions as well: the Tōkai coast to the south, and interior Shinano to the north. These four population clusters—two proximate, two more distant—were the most important points on Shimoina's economic compass.

RELATIONS WITH THE SOUTHWEST

The most powerful pole of Shimoina's economic field was located to the south and west, in the thriving community centered on the Ise and Mikawa bays (map 3). By far the largest city in the region was Nagoya, castle town of the Owari domain.[3] Overseeing one of Japan's largest fiefs (estimated officially at over 600,000 koku), Nagoya was home to nearly 70,000 inhabitants in 1700, and to over 90,000 by the end of the Tokugawa period. Its sizable concentration of warriors, artisans, and merchants constituted an important magnet for foodstuffs and consumer goods, making Nagoya the core of an economic region that extended throughout the adjacent Nōbi plain, into the surrounding mountains, and southward around the Ise and Mikawa bay shores.

Nagoya's importance as a center of political power and consumer demand was not, however, matched by any significant wholesaling or shipping role. Located on high ground several kilometers inland, with superior ports to both east and west, Nagoya was poorly positioned to dominate exchanges within the larger transbay community. Instead, those interactions assumed the form of a complex web of multilateral ties, sustained by a shipping network that eluded the domination of any single port. Rather than being a sharply focused nodal region, the trading world of greater Nōbi was a loose confederation of differently endowed subregions, tied together by a latticework of roads and shipping lines that crisscrossed the Nōbi plain and the protected body of water on which it fronted.[4]

Reflecting this economic structure, Shimoina's trade with the region was oriented not solely to Nagoya, but also to two smaller ports on nearby Mikawa Bay: Yoshida (on the Toyo River) and Okazaki (on the Yahagi). Nagoya absorbed the bulk of Iida's southbound exports and

3. The Owari daimyo were a collateral branch of the Tokugawa house. Ieyasu, founder of the shogunate, had his third son installed at Nagoya castle in order to secure control of this strategic area, which included a vital shipping center in Ise Bay, the fertile Nōbi plain, and the prime forests of the Kiso valley.

4. For a more extensive discussion of trade and transport in greater Nōbi, see Wigen n.d.

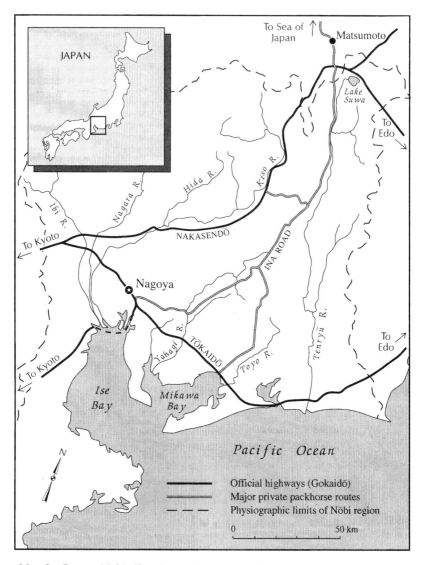

Map 3. Greater Nōbi, Showing Major Rivers and Roads.

served as its primary point of contact with Kyoto and Osaka, but Yoshida and Okazaki supplied most of its coastal trade. With a commoner population of roughly 6,500 in 1833, Okazaki occupied the center of Mikawa's cotton-growing district; the slightly larger Yoshida (population 7,000 in 1710)—later renamed Toyohashi—was the center of the region's weaving industry. Both were important destinations for the pack trains connecting Shimoina with the Pacific coast.

Finally, through these port towns, Iida and its hinterland were connected to the most important cities of western Japan, Kyoto and Osaka. Although the Tokugawa political order relegated the imperial court to a purely ceremonial role, Kyoto remained an important cultural and religious center throughout the Edo period, housing some 400,000 persons. It also supported a host of craft industries, including the country's largest center for the production of silk textiles. But it was the nearby port of Osaka, home to over 400,000, that dominated the economy of western Japan, growing rich as a warehousing and shipping center for domainal tax rice and other commodities, and as the focal point of the nation's financial system.

Lying between these powerhouses to the west and the shogun's capital to the east put the Nōbi trading community in an enviable position. All vessels making the passage between Osaka and Edo were forced to anchor in one of Ise Bay's many harbors for at least a night. In addition, Ise shippers capitalized on their long history of maritime expertise to seize part of the lucrative Osaka-Edo carrying trade, whose volume far exceeded that of any other coastal shipping in Japan. Through these connections, the Ise Bay community maintained regular ties with points west. The volume of exchange between Kyoto or Osaka and a remote outpost like Iida was relatively meager, but it would nonetheless prove significant for Shimoina's development.

RELATIONS WITH THE NORTHEAST

The second important pole of the Shimoina economy was interior Shinano, and through it the shogun's headquarters of Edo on the Kantō plain farther east. It was as a southern gateway into the highland area of central Honshu that the Ina valley found its identity within Shinano province. Ina was not, however, the only southern gateway available. The parallel Kiso valley to the west offered an essentially equivalent path between the interior of the province and areas to the southwest (see map 3). As a result, residents of Kiso and Ina jockeyed repeatedly for primacy as trade intermediary, competing to carry the cargo that tied central Shinano to the Tōkai coast. During the Heian period (794–1185), the official road between Kyoto and the north, the Tōsandō, had followed the upper Tenryū River. But its successor in the Tokugawa period, the Nakasendō, shifted official traffic into the Kiso valley—a shift doubtless motivated in part by the strategic importance of Kiso's rich forests.[5]

To serve travelers along the Nakasendō, the Bakufu established eleven post stations (*shukueki*) in Kiso, whose combined population by the end of the Edo period was 15,000.[6] Yet the Kiso valley was unable to support this swollen human presence. Unlike the broader tectonic basin of Ina, Kiso was only a narrow canyon, its sides too steep and rocky to support more than scattered patches of arable land. As a result, in order to sustain both its sizable resident populace and the frequent travelers on the Nakasendō, Kiso was licensed to requisition large quantities of grain from its neighbors. The lower Ina valley, in lieu of supplying grain, contributed corvée labor. Some twenty-four villages south of Iida were designated as helper villages (*sukegō mura*) for four post stations in the lower Kiso valley, obliging them to provide men and horses each spring to assist with the annual daimyo processions along the Nakasendō.[7]

While Kiso was thus an important rival (and, to some extent, a drain on Shimoina's resources), it was not a major trading partner for the region. More important in the latter capacity were the Suwa and Matsumoto basins, the Ina valley's neighbors to the north. The area around Lake Suwa, home to the smallest but most centrally located of the five population centers in Shinano province, was politically unified during the Edo period under the Suwa clan, which held sway over a 30,000-*koku* fief from the castle town of Takashima. Mid-Tokugawa income from Suwa's productive rice lands (including considerable reclaimed acreage along the lake shores) was augmented by a lucrative lacustrine fishing industry and a number of specialty products, including timber, lacquerware, and processed cotton. North and west of Suwa was the larger Matsumoto basin, home of the powerful Toda daimyo. From a castle town at the southeastern edge of the basin, Toda commanded an expansive 60,000-*koku* fiefdom, the second largest in Shinano. In addition to its grain exports, Matsumoto was famous for several excellent varieties of tobacco.

5. Management of the Kiso watershed was assigned to the collateral Tokugawa line installed in Nagoya castle, making it part of Owari domain. On the importance (and difficulties) of Kiso forestry, see Totman 1983:185–90; Totman 1989:61–63.

6. Post stations were settlements along the major turnpikes that were designated to provide relay horses and porters for passengers, mail, and cargo; see Vaporis 1994. Brayshay 1991 provides suggestive details of a comparable post-horse network in sixteenth-century England and Wales.

7. On the tribulations of "helper villages" (*sukegō mura*), see Vaporis 1986.

Both Takashima and Matsumoto were midsize castle towns, the latter approaching 10,000 inhabitants by the middle of the Tokugawa era. Both accordingly constituted important markets for such coastal products as salt and dried fish. Yet both basins were closer to coastal lowlands other than the Ise and Mikawa bays, and the bulk of their trade accordingly passed through avenues other than the Ina valley. As map 4 suggests, Takashima enjoyed superior access to the Pacific coast through the Kōfu basin; Matsumoto, on the other hand, was oriented primarily to the Sea of Japan. And both basins were close enough to Edo, via the long tentacle of the Tone River, to fall increasingly under that city's sphere of influence during the eighteenth century. By 1750, Edo housed close to one million inhabitants (an order of magnitude more than Nagoya), and its inland waterway to the west was navigable for well over 100 kilometers. Consequently, as close as they were in physical space, Suwa and Matsumoto lay in a fundamentally different economic realm from that of Shimoina. What tied them together was primarily their respective specialty trades with the Pacific coast.

Such were the sinews that bound Shimoina to the larger Japanese economy in the middle of the Edo period. Its position along an axis between Ise Bay and central Shinano provided the Ina valley not only with a convenient source of goods from both the coast and the interior, but with a significant carrying trade as well. It is to the infrastructure of that trade, and the conflict surrounding it, that we now turn.

The Thwarting of River Transport

Given its location between the mountainous interior and the coast, Shimoina's domain lords and merchants alike had powerful incentives to develop the valley's riparian capacity. The Tenryū is one of Japan's longest rivers, and in Tokugawa Japan, as in early modern Europe, water routes constituted the true arteries of the circulation system.[8] Overland transport was at least twice as costly as maritime or potamic shipping, even in the best of circumstances; moreover, water

8. Despite their prominence on maps of early modern Japan, the five national turnpikes known collectively as the Gokaidō did not serve this function. The Gokaidō was essentially a passenger system, designed with an eye to the political needs of the shogunate rather than the economic imperatives of trade.

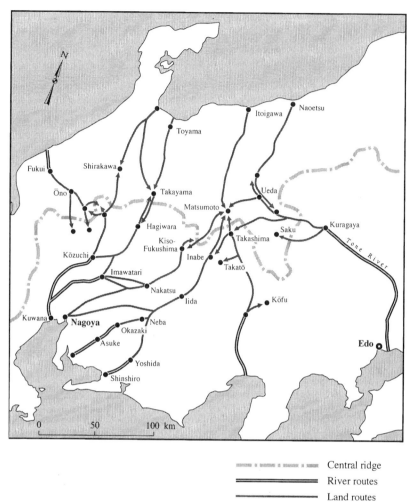

━ ▨ ▨ ▨ ▨ ▨ ▨ ━   Central ridge
═══════════   River routes
━━━━━━━   Land routes

Map 4. Primary Trade Routes for Salt in Central Japan, Late Tokugawa to
Early Meiji. (Adapted from Tomioka 1978.)

carriage offered great advantages in load capacity.[9] In fact, the costs and
labor associated with pack animals were sufficient to induce heroic labors
of dredging each year so that goods might be ferried inland as far as possible. Accordingly, one might expect the residents of the Tenryū valley

9. In assessing the contribution of canals to British industrialization, Michael Freeman observes: "Water carriage cost, on average, half that by road. But it was in its release
of road transport's capacity constraints that water carriage served above all" (1986:86).
Both differentials were undoubtedly higher in Japan, where overland transport was fur-

to have exerted similar efforts to establish riparian access to the Pacific coast. For a combination of physical and political reasons, however, the Tenryū's transport potential remained undeveloped until the Meiji era.

The physical obstacles to navigation on the Tenryū were based in the area's complex terrain. As its technical name suggests, the Ina graben was formed tectonically rather than by erosion. In other words, the Tenryū River, which drains the valley, did not carve it out. Rather, the dramatic contours of the present basin were created by a combination of subsidence in the valley floor and uplift in the mountains that surround it on three sides. These include not only the Ina and Akaishi ranges to the east and the Kiso range to the west, but a jumbled highland known as the Nanbu uplands to the south (map 5). The latter formed a serious impediment to navigation along the middle reaches of the Tenryū. With falls and boulders blocking the southern entrance to the graben, the Tenryū proved less help than hindrance in Shimoina's long-distance trade contacts.

Severe fluctuations in water level were an additional obstacle to navigation on the river. Its source is Lake Suwa, a shallow body of water that fills the floor of the Suwa basin. Fed by numerous mountain streams, and with an average depth of only 4.4 meters (despite a diameter of over four kilometers), the lake regularly flooded its banks following heavy storms in early modern times.[10] This wreaked havoc not only for dwellers along the nearby shores but for those in the Ina valley as well. As Lake Suwa's sole outlet, the Tenryū River saw its water levels surge erratically.[11]

Such temporary surges during the rainy season contrasted with low water levels in the dry winter months, when numerous obstacles in the river bed were exposed.[12] By the eighteenth century, erosion and deforestation in the mountainous terrain upstream aggravated the difficulty, filling the Tenryū each spring and summer with great loads of sediment. Rocks, gravel, and mud carried down from the mountains settled out

---

ther constrained by restrictions on wheeled vehicles. For a discussion of transport modes and costs under similarly difficult conditions in medieval Europe, see Postan 1952:149–55.

10. Annual precipitation averages 1,600–2,000 mm (65–80 inches) on the valley floor, and from 2,000 to 2,700 mm (80 to 110 inches) in the mountains.

11. During the first century of Tokugawa rule, Shimoina suffered some degree of flooding in one year out of three. Shimohisakata-mura 1973: 470–71.

12. See Ichikawa Masami et al. 1980. Japan's steep terrain and seasonally erratic water supply not only created obstacles to the use of rivers for commercial navigation; it also impeded the development of canals, which played a pivotal role in the economic integration of England (Mantoux 1961:120–32; Langton 1984) and the European continent (Vance 1986) during the early modern period.

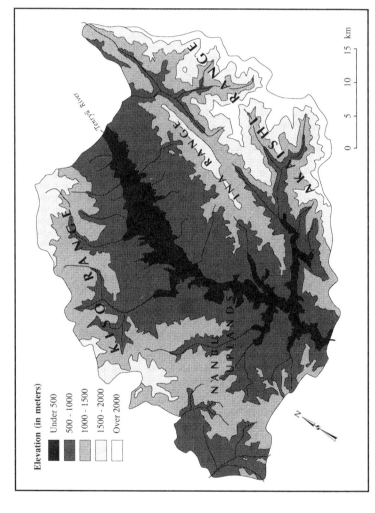

Elevation (in meters)

- Under 500
- 500 - 1000
- 1000 - 1500
- 1500 - 2000
- Over 2000

KISO RANGE

TINA RANGE

AKFISHI RANGE

NANBU UPLANDS

Tenryū River

0   5   10   15 km

N

Map 5. The Topography of the Shimoina Region.

where the river debouched onto the lowland plains, creating a morass of braided, debris-choked channels along its lower reaches.[13]

This combination of erratic discharge levels and irregular terrain clearly obstructed navigation on the Tenryū. While regular boat traffic did develop along short stretches of the river, no unified shipping service was created until after the Tokugawa era. Yet as difficult as the challenges posed by the river may have been, it was not solely technological limitations that kept the Ina valley dependent on packhorse transport until Meiji. Political intervention played a role as well.

By mid Edo times, both capital and technology were at hand to expand and coordinate transportation on the Tenryū. As early as the 1770s, an Iida merchant gambled his personal fortune and a decade of his life on the possibility of opening a commercial shipping business to run the length of the river. A surveyor in his hire counted sixty places where the riverbed would require work or where portages would have to be improved, estimating that 2,200 days' labor by skilled stone masons, and an additional 6,525 days' input from manual laborers, would be required for the job. Undaunted, the merchant petitioned the Bakufu for years for permission to undertake the work.

But the prospect of competition along a river route that paralleled their primary north-south artery threatened the Shinano pack trains, which were a well-established force by this time. Moreover, the prospect of confronting the drivers was unsettling to the Bakufu. The timing could not have been less propitious: the Tokugawa courts had just settled the 1764 suit legitimizing the private network. As a result, when the Bakufu finally gave permission for Tenryū River shipping in 1803, it appended enough regulations to ensure that the venture would not upset the packhorse drivers. The business could employ no more than one hundred boats, could not operate north of the Tenryū Gorge (where the river closely paralleled the Ina road), and, most devastatingly, could carry no cargo for which packhorses might compete. As this excluded essentially everything but long timbers, the enterprise quickly folded.

Forty years after the original effort, in 1823, a merchant of Matsumoto resurrected the issue, this time proposing a shipping service all the way from the Tōkai coast to Lake Suwa. Before acting on his petition, Matsumoto officials solicited responses from the affected areas, including Suwa, Matsumoto, Takatō, and Kiso as well as the lower Ina valley. The result was unanimous opposition. Packhorse drivers were

13. Trewartha 1965:471.

not the only ones fearing for their livelihoods: villagers around the lake worried about their land being taken up by docking facilities, and about the burden of dredging the lake mouth; Suwa domain officials were anxious to protect tax income from lakeside fields; and the post-station operators in Kiso were alarmed at the prospect of rice from upper Ina being shipped downriver rather than hauled overland to the Kiso valley. When the plan was scaled down to mollify all parties, it again left the would-be operator with nothing but long timbers to ship. By 1835, the last Tokugawa-period attempt to open a pan-Tenryū shipping enterprise had collapsed.[14]

In short, while physical obstacles impeded transport on the Tenryū during the first half of the Edo era, political machinations were essential to preserving the status quo after 1764. It was this intersection of environmental and social constraints that kept the Ina valley dependent on overland transport—a dependence that, in turn, set the stage for the Ina valley pack trains.

The Political Geography of Overland Cargo Transport

The thwarting of navigation on the Tenryū meant that Shimoina's sole connections with the outside world were a series of steep, narrow overland passages. Travelers entering or leaving the region in any direction other than toward the north had to cross one or another of the surrounding ranges at passes of over 1,000 meters elevation. Moreover, throughout the Tokugawa era, people and goods traversing the mountainous terrain of the area had to make their way exclusively on foot or horseback, without the benefit of wheeled vehicles.[15] Under these circumstances, two and a half to three days were required simply to complete a journey from one end of the present-day county to the other. Finally, all travelers had to pass armed guards at barriers (*sekisho*)

14. Kusakabe 1987:26; Misawa 1971:30–31, 39.
15. During the Tokugawa period, the use of wheeled vehicles was sharply restricted; in rural areas, they were virtually unknown (cf. Rein 1889:32). Given the wet climate and the poor condition of most of the country's roads, provincial merchants may well have preferred packhorses to carts in any case, as the latter were slow, expensive, and impractical except in paved areas; Mantoux (1961:114), at any rate, comes to a similar conclusion concerning early-eighteenth-century British merchants.

and inspection stations (*bansho*) strategically positioned on all southern and western approaches to the basin, including the Tenryū River.

Despite these obstacles, three major throughways entered the valley from the south, converging in the central castle town of Iida to form the Ina road (map 6). On the east bank of the Tenryū was the Akiha (Akiba) road, a pilgrimage route to the shrine of the same name located in Tōtōmi province (south of Shimoina). A steep mountain path through sparsely inhabited forestland, the Akiha road was traversed primarily by pilgrims and local inhabitants. More heavily trafficked was the Enshū or Shimo Kaidō (lower road), closely paralleling the Tenryū along its western bank. This route carried roughly a quarter of Shimoina's packhorse trade with the south and served dozens of small hamlets scattered through the Nanbu uplands as well. But the bulk of the area's commerce traversed the Sanshū or Kami Kaidō (upper road), the farthest west of the three. Also known as the Ina road or Chūma Kaidō (packhorse road), it was this corridor through Shimoina that constituted the true trunk line of the region's transport system.

The Ina road had functioned for centuries as a national highway, having been designated in the Heian period as the official route between the capital and the interior. Under the Tokugawa regime, however, it was downgraded to a back road, displaced in the national turnpike system by the parallel Kiso road (part of the Nakasendō). Yet it seems unlikely that this usurpation of the valley's official transport function would have been mourned at the time. On the one hand, scholars believe that the Bakufu was merely sanctioning a de facto shift in travel patterns that had sent increasing traffic onto the Kiso road over the intervening centuries in any case. On the other hand, being demoted to a lower level of the transportation hierarchy appears in retrospect to have been more blessing than curse. With its primary roadway designated a mere auxiliary route (*wakiōkan*) in Edo's transportation taxonomy, the Ina valley was spared the onerous burden of underwriting official transport on an extensive scale.[16] Paradoxically, its designation as an auxiliary route was precisely what allowed carriers using the Ina road to best their Kiso counterparts in the contest for commercial traffic.

16. Vaporis 1986 testifies powerfully to the extent of this burden for residents of the Tōkaidō corridor. Although numerous villages in Shimoina were later drafted to serve periodically as "helper villages" (*sukegō mura*) for the Nakasendō, Ina residents were far enough removed from the official turnpike to avoid most of the burden of subsidizing official traffic.

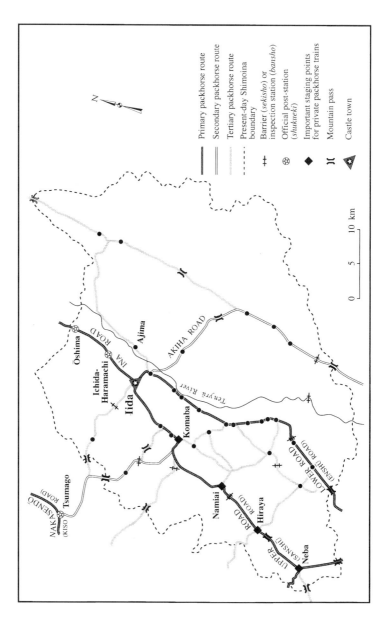

Map 6. Transport in Lower Ina *Gun*, Mid Tokugawa.

Legend:

— Primary packhorse route
— Secondary packhorse route
— Tertiary packhorse route
--- Present-day Shimoina boundary
‡ Barrier (*sekisho*) or inspection station (*bansho*)
⊗ Official post-station (*shukueki*)
◆ Important staging points for private packhorse trains
)(Mountain pass
◮ Castle town

0   5   10 km

Labels on map:

Oshima
Ichida-Haramachi
Iida
Ajima
Komaba
INA ROAD
AKIHA ROAD
Tenryū River
Tsumago
NAKASENDŌ (KISO) ROAD
Namiai
Hiraya
Neba
UPPER ROAD (SANSHŪ)
LOWER ROAD (ENSHŪ ROAD)

N

THE OFFICIAL POST-TOWN SYSTEM
ON THE INA ROAD

For the most part, the contest between Ina and Kiso was not fought directly. The Ina valley's success in securing the bulk of the trade through the Japanese Alps was rather a by-product of an intraregional conflict, pitting one local transport network against another. It was primarily against post stations within the Ina valley, rather than outside it, that the area's private pack trains struggled for recognition.

Escaping designation as a Bakufu highway had not exempted residents along the Ina road from official service altogether. Still the primary throughway for the Tenryū valley, it became the route of choice for a locally based post-station system, similar to the Nakasendō's in operation if more modest in size (map 7).[17] In completed form, the Ina road network boasted fourteen *shukueki*, or official relay stations, between Iida and Matsumoto (inclusive). In this short distance (less than 100 kilometers), the road traversed the territory of at least ten separate local authorities. Despite being supervised piecemeal, the network functioned effectively as an interconnected whole, duly seeing official travelers and their cargo from one end of the valley to the other.

Since travelers on official business were privileged to use post-station facilities at below cost, economic support for the system had to come from other sources. In Ina, as on the five national turnpikes (the Gokaidō), such support came primarily from two sources. One part of the solution was to exempt local lands from crop taxes. Thus, in the 1640s, all official relay stations on the Ina road were declared tax-exempt by their respective overlords. At the same time, post stations were given a legal monopoly on all commercial cargo transport along the road. This arrangement was designed to allow them to set fees on merchants' goods at high enough levels to offset the cost of the fixed-rate work they had to perform for the rulers.

In the limited context within which the post-station system had been conceived, it ought by rights to have worked. If the feudal authorities had imposed the burden of transporting official retinues, they had compensated at least in part by granting the post-station system broad monopoly powers over the increasing commercial trade, which should have proven a lucrative source of income. But by the turn of the eighteenth

---

17. In Ina *gun*, fragments of such a relay system predated the Tokugawa peace. Furushima 1944:79; Furushima 1974a:368. Nor was post-horse duty confined to the Ina road; even local paths between villages were the site of periodic visits by officials, and where an official went, an entourage inevitably followed. Temporary relays were thus occasionally established in the smallest of villages. Hirasawa 1959b; Hirasawa 1965b.

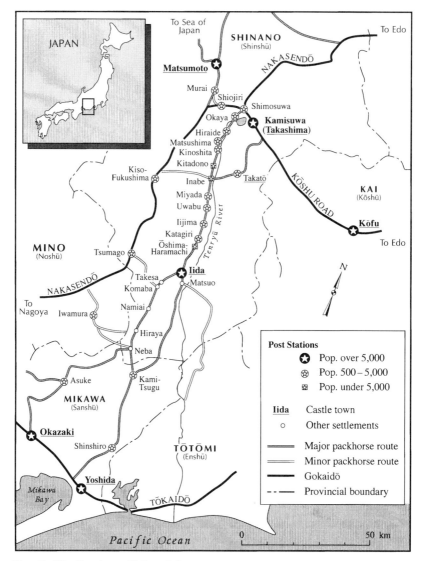

Map 7. The Southern *Chūma* Sphere.

century, both of the initial sources of support were in jeopardy. On the one hand, political authorities anxious to raise their own revenues were beginning to revoke the post stations' tax-exempt status. Such was the fate of four villages north of Iida (Uwabu, Iijima, Katagiri, and Ōshima-Haramachi) (see map 7), which had their exemptions repealed in 1700. Desperate, the four petitioned for the right at least to charge tolls (*kōsen*) on the private carriers who passed through their checkpoints, but this proved politically impractical.[18] At the same time, moreover, peasants from neighboring villages were becoming increasingly bold in wresting away the vital commercial shipping business. Although the villages that had lost their tax exemptions in 1700 would lead a protracted fight against these private upstarts, they would have to confront a powerful merchant block in the castle town, able to argue that the post-station monopoly was no longer in the best interests of the domain lords who had established it.

THE RISE OF THE PEASANT TRANSPORTERS

The alternative favored by area merchants was a private packhorse network. Pack trains from villages outside the official system, often called "rural horses" (*zaigō ma*), originated with peasants carrying their own produce to market. Such independent operators were ubiquitous in Japan during the Edo period. But in Shinano Province, "rural horses" evolved over time into an organized, specialized private packhorse system (*chūma seido*), a development unique to this area.[19] The Ina road was by no means the sole artery of the Shinano packhorse network, but it was both cradle and core of the system; over time, the route along the Tenryū came to be called simply "the packhorse road" (*chūma kaidō*). What allowed private carriers to flourish in Ina was the relative weakness of the valley's official post towns. For two reasons, the local relay stations were weaker than their counterparts elsewhere.

First, unlike the Nakasendō and Kōshū roads, the Ina road was not under direct Bakufu control. On the contrary, it ran through politically fragmented territory, a patchwork of minor domains and exclaves interspersed with Bakufu holdings. As a result, although the valley's post stations had been established for the use of local daimyo, they were not protected by a force powerful enough to undermine private packhorse trains as they arose. Secondly, the scale of official use was

18. Only two stations on the Ina road, Murai and Shiojiri, were granted permission to collect fees, at the level of 4 *mon* per load, for goods other than grain that were in transit by private carriers. Furushima 1974a:374, 400–401.

19. The word *chūma* first appears in the documentary record in 1694. Several theories have been propounded as to the origins of the term, the characters for which literally mean "middle horse," but Furushima Toshio, the foremost scholar of the *chūma* system, makes a persuasive case for the characters having been chosen phonetically to fit a spoken slurring of *teuma* (literally "hand horses"). The latter in turn may have been an abbreviation of *temae uma*, a common alternative term for *zaigō ma*, which suggests that—in con-

much lower here than on either the Nakasendō or the Kōshū road. The latter turnpikes, although they would eventually become secondary packhorse thoroughfares,[20] also bore heavy official traffic that required support.[21] The only official traffic on the Ina road, by contrast, was both small in scale and local in origin. South of Iida in particular, no daimyo traversed the Ina road en route to or from Edo. As a result, this natural corridor between the Tōkai coast and the interior of Shinano was but lightly regulated, a circumstance that facilitated early and vigorous development of private packhorse transport.[22]

The drivers who organized to take advantage of this situation included three major types of operators. Some worked primarily as peasants in nearby villages, hauling their own goods or those of fellow villagers to market. Others were part-time transporters from the interior, who, to fill orders from Iida merchants, purchased goods in remote villages and brought them to the castle town. Lastly, there were professional traders who purchased local goods on their own initiative without requisitions from merchants in the castle town. Upon arriving in Iida, these traders would seek out interested parties and barter the goods they had brought for items they could sell back home.[23]

Despite these differences in their mode of operation, the Ina valley pack drivers came together during the later seventeenth century to protect their collective role in the trade. Galvanized by the concerted efforts of the post stations against them, operators from seventy-three villages in northern and southern Ina banded together in the 1690s to create a trade association (*kumimura* or *chūma nakama*) that, like the Ina road itself, transcended a jigsaw puzzle of fragmented political territories. For

---

trast to the reserve teams of the post towns—peasants simply used horses that were "at hand." Kikuchi 1981:140; Furushima 1974a:408; Furushima 1974b:8–9.

20. By the 1740s, some five to six hundred horses per day were traversing the Kōshū road between Suwa and Edo. Nonetheless, the volume of commercial cargo transported on the Ina road appears to have been much greater than that on either the Nakasendō or the Kōshū road. For estimates, see Furushima 1944:61, 125.

21. Some thirty-four daimyo used the Nakasendō when commuting to and from Edo; six of these possessed large domains of over 100,000 *koku*, an indication of the extent of their retinues. The Kōshū road was traversed only by daimyo from three minor Shinano domains (Takashima, Takatō, and Iida), yet even this modest public corvée strengthened the official post stations there. Furushima 1974a:401.

22. An additional spur to packhorse development may have been the relative paucity of agricultural opportunities in the area; see Yamauchi 1980:1–2.

23. The source of this typology is a document submitted to Iida domain officials by castle-town merchants; Furushima (1974a:381–82) surmises that the actual forms may have been still more diverse than indicated here.

the next century and a half, this organization coordinated numerous activities on behalf of the drivers: raising funds and writing petitions for legal confrontations, limiting the number of horses competitors could lead, securing a favorable division of cargo at the Tōkai terminals, and sharply curtailing the rights of unaffiliated carriers. Members who broke the organization's rules were punished by expulsion.[24]

At the height of the trade, according to a saying widely quoted in Iida, a thousand horses a day passed in and out of the castle town (*irini senda, deni senda*). This figure, recited like a litany by local historians, no doubt contains a margin of hyperbole. But while the 1763 survey found only one-fifth as much traffic, it is clear that subsequent years saw considerable growth in the trade.[25] Counts of active drivers and pack animals must also be interpreted critically, but the settlement of 1764 (also known as the Meiwa decision) found 678 villages throughout Shinano province to be involved in the trade, and permitted over 18,000 animals within them (including both horses and oxen) to operate in the pack trains. The highest density was found in Ina *gun*, where the shogunate licensed nearly 8,000 pack animals in 163 villages—almost 50 per village, or twice the provincial average (map 8; table 1).[26]

FEATURES OF THE PRIVATE PACKHORSE OPERATION

As a private cargo-shipping service, the *chūma* system differed in several particulars from the official post-station system with which it competed. A first, essential difference was that, whereas all cargo traveling in the official system had to be unloaded at each post station

24. Similar trade associations flourished in the mid to late Tokugawa period throughout Japan; for a general discussion of *nakama*, see Hauser 1974:16–20. On the *chūma nakama*, see Furushima 1944:287–95.

25. Furushima (1944:385–86), for instance, estimates that by 1815, southbound exports from Iida had tripled over their level of fifty years earlier. Iwashima (1967:44) quotes the saying "1,000 loads in, 1,000 loads out" (*irini senda, deni senda*) with regard to the Tokugawa period in general.

26. The complete listing of villages and animal counts may be found in volume 1 of the *Chūma ikken kirokushū* (a collection of documents transcribed by the Kinsei Ina Shiryō Kankōkai and published in 1953), pp. 23–38. Almost all of the pack trade in Ina was found to be concentrated west of the Tenryū River, along the Ina road; numerous villages on the east bank were not listed at all. For a more extended discussion, see Miyashita 1980:106.

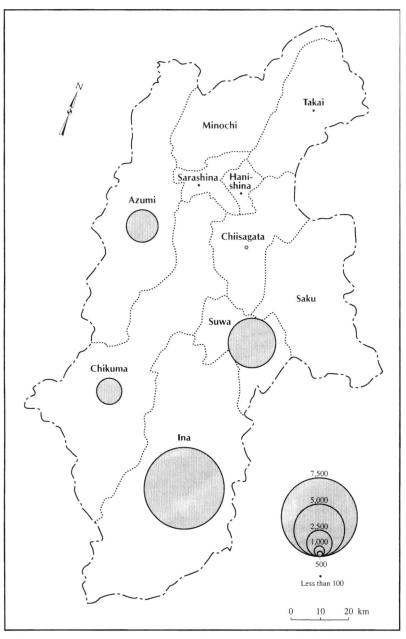

Map 8. The Distribution of Pack Animals Licensed to Operate in the *Chūma* Network of Shinano Province in 1764, by District (*Gun*). (Data from Kinsei Ina Shiryō Kankōkai 1953, 1:28–38.)

Table 1 *Villages and Pack Animals Licensed to Operate in the Shinano Pack-*
horse Network in 1764, by District (gun)

District	Villages	Animals	Animals/Village
Ina	163	7,849	48
Suwa	123	4,689	38
Azumi	180	3,178	18
Chikuma	159	2,525	16
Chiisagata	35	321	9
Takai	8	81	10
Hanishina	6	76	13
Sarashina	5	58	12
TOTAL	679	18,768	28

SOURCE: Kinsei Ina Shiryō Kankōkai 1953, 1:28–38.

NOTE: Substantially fewer animals may have participated in cargo transport than the
numbers licensed; for more extended discussion, see Wigen 1990:106–7.

and reloaded onto a fresh horse, private operators were under no such
obligation. Not only did the drivers carry their cargo over greater distances between stops, but the same driver and team would typically see
a load through from origin to destination. The difference was significant. While cargo in the official system had to be transferred every 8–10
kilometers along the Ina road, private operators could go 32 to 36 kilometers in a single day; on long-distance hauls, they would commonly
travel 80 to 120 kilometers without reloading. In recognition of this
central difference, the private operators were often called "through-
horses" (*tōshi-uma*), to distinguish them from the "relay horses" (*tsugi-
uma*) of the official post-station system.[27]

Going straight through gave the packhorse trains a number of advantages. For one, it increased their speed. In supporting documents submitted at the time of the investigation of 1763, Iida's merchant community claimed that a journey that would take ten days by post horse
could be accomplished in a single day by *chūma*. Although this was a
gross exaggeration, the general point was well taken. Peasant drivers
could make the trip from Iida to Nagoya, for instance, in about six days
each way; for that from Iida to Matsumoto, a trip of twenty-five *ri* (100
kilometers), four days was considered the outer limit. Delivery times by
post horse, on the other hand, were both longer and unpredictable. Any
official demand for the facilities could displace merchant cargo and its

27. On the geography of the local "through-horse" system, see Mukaiyama 1969:245.

handlers, adding further delays to an already cumbersome system. Such delays exacerbated the risk of cargo damage or spoilage, a recurrent problem with the constant loading and unloading under the relay system.[28] But speed was not the only measure differentiating the two systems. A second and equally important index was labor efficiency. While individual post horses could be saddled with slightly larger loads (up to 150 kilograms, compared to 120 to 135 kilograms for the private pack trains), the *chūma* system operated with a lower ratio of drivers to animals. In the post-station system, the firm rule was one man per horse.[29] But peasant operators developed techniques for managing multiple pack animals, typically three to four horses or as many as five oxen, by driving them from behind.[30] As the Shinano packhorse network coalesced into a trade association, the drivers began to treat their techniques for leading multiple pack trains as a form of trade property, defending as their prerogative and theirs alone the right to lead four animals. In fact, a whole generation of legal battles turned largely on this question, and in a decisive court case of the early 1800s, the Shinano packhorse drivers' claim to a monopoly on this practice was upheld.[31]

A third striking characteristic of the private packhorse system was the freedom of its operators to choose from multiple routes. Unlike the official turnpikes, which comprised strictly demarcated stretches of road, punctuated by mandatory stopping points, the Ina road was an informal local route with multiple channels, and drivers could and did choose for themselves which route to take. In negotiations with the post-station managers, Iida merchants consistently referred to peasant operators as "*okabune*," or "ships of the interior," a pointed reminder of the pack drivers' right to go anywhere they pleased.[32]

All of these differences with the official system translated into price differentials. Having to stop at each post station, the relay horses incurred indirect as well as direct expenses in the form of cargo transfer fees, bills for food and lodging, and damage to cargo. The high ratio of

28. Fragile cargo, for which this was particularly important, included dried persimmons, cocoons, and lacquerware—three prominent local products exported from Shimoina.
29. The gendered referent is used advisedly here. Although women are known to have worked as porters, leading pack animals appears to have been an exclusively male occupation.
30. The drivers in the *chūma* system were aptly called *umaoi*, literally "horse-followers," for they drove their animals from behind (controlling them primarily through spoken commands). Andō and Yamori 1972:80.
31. For details, see Wigen 1990, ch. 3.
32. Furushima 1974a:405–7.

drivers to animals compounded these expenses. Furthermore, all goods traveling by post horse were susceptible to toll surcharges. And whereas the fee schedule in the relay system was nonnegotiable, merchants could bargain with the peasant drivers. Overall, the cost of sending goods by private pack train worked out to only about half the cost of using the official counterpart. The difference was significant enough that the post stations objected to carrying merchant cargo at all if it meant they would have to carry it at *chūma* rates.[33] That profit-minded merchants sought out these private carriers is not hard to understand.

Cargo Brokers: The Marketing Nexus of the Network

Among the merchants who played a critical advocacy role in the packhorse drivers' suit, none were more important than the cargo brokers. These specialized shipping agents evolved in tandem with the packhorse network itself. In the early 1600s, packhorse drivers had operated not only as transporters but as marketers too, unloading their wares and hawking them directly on the streets in various commercial wards of Iida. During the Kan'ei period (1624–44), for instance, drivers were ordered to go to designated wards, specified according to the origin and nature of their cargo, so that each section of the castle town might share in the market prosperity they brought. By the time they began to be called *chūma*, however, the peasant packhorse drivers of the Ina road had become transport specialists, shedding their earlier marketing function. In legal documents beginning in the 1720s, *chūma* were defined in part by their relationship with cargo brokers, or *nidon'ya*.[34]

The word *ton'ya* (and its variants, *toiya* and *don'ya*) designated a broad and diverse class of merchants in Tokugawa Japan, usually with

33. Post-horse fees averaged 1.7 to 1.8 times the level of private packhorse fees for the same distance, although there was some variation (since *chūma* fares were negotiable). For sample fares, see Furushima 1974a:414.

34. Furushima 1974a:372ff., 409. A 1769 petition penned by the merchants of Iida explicitly differentiated *chūma* drivers from intermediary merchants (*nakagai shōnin*), and other records confirm the point that the drivers did not typically buy and sell goods on their own account.

some degree of monopoly privileges. Some operated as wholesalers, consignment agents, or receiving agents; others supervised processing industries; still others oversaw shipping concerns.[35] It is the latter that are of interest here: brokerage houses engaged in cargo shipping, known locally as horse or cargo *ton'ya* (*umadon'ya* or *nidon'ya*).[36] Like their counterparts in wholesaling, cargo *ton'ya* were typically licensed by the local authorities to collect tolls (*kōsen*) on each transaction, in exchange for which they usually paid licensing fees (*unjōkin*) to local officials.[37] Brokerage firms operating in this way could be found in all the main *chūma* stopping points along the Ina road. They were particularly numerous in the villages where packhorse traffic was heavy, and in and around the castle town. By the eighteenth century, five full-time cargo shippers had set up shop in Iida, while numerous castle-town merchants handled shipping work on the side. Their rural counterparts were clustered in villages to the south, including Namiai, Hiraya, and Neba, where cargo was collected (either from local merchants or from incoming drivers) and redistributed for posting to its final destination.

Interviews with men who drove pack trains in the Meiji period leave us most of what we know of the day-to-day operations of the cargo shippers. Their tales are consistent in picturing these brokers as among the wealthiest members of their villages. Minimum requirements included a large house, a wide courtyard, and storage space for cargo, all with street frontage. The home of the central broker in Namiai, for instance, was an imposing structure of three stories. The gate was wide enough to allow the pack drivers to walk their horses straight into the courtyard (*doma*), where there were seats for a dozen or more men to rest and eat. Employees would serve food to the waiting drivers, while others attended to the horses and cargo. Throughout the day, the head of the household would sit on a dais in the courtyard, paying for freight delivered, receiving deposits for goods shipped out, drawing up bills of lad-

35. Hauser 1974:13. This roster of roles corresponds closely to James Vance's broad typology of European wholesalers; see Vance 1970:29–33.

36. Because of the breadth of the term and the importance ascribed to *ton'ya* merchants in the debates on early Japanese capitalism (usefully dissected in Hoston 1986, ch. 5), it is essential to be clear about the type of *ton'ya* in question here. Hirasawa Kiyoto (1953b) distinguishes three discrete types: station managers (*shuku ton'ya*), wholesalers in specific commodities (*shōhin ton'ya*), and cargo brokerage firms (*umadon'ya* or *nidon'ya*). Packhorse drivers were most closely connected with the latter. On the emergence and functions of *shuku ton'ya*, see Watanabe 1979.

37. The *ton'ya* in Namiai, for instance, were permitted to charge twenty-four *mon* on each load that they handled, in return for which privilege they paid ten *ryō* per year to the village officials. Andō and Yamori 1972:86.

ing, and keeping accounts of all transactions. The brokers usually sold grain and straw for the animals as well, and some drivers, lacking the cash to pay on the spot, were chronically in their debt.[38]

In contrast to the porters and drivers working for the official post towns, who had fixed arrangements with specific station managers, private packhorse drivers were free agents. Their relationships with brokers at both the shipping and the receiving end were continually open to negotiation and had to be contracted anew for each job. This fluid situation created risks for those whose goods were being entrusted to the drivers, an uncertainty that merchants and shippers hedged in two important ways. Cargo brokers, on the one hand, drew up bills of lading (*okurijō*) for each shipment they sent out. The owners of the cargo, for their part, insisted on collecting deposits (*shikikin*) from the drivers when entrusting goods to them. These deposits commonly amounted to 70 or 80 percent, and sometimes ran as high as 90 percent, of the value of the cargo. They would be collectible, along with the drivers' fares, upon delivery of the goods, when the receiving broker would pay the full value of the shipment. Some packhorse operators also functioned as advance purchasing agents for the cargo brokers.[39]

It was a great advantage for those placing orders that, unlike the relay operators, *chūma* drivers did not insist on prepayment for their transport services. Taken together, advance deposits and the system of payment on delivery in effect functioned to provide short-term financing for castle-town merchants, particularly small-scale operators who could not always pay the full cost of their purchases in advance.[40] Who ultimately controlled the capital dispensed by the *chūma* is a critical question that has not been adequately answered to date; it is highly unlikely that a common driver could command cash worth 90 percent of the

38. Nagano-ken Kyōiku Iinkai 1959:87.

39. This involved forwarding cash deposits on the brokers' behalf for goods requested from the countryside, as well as for merchants' goods acquired from other *nidon'ya*. Furushima 1944:223–26.

40. In fact, some evidence suggests that deposits from the drivers, while originating as advances for cargo, may have gone beyond this into general-purpose moneylending. The document mentioned above, for instance, indicates that merchants obtained money-changing services from the drivers (*chūmakata yori kawasekin uketori*). This suggested to Furushima that the *chūma* drivers may have used their unusual mobility—passing through two, three, or even four commercial centers at regular intervals—to develop a financing network of sorts. In other words, in addition to providing transport services for cargo, they may have provided "transport services for capital" (*shiharaikin no unsō*). Furushima 1974a:409–12.

value of the cargo that four beasts of burden could carry.[41] In any case, however, this feature of the packhorse system was vital to the economic well-being of the castle town as a whole, prompting not only cargo brokers but other merchants as well to rally to the packhorse drivers' defense in their numerous legal battles.

SOCIAL CONFLICT OVER CONTROL
OF THE TRADE

Of keen interest to the merchant community was the contest between the official post-station system, which had theoretically been given a monopoly on cargo transport, and peasant packhorse drivers from other villages along the road. This conflict surfaced early in the seventeenth century, as individual peasants tried to transport their own goods to local markets. No sooner was that reasonable request granted, however, than the post stations began to complain that rural drivers were illegally mixing commercial cargo with their own produce. But efforts to halt this practice were to little avail. The trend in a long series of suits was for the balance to be tipped fitfully but progressively in favor of the *chūma*.[42] Meanwhile, the scope of packhorse activity steadily widened. By the turn of the eighteenth century, private operators were openly challenging the post-station monopoly even to carry commercial cargo to distant markets. As a result, the scope of the disputes also widened. By the mid eighteenth century, what had started as an intradomainal conflict had spread throughout Shinano and into neighboring Kai province, later to extend south to include operators from Mikawa as well.[43]

In most of these suits, the *chūma* drivers were forcefully backed by the merchants of Iida. The latter repeatedly submitted strongly worded testimony to argue that the private packhorse network was vital to the

41. Furushima, personal communication, May 27, 1990. Masaki Keiji suggests that a driver who could not front all the capital for the merchandise he handled might borrow from a *sake* maker or other rich moneylender in his village (1978:53). But whether this was in fact a general practice is not clear. Both the origin of the capital fronted to make deposits on goods the *chūma* carried and the relationship between the drivers and the owners of that capital (presumably wealthy peasants in the drivers' villages) remain unanswered questions.

42. For details on these early conflicts, see Murase 1984.

43. In battles pitting private operators in one region against their counterparts from other areas, the loyalties of the castle-town merchants were less clearly on the side of the Shinano *chūma*. For a timeline and summary of major suits between private carriers from different regions, see Onogi 1968.

well-being, not only of the merchant community, but of the castle town and indeed of the domain as a whole. In 1718, for instance, Iida merchants testified that goods imported from nearby areas, as well as from the interior of Shinano, would become scarcer and more expensive if peasant drivers were not given free rein. Their case centered on three assertions: (1) the castle town's supply of rice and other foodstuffs from nearby villages might decrease if peasants were prevented from carrying commercial cargo on their return trip; (2) if drivers from farther in the interior found nothing to take back with them, the price of the goods they brought into the city would have to rise to compensate for their lost income; and (3) if private carriers were charged for using the facilities of the castle town, most would bypass the town altogether, spelling ruin for Iida as a trade entrepôt.[44]

This presented the domain with a serious dilemma. The lord of Iida and his retainers still needed the post-station system to support their official travel. Yet the castle-town economy had become equally dependent on an alternative shipping system that threatened to undermine the post stations altogether. The domainal courts would have to broker some sort of settlement that would permit peaceful coexistence of the two systems. The trick was finding a compromise that would stick.

In one early pass at resolving the problem, Iida in 1727 handed down a "post-horse first" policy. Under this ruling, private carriers would be allowed to take commercial loads, but only after all the available post horses had been employed. Just as the merchants had feared, however, their business was hurt. Not only did the local peasants' incoming cargo decrease, but its cost to the merchant houses increased. Furthermore, drivers from the coastal provinces of Mikawa and Owari began to avoid the castle town to do business in smaller settlements nearby, where they could freely acquire return cargo. The merchants lost no time in demanding that the policy be overturned. In response, the domain compromised further later in the same year, giving the private operators wider latitude. The new arrangement quieted things for a time, but the conflict continued to simmer.[45]

The post stations' chronic frustration finally broke out into the open in the 1750s, prompting the landmark suit that capped the first generation of litigation. This was an affair of such central importance as to be referred to in the literature simply as "the packhorse incident" (*chūma*

44. Furushima 1974a:378.
45. For a more extended discussion, see Wigen 1990:121ff.

ikken). It began as a dispute between three post towns in upper Ina (Matsushima, Kitadono, and Miyagi) and eighty-two nearby villages. The initial provocation was narrow but symbolic: the three post towns insisted that all goods carried into their jurisdictions on private pack animals be transferred to post horses before being allowed to continue. The peasant drivers protested, citing precedents that established the right of merchants and cargo brokers to decide for themselves whether to use post horses or their private competitors, and a legal suit was begun. Within months, the merchants of Matsumoto, as well as thirty-eight *chūma* villages from lower Ina, had joined in, and in time the merchants of Iida entered the fray as well, turning the case into a major regionwide battle. Over the next half decade, the suit worked its way through the Bakufu's courts, leaving behind a prodigious set of records that have been central to every subsequent analysis of the packhorse network.

With the merchants' support, the *chūma* won a highly favorable decision in 1764 (the first year of a new era, Meiwa 1). The Bakufu chose to resolve the dispute by allocating loads throughout Shinano between the post stations and the private carriers, and by standardizing the fare structure between the two. On both counts, however, the Ina road villages were treated notably differently from their counterparts elsewhere in the province. While on other roads the fees for both *chūma* and post horses were set at the higher post-station rates, those on the Ina road were fixed at the lower prevailing packhorse rates. Ina received similarly special treatment in the matter of cargo allocation, leaving the local packhorse drivers unrestricted as to the content of their cargo.[46]

These provisions were of signal importance. The once-illicit private pack trains had won recognition as a legal transport network, with guaranteed rights to the increasingly lucrative cargo trade in the mountains. More important, although the battle had essentially been fought between private and official carriers along the Ina road, its outcome would have supraregional implications. With its fares now fixed at lower rates than those of its neighbors, and its drivers freer with respect to cargo, the Ina valley carriers as a whole had secured an important advantage against their competitors in the Kiso valley.

46. In most of Shinano, merchants were ordered to divide the available cargo between the two types of drivers on the basis of content, with some goods specified as the prerogative of the post-horse system alone. Along the Ina road, on the other hand—owing both to the combined force of merchant and pack-driver resistance and to the relative weakness of the area's post stations—the *chūma* were not limited in any way as to the content of their cargo. Here the split was rather determined in load amounts, with approximately one-fifth of all traffic to be carried through the official relay system. Furushima 1974a:399.

For all its significance, however, it is important to see the Meiwa decision in context. Legally, it did not represent the end of all conflict over
control of the trade; new skirmishes would continue to be fought
throughout the remainder of the Tokugawa period.[47] And in practical
terms, it is unclear how much choice the shogun's courts really had. In
sanctioning the Ina valley pack trains, the Bakufu may merely have been
extending its blessing to a fait accompli, for if the trade survey of 1763
is to be believed, the Ina valley had already secured the bulk of the trade.
It is to the details of that comprehensive survey that we now turn.

A Synchronic View of Trade Patterns in 1763

Investigations of cargo flows were relatively rare in Tokugawa Japan. Even in the Ina valley, where transport rights were the subject of litigation for centuries, few surviving documents afford a quantitative picture of cargo circulation in the mountains. The pair of surveys
undertaken at the Bakufu's behest in 1763, on the eve of the Meiwa settlement that legitimized the *chūma* network, accordingly provide an unusual and valuable glimpse into Tokugawa economic geography.

The survey results were reported in two documents. The first, entitled the "Hōreki 13 [1763] Report of Incoming Cargo from Various
Places" (*Hōreki jūsannen tokoro-dokoro yori irinimotsu mokuroku*),
records Iida's imports; the second, "Record of Cargo Entrusted to
Packhorse Drivers by the Merchants of Iida City, Ina District, Shinano
Province" (*Shinshū Inagun Iidamachi shōnin chūma e aiwatashisōrō ni-
motsu kakinukichō*), documents the city's exports.[48] Both take as their
locus the castle town of Iida, cataloging trade goods by origin and

47. Over the next century, competition from private operators headquartered south
of the Shinano border led to a second round of suits, followed in turn by a third round of
litigation directed against local peasant part-timers. For details of these cases, see Wigen
1990:124–29.

48. The original documents, along with others pertaining to the packhorse trade, were
transcribed by hand and published in 1953 in the three-volume *Chūma ikken kirokushū*
(Kinsei Ina Shiryō Kankōkai, ed.). More recently, both documents have been reprinted by
Nagano prefecture as part of a massive collection pertaining to the history of Shimoina
during the early modern period (Nagano-ken 1983; see documents nos. 1861 and 1870
esp. pp. 223–29 and 268–71, respectively). Data from these mid-eighteenth-century surveys, adjusted where necessary to correct inconsistencies in the original, have been drawn
on throughout the discussion here, which has also benefited from a prior analysis by Masaki
Keiji (1978, ch. 2).

Table 2 *The Proximate Origin of Goods Imported to Iida from the South*

Origin	Loads	Percentage of Total
Mikawa[a]	14,280	60
Nagoya	6,766	28
Tōtōmi	2,726	11
Mino	190	1
TOTAL	23,962	100

[a]Okazaki, Yoshida, and Shinshiro.

destination. Although they cover only cargo carried by licensed packhorse drivers, omitting that handled by independent operators (*teuma*) as well as by carriers from Mikawa Province (known as *tsugi-uma*), the pattern they reveal is believed to provide a generally credible model of trade flows in the southern packhorse sphere during the mid Edo period.

According to these records, the total volume of traffic passing into and out of the town exceeded 73,000 animal loads. Of that total, imports accounted for more than 52,000. Goods leaving Iida, by contrast—including both locally produced items and re-exports—amounted to only 21,000 loads, or less than half the level of imports. More significant, however, are the sharp differences that can be seen in the nature of Iida's exchanges with the coast and the interior. Whereas Shimoina appears to have been little more than a consumption and distribution center for goods from the coast, imports from the interior penetrated more deeply into the local economy, in some cases providing important raw materials for local manufacturing. The following account analyzes these and other patterns in the eighteenth-century trade figures, beginning (like the surveys themselves) with the larger volume of incoming trade.

IIDA'S IMPORTS

Forty-five percent of goods brought into Iida in 1763 originated south and west of the castle town, primarily along the Tōkai coast. A breakdown of this traffic from the south shows a preponderance of Mikawa Bay over Ise Bay connections. Over two-thirds of Iida's Tōkai traffic was routed through the former area, with 60 percent of southern imports coming through Mikawa province's ports of Okazaki, Yoshida, and Shinshiro, 11 percent through Tōtōmi province, and less than 30 percent by way of Nagoya (table 2).

Table 3 *Goods Imported into Iida from the South*

Item	Origin	Loads
Tea	Mikawa and Tōtōmi	10,756
Salt	Inland Sea	4,600
Dried fish	Kumano, Owari, Echigo	2,485
Ginned and raw cotton	Kinai, Mikawa, Owari	2,220
Used clothing	Nagoya, Okazaki, Yoshida	228
Cotton cloth	Osaka, Sakai, Owari	303
Silk cloth	Kyoto	40
Indigo	Okazaki, Yoshida, Shinshiro	350
Stenciling paper	Ise	3
Mandarin oranges	Kii, Mikawa	450
Pharmacopoeia	Kyoto, Osaka, Owari	225
Vegetables	Owari	20
Iron and copper goods[a]	Ise, Echizen, Osaka	547
Tatami mat facing	Osaka, Higō, Owari	250
Tatami fiber	Nagoya	5
Lacquer	Izumi	20
Pottery	Mino	190
Earthenware	Hizen	70
Bleached wax	Kyoto, Osaka	130
Unbleached wax	Ōmi	110
Mosquito netting	Ōmi	5
Whetstones	Ōmi	18
Hairdress ties	Nagoya	20
Notions	Kyoto, Osaka, Owari	230
Hair and lamp oil	Nagoya	240
Sedge hats	Ise	90
Garden plants	Kyoto, Osaka	7
Miscellaneous[b]	Kyoto, Osaka, Owari	350
TOTAL		23,962

[a]Includes pots, sickles, and nails.
[b]Timber, paintbrushes, incense, sugar, laundry starch, glue, ink, lead, cosmetics, dyestuffs, shells, etc.

The bulk of freight from these coastal cities consisted of textiles, salt, dried fish, mandarin oranges, and tea, the last of these alone constituting 45 percent of the total. All of these goods, and many lesser imports, were procured primarily from production sites around the shores of the Ise and Mikawa bays. Yet, as may be seen from table 3, a number of items from more distant areas found their way to Iida through these ports as well. Prominent

Table 4 *Origin of Imports into Iida from the North*

Origin	Loads	Percentage of Total
Beyond the Ina Valley		
Edo and the Kantō plain	205	1
Matsumoto	2,665	9
Suwa basin (Shimosuwa)	72	<1
Sea of Japan coast	394	1
Kiso	70	<1
SUBTOTAL	3,406	4
Within the Ina Valley		
Takatō	1,452	5
Upper Ina valley	16,286	57
East bank of Tenryū River	5,700	20
Iida vicinity	1,784	6
SUBTOTAL	25,222	88
TOTAL	28,628	100

SOURCE: NSTI:72–73.

here were manufactured goods and specialty products from the Inland Sea, but the list also included dried fish from as far away as Echigo on the coast of the Sea of Japan, brought round the western end of Honshu by ship.

The reverse flow of goods entering Iida from the north constituted the majority (55 percent) of the castle town's imports. In origin, this trade was heavily weighted toward products of the upper Ina valley and interior Shinano. Some 15,900 loads of grains and pulses (including rice, barley, and beans) made up the largest single item on the list. These staples are recorded as having come predominantly from the agricultural settlements north of Iida in the central Ina valley, with a lesser share (2,300 loads) deriving from Matsumoto.[49] Apportioning this item accordingly in the origin-specific breakdown in table 4 makes it clear that only a small fraction of Iida's trade goods from the north originated outside of the Ina valley.

The bulk of imports from the north consisted of agricultural and forest resources for consumption in Iida or for processing in the area's craft

49. The upper Ina valley's share of this line item (i.e., the remaining 13,600 loads of grains and pulses) was distributed as follows: 450 loads were recorded as having come from Takatō, 2,650 from the east bank of the Tenryū River, and 10,500 from villages along the Ina road between Iida and Matsumoto.

industries. In addition to provisions, the northern Ina valley supplied Iida with tobacco (much of which was blended in the castle town for re-export) and cooperage (especially bamboo hoops used to secure barrels and casks). A detailed accounting is presented in table 5.

Independent sources, it should be noted, suggest that a much larger volume of tobacco was brought south from central Shinano through the Ina valley than is recorded here. These imports escaped the census tak-ers simply because they were handled by carriers other than the *chūma* organization. The heart of Shinano tobacco production at the time was the Ōi watershed in Chikuma *gun*, north of the central ridge. In 1763, virtually all of this area's tobacco was shipped south; 90 percent went to Nagoya, most of it over the Ina road (table 6).[50]

IIDA'S EXPORTS

The total volume of goods dispatched from Iida during the year came to 21,453 animal loads, or roughly two-fifths the level of imports. By bulk, the division of these exports was decidedly unbalanced. Only a quarter of the total was directed toward the Tōkai terminals; the remainder was bound for the interior (map 9).

It is significant that most of this northbound cargo—more than 12,000 loads' worth, or half of all items dispatched from the castle town—consisted of goods that had originated along the Ise-Mikawa coast. This figure testifies to the importance of the through-trade for the packhorse network. In fact, only half of all cargo brought into the val-ley from the Tōkai area was consumed within Iida. While the domestic market absorbed nearly all of the imported salt and dried fish (which the interior basins of Shinano could readily procure from other sources), many specialty products of the coastal region merely passed through Iida en route to the interior. This was especially true of tea and cotton. Of the 12,000-odd loads of re-exported cargo carried by the pack trains into interior Shinano, more than 11,000 consisted entirely of these two specialty products of the Pacific coast. More than 85 percent of the recorded import levels for both goods went to satisfy demand north of the lower Ina valley (table 7).

In addition to those items that passed into and through Iida from outside, the Hōreki documents identify cargo originating in Iida as well.

50. The geography of the tobacco trade proved highly responsive to later price shifts, however; after the turn of the century, the bulk of Shinshū tobacco was rerouted to Edo. Miyagawa 1960:158–59.

Table 5 *Goods Imported into Iida from the North*

Item	Origin	Loads
Grains and pulses	Matsumoto, Ina valley	15,900
Silk thread	Ina valley	10
Cocoons	Matsumoto, Ina valley	80
Silk cloth	Takatō	17
Hemp thread	Takatō, Ina valley	60
Raw hemp	Matsumoto, Shimosuwa	120
Dried gourd shavings	Matsumoto	20
Perilla seeds	Matsumoto, Takatō, Ina valley	1,140
Dried fish	Sea of Japan coast	394
Hair and lamp oil	Ina valley	60
Walnuts	Ina valley	70
Chestnuts	East Tenryū	100
Dried persimmons	Iida area	900
Shōchū lees	Takatō	500
Rapeseed wastes	Matsumoto, Takatō	270
Lime	Ina valley	150
Lacquerware	Kiso	70
Flint	Edo	5
Kettles	Edo	20
Used metalwares	Ina valley	60
Bamboo cooperage	East Tenryū	510
Wooden ladles	Iida area	30
Tobacco	Matsumoto, Kai province	55
Tobacco	East Tenryū, Ina valley	3,900
Ash-based bleach	Ina valley	180
Whetstones	Kōzuke	180
Pharmacopoeia	Ina valley, East Tenryū	170
Cryptomeria bark	East Tenryū	80
Dyestuffs	Iida area	42
Wooden combs	Iida area	12
Rope	Shimosuwa	12
Wooden clogs	East Tenryū	340
Freeze-dried tofu	Matsumoto	40
Paper mulberry	Takatō	150
Hardwoods for crafts	Takatō, Iida area, East Tenryū	2,050
Other (not specified)	Takatō, Ina valley	926
TOTAL		28,628

Table 6 *Exports of Chikuma Tobacco, 1763*

Destination	Loads	Percentage
Nagoya, via Ina road	5,400	75
Nagoya, via Kiso road	1,000	14
Okazaki	250	3
Shinshiro	250	3
Yoshida	100	1
Suwa	130	1
Takatō	50	1
TOTAL	7,200	100

SOURCE: Miyagawa 1960:159.

The latter consisted primarily of semiprocessed agricultural goods, notably dried persimmons and tobacco, supplemented by a small quantity of high-value local protoindustrial goods, notably paper hair ornaments, silk thread, lacquerware, and small wooden craft objects.

Tobacco is a difficult item to categorize in these terms. Ninety-nine percent of the leaves carried by the *chūma* were grown, dried, and shredded in the northern Ina valley, beyond the borders of the Shimoina economy. The only processing done in the castle town involved blending the different varieties and repackaging them for further shipment. If one considers this sufficient to call tobacco a local product, it dwarfs all others, accounting for roughly 3,800 of 8,000 loads of "local" goods dispatched from the castle town. It also biases the export trade heavily toward the Tōkai, since all but a single load of the tobacco that passed through Iida was bound for the Tōkai coast. Given the minimal processing that went on in Iida, however, I find it more accurate to designate tobacco an item of through-trade. Removing tobacco from the list of locally produced goods reveals that the markets for the remaining items of local manufacture were roughly balanced between north and south. In fact, of the remaining 5,028 horse loads of local products exported in 1763, slightly *fewer* than half (2,082) were destined for the Tōkai coast. All of Iida's silk and unlacquered bowls, however, were shipped in that direction (table 8).

The remaining three-fifths of local products, including some of Iida's most valuable protoindustrial goods, were exported to the north and east. All of the area's paper parasols and hairdress ties, 90 percent of its finished lacquerware, and 70 percent of its dried persimmons were dispatched toward central Shinano. It is unclear how much of the north-

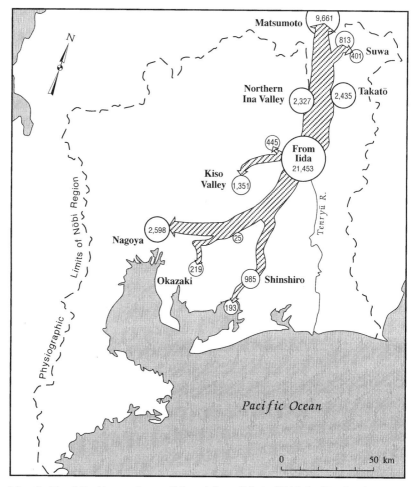

Map 9. Total Packhorse Cargo Shipped from Iida in 1763, by Destination.
Unit: packhorse loads (*da*). (Adapted from Nishioka and Hattori
1956:214.)

bound total was destined ultimately for the wealthy markets of the
Kantō, but at least a portion of Iida's hairdress ties were specified as Edo-
bound. A detailed breakdown is provided in table 9.

ANALYSIS

The data in tables 1–9 offer numerous suggestions about
Iida's orientation in the wider Tokugawa economy. At the outset of this
chapter, Shimoina was confidently identified as belonging within the

Table 7 *Goods Passing through Iida en Route from the Tōkai Coast*
       to the Interior

Item	Destination	Loads
Tea	Matsumoto, Takatō, Ina road, Kamisuwa	9,664
Raw cotton	Matsumoto, Takatō, Ina road, Kamisuwa	1,877
Salt	Ina road	417
Dried fish	Matsumoto, Takatō, Ina road, Kamisuwa	197
Cotton cloth	Matsumoto, Takatō, Ina road, Kamisuwa	165
Used clothing	Shimosuwa, Takatō, Ina road	28
Notions	Shimosuwa, Takatō, Ina road	85
Indigo	Ina road	55
Mandarin oranges	All destinations	21
Iron goods	All destinations	92
Silk cloth	Suwa, Takatō	13
Ceramics	Ina road	4
Sedge hats	Takatō, Ina road	8
Pharmacopoeia	All destinations	55
Whetstones	Takatō	4
TOTAL		12,685

greater Nōbi trading sphere. The content and volume of the packhorse trade allow us to give substance to this assertion; they also indicate that Shimoina's ties were somewhat more complicated than this simple formula suggests.

On the one hand, the differential volume of goods coming into the castle town clearly confirms the assessment that the area's principal ties were with the coast. Although the roughly 24,000 loads originating to the south were overshadowed by nearly 29,000 loads entering Iida from the north, almost all of the latter (88 percent) came from within the Ina valley. In contrast to the steady stream of goods originating more than three days' distance to the south, in other words, Iida's import trade from comparably distant areas to the north and east was a mere trickle.

On the other hand, if one cuts the data in a different way, separating goods consumed or processed locally from those that were merely warehoused in the castle town, a rather different set of patterns is revealed. Most conspicuous is a curious imbalance in the through-trade. The volume of coastal cargo passing through Iida to points north, approximately 12,000 loads, was roughly triple that of through-traffic to the south, which included 3,800 loads of tobacco but little else. Even more interesting than

Table 8 *Goods Processed or Produced in the Iida Area for Export to Nagoya, Okazaki, Yoshida, Mino, and Kiso*

Item	Destination	Loads
Unlacquered bowls	Shinshiro, Yoshida, Nagoya	407
Lacquered bowls	Shinshiro, Yoshida, Nagoya	132
Dried persimmons	Nagoya, Okazaki, Iwamura	314
Silk cloth	Nagoya, Okazaki	100
Silk wadding	Yoshida	1
Silk floss	Tsumago (Kiso)	81
Cocoons	Iwamura (Kiso)	140
Semirefined oil (*mizu-abura*)	Nagoya, Kiso	80
Paper	Nagoya, Mikawa, Mino	188
Pharmacopoeia	Nagoya, Shinshiro	62
Hemp thread	Nagoya, Yoshida, Shinshiro	268
Cryptomeria bark	Shinshiro	72
Soy sauce	Mikawa, Kiso	24
Tamari sauce	Mikawa, Kiso	47
Combs	Shinshiro	11
Lacquerware	Shinshiro	7
Women's clogs	Iwamura (Kiso)	138
Camphor	Nagoya, Shinshiro	5
Miscellaneous	—	5
TOTAL		2,082

this quantitative difference is a striking qualitative difference in Iida's trade with the two main poles of its economic compass. The contrast is simply stated: goods from the north, unlike those from the south, passed not only through Iida's markets but into its commercial production circuits as well.

In relation to the Tōkai coast, Shimoina played primarily a passive role as a consumption and distribution point. As we have seen, fully half of the goods imported from the Nōbi region were merely warehoused in Iida before being passed on, in an unaltered state, to markets farther inland. The remaining half, which was destined for local markets, consisted primarily of ready-to-eat specialty foods (salt, tea, mandarin oranges, fish, vegetables) and manufactured items (silk, textiles, used clothing, refined hair oils, tatami mat facings, pottery, and sundries). There were, to be sure, two important exceptions to this pattern: ginned cotton and indigo. Both required further processing before they could be used. Yet in eighteenth-century Japan, these two items were usually processed *at the point of consumption* (i.e., within the peasant house-

Table 9 *Goods Processed or Produced in Iida for Export to Interior Shinano*

Item	Destination	Loads
Lacquered bowls	Matsumoto, Suwa, Takatō, Ina valley	1,173
Hairdress ties	Matsumoto, Suwa, Takatō, Ina valley	444
Dried persimmons	Matsumoto, Suwa, Takatō, Ina valley	779
Bamboo cooperage	Matsumoto, Takatō, Ina valley	306
Vinegar	Takatō, Ina valley	62
Zumikawa (a yellow dye)	Takatō	37
Kimono	Takatō	7
Tobacco	Takatō	1
Paper	Matsumoto, Suwa, Takatō	20
Paper parasols	Matsumoto	19
Candles	Matsumoto, Takatō	28
Wooden combs	Matsumoto, Takatō	6
Soy sauce and *tamari*	Ina valley	14
Refined oil	Suwa	48
Ladles	Ina valley	2
TOTAL		2,946

hold), so that, like the other goods from the coast, they were primarily resold in the form in which they arrived.[51]

The same cannot be said of goods from the interior. With the exception of tobacco, the latter consisted mainly of grains and pulses, nuts, fertilizers (*shōchū* lees, rapeseed wastes, and lime), and a host of raw or barely processed materials: oil seeds, mulberry bark, hardwoods, cocoons, raw hemp, dyestuffs, and the like. After they entered Iida, the path of those goods differed from that of their coastal counterparts in two respects. First, a much higher percentage of the influx from the interior was consumed locally; second, a significant proportion of the remainder was processed by Iida residents for re-export to the south *in more refined forms*. This pattern is exemplified by the case of textiles, where the castle town imported cocoons and raw hemp from the north but exported hemp thread, silk floss, and a small amount of specialty silk cloth to the south. Similar patterns prevailed in four additional industrial sectors: paper, oils, hardwood crafts, and soy products (soy sauce and *tamari*).

In short, Iida's role in relation to northbound goods was significantly different from the role it played in the southbound trade. In economic terms, the northern reaches of Iida's packhorse sphere constituted a re-

51. On the growth of a national market for cotton, see Hauser 1974:59.

source hinterland for the castle town, whereas the coastal areas did not. This conclusion is reinforced when the focus is narrowed to processed goods alone, for the Nōbi littoral may be clearly distinguished from interior Shinano on the basis of the kinds of locally processed items each market absorbed. While the Tōkai region mainly purchased its semi-processed goods, Iida was able to export numerous finished goods to both interior Shinano *and Edo*. For instance, all of Iida's silk thread and unlacquered bowls—both of which required further processing before they reached the final user—were shipped out to the south; by contrast, all or most of the area's paper parasols, paper hairdress ties, and finished lacquerware were sent northward toward central Shinano, with some going on to Edo. Again, this suggests that, in its relations with the coast, Iida essentially played the role of a periphery—importing more advanced or more fully processed goods than it exported—whereas in its trade with the interior, the roles were reversed. These are among the most important implications of the 1763 data for Shimoina's orientation in the Tokugawa economic landscape.

Conclusion

In addition to the specific findings about Shimoina's economic role, the foregoing discussion has suggested a number of more general conclusions about the organization of the Tokugawa space-economy at the middle of the eighteenth century. Three of these findings are of central interest for the present study.

First, despite concerted attempts by various levels of government to stabilize it, the infrastructure of trade proved highly elusive of regulation. We have seen this most clearly in the rise of an illicit packhorse network, which secured a foothold by providing merchants with an advantageous system for carrying freight in and out of the mountains. Individual drivers traveled straight through to distant destinations rather than transferring their loads en route, arriving faster and with less damage to cargo; they exercised free choice of routing, being able not only to bypass the post stations but to take shortcuts or detours at will; and each driver led multiple animals, reducing the labor inputs necessary to transport the same volume of freight. Such a level of innovation was unmatched elsewhere in the transport developments surveyed here. Yet some degree of flexibility characterized all major elements of the circulation infrastructure, including its nodes (towns and marketplaces) and channels (roads and shipping lines) as well as the actual carriers that shuttled between them. Castle towns, for instance, were officially granted significant roles in the cargo trade, as both wholesaling centers and distribution points. But they too had to fight to preserve their prerogatives. Just as the official relay stations saw their carrying business usurped by peasant packhorse teams, so castle-town merchants saw their wholesaling trade threatened by rural brokers. In Iida, this lent the merchants' position a certain paradoxical, if not hypocritical, quality. While they were happy enough to see upstarts displacing official transporters, they lost no time rallying against precisely the same sort of development when it threatened their own domain.[52]

No amount of protective legislation, however, could guarantee the castle towns' position as the most important nodes in the growing commercial trade. Under provocation of surcharges or simply unfavorable terms of trade, drivers repeatedly took the risk of circumventing designated markets to negotiate more profitable deals in rural satellites. And in extreme cases, as they had done in sanctioning the packhorse network itself, the authorities eventually recognized these changes by revoking the privileges they had ordained in the first place. This finding is perhaps not surprising, given evidence of comparable movement away from early Tokugawa norms in other areas of economic and social life.[53] But it does provide additional confirmation for characterizing the regime's attempts at economic regulation as more "flamboyant" than effectual.[54]

The second general finding of the chapter concerns the extent to which the Tokugawa space-economy was a political creation. We have seen this most clearly in the fight between the post stations and the packhorses over legalization of the latter network. In considering the survival of that network, however, it is also important to recall the efforts expended to prevent the Tenryū River from being developed for shipping. The importance of the Bakufu's intervention in blocking development of the proposed navigation system on the Tenryū cannot be overesti-

52. This struggle is discussed more fully in chapter 4.

53. Examples range from domain finances to fertility control, rural-urban migration, the sapping of commercial production in castle towns, and chronic infringements of sumptuary legislation. For a survey of strains that had developed by the early nineteenth century, see Jansen 1989a.

54. The source of this term is Philip C. Brown (1991), who defines the flamboyant state as one that is nominally granted wider authority than it is actually able to exercise. See also Brown 1993. For a related discussion contrasting the Tokugawa shoguns' broad claims with their more limited de facto powers, see White 1988.

mated. The code governing packhorse activities restricted loads to roughly one hundred kilograms per animal on level ground; in difficult terrain, the legal limit was reduced to between fifty and eighty kilograms. Since the capacity of a riverboat was twenty to forty times greater, it was decisive for the packhorse system's survival that the middle reaches of the Tenryū River were largely unusable for long-distance shipping until the Meiji period.[55]

The thwarting of riparian transport thus represents one more instance where politics decisively shaped the Ina valley's regional economic infrastructure. It deserves recounting in the present context because it was an essential backdrop to Shimoina's development, and a necessary, but by no means given, condition for the survival of the packhorse network. Terrain and technology alone did not determine the geography of transportation in the southern Japanese Alps during the Tokugawa period. Rather, the routing of cargo flows was a matter of policy, openly contested by interested parties. Just as the recognition of private packhorse drivers along the Ina road was a political act, so the failure to exploit the Tenryū River was ultimately a political decision.

Finally, the data presented here suggest that constraints on transport capacity—whatever their origin—affected local economic development in critical ways. Beyond the limits of navigation, hauling high-volume, low-value goods such as grain and fuel more than one to two days' journey from their source of origin was simply not practical. On the one hand, this sharply limited the scale of urban development beyond the coastal littoral (and its inland extension onto a handful of alluvial plains). Equally important, it constrained the kinds of commodities that the larger lowland cities could profitably extract from beyond this pale as well. With the important exception of lumber, which could usually be floated or rafted downstream from high in the mountains (even on rivers like the Tenryū that were otherwise not navigable),[56] highland areas of Japan simply could not be exploited for bulky commodities until the late nineteenth century.

It is essential to note that this constraint was not entirely a negative force in the development of the Japanese uplands. To the extent that the country's interior regions developed exports at all—and the com-

55. Figures on load limits are from Tomioka 1978:86, 153; Mukaiyama 1969: 250–53.

56. Sources on Tenryū logging and riparian navigation include Hayashi Tōmito 1987, Ishikawa 1980, Kusakabe 1987, Misawa 1971, and Toyoda et al. 1978. As Conrad Totman (1989:71–74) persuasively documents, the losses—here as on other rivers used for transporting lumber—were often exorbitant.

petitive mercantilism of domain ideology dictated that all feudal lords at least make this attempt—the expense of transport dictated that those exports be lightweight, high-value goods. This principle is borne out in the cargo manifests identifying the goods exported from Iida to the large metropolises that surrounded it: highly processed agricultural goods (dried persimmons and tobacco), paper and paper wares, silk thread, and lacquerware top the list. In sum, being relegated to the inland transport periphery in Tokugawa Japan in some ways encouraged precisely the kind of economic development that we have come to think of as characterizing advanced areas.

This represents one way in which the composition of trade is as enlightening about patterns of production as about patterns of exchange. Of the data provided in the 1763 trade surveys, the figures on Iida's own exports are particular revealing about the local economy; between them, tables 7 and 8 constitute in effect an X ray of the castle town's industrial base. To flesh out that skeletal image, our focus must now shift explicitly from circulation to production: from the politically charged history of the region's transport infrastructure to the equally political geography of its territorial division of labor.

CHAPTER THREE

The Landscape of Protoindustrial Production as Contested Terrain

By the time of the 1764 decision, the Ina valley had clearly become an important conduit for exchange between the mountains and the sea. Local peasants and merchants had managed to take control of the bulk of cargo traffic in the southern Japanese Alps, effectively reconfiguring the map of Tokugawa transport. But the significance of the Ina valley pack trains extended beyond the sphere of circulation, exercising a far-reaching influence over the geography of production as well. Long before they achieved legal recognition, the valley's *chūma* drivers were calling forth new forms of economic integration across the region. The catalyst for this integration was the accessibility of urban markets for local commodities.[1]

As the 1763 survey revealed, by the mid eighteenth century, Shimoina was already shipping a number of specialty items to distant urban centers. Three clusters of local goods were particularly prominent in the cargo manifests: hardwood wares, silk products, and paper crafts. All were relatively lightweight luxury goods, accessible only to the nation's

1. Compare Gay Gullickson's (1986) observation that the geography of protoindustrial development in France hinged largely on two variables: the ease of transport between a region and its potential markets, and the complementarity of craft work with rural labor demands. Shimoina was well positioned in both senses.

wealthier consumers. Yet the manufacture of these commodities worked essential changes on the Iida area landscape, drawing on resources from a score or more of villages around the castle town. By the middle of the eighteenth century, the overlapping circuits of labor and finance that supported their production had effectively knit central Shimoina into a cohesive economic region.

The present chapter explores the nature of territorial differentiation and integration under this nascent protoindustrial regime. An essential aim is to demonstrate that the landscape of Tokugawa commodity production was shaped by a surprisingly intricate locational calculus. Early modern industrial processes depended heavily on place-specific resources of several kinds: environmental, commercial, and political. It is my first contention that we cannot understand Japanese protoindustrialization without exploring in detail the geographical restrictions within which it operated.

The second claim is a bolder and perhaps less familiar one. Most historians of the Tokugawa era would agree that, while feudal authorities had a heavy hand in fostering Japanese protoindustry, the revenues generated in this new sphere of production largely eluded their grasp. I would go further to suggest that this crucial political limitation, too, was ultimately grounded in geography. Commercial ventures followed their own spatial logic, which was not that of the Tokugawa feudal settlement; economic regions coexisted with political regions, but did not conform to them. Although complexly intertwined, the political and commercial orders were not congruent; in southern Shinano, at least, their boundaries showed a significant degree of slippage. The sheer geographical mismatch between the bounds of local political authority and the extent of the new commercial networks precluded the former not only from controlling protoindustry but even from taxing it effectively.

To lay the groundwork for this twofold argument, the chapter opens with a broad overview of Shimoina's landforms, settlement morphology, and administrative divisions in the mid Tokugawa period. The remainder of the chapter is devoted to tracing the origins and organization of Shimoina's major protoindustries. A brief discussion of the lacquerware and textile complexes is followed by a more extended analysis of papermaking and associated crafts, the primary focus of domainal patronage in the area. The paper sector also became the most conspicuous arena of social conflict, partly because its spatial order repeatedly eluded regulatory controls. The final section traces one decisive battle over taxation rights, demonstrating how the attempt to impose such controls could backfire—and underscoring the extent to which Tokugawa Japan's protoindustrial landscapes were contested terrain.

Internal Configurations

The preceding chapter situated Shimoina in relation to a constellation of larger economic spheres, a universe in which the valley occupied an integral, if interstitial, place. But the productive complex of Iida and its hinterland took shape in a highly specific local configuration as well. Three landscape features in particular bore directly on the geography of production: landforms, settlement distribution, and political jurisdictions. A careful look at topographical and historical maps reveals the essential contours of each.

TOPOGRAPHY AND SETTLEMENT FORM

The first and most fundamental constraint on the spatial structure of the Shimoina economy was the area's physical environment.[2] In the rugged, mountainous country of southern Shinano, the largest expanse of level land was a steplike series of terrace surfaces along the Tenryū River, in the landform known as the Ina graben, or trench. This elongated basin constituted the economic as well as the physical core of the region (see map 5).

By the mid Edo era, Shimoina's rural population was concentrated along these terrace surfaces on either side of the Tenryū River. As can be seen by comparing map 10 with map 5, most of Shimoina's 160-odd villages in the later Tokugawa period—and all of the largest—were located in the basin, between 400 and 600 meters in elevation. Scattered upland settlements were to be found through the extensive mountainous hinterland as well (the highest during the Tokugawa era being the now-abandoned hamlet of Ōdaira, at over 1,100 meters). The more extensive terrace belt having developed along the western scarp (fig. 1), both arable land and population were disproportionately distributed west of the river.

2. The best source on the geomorphology of the Ina valley is Matsushima 1987. See also Shimoina Kyōikukai 1976.

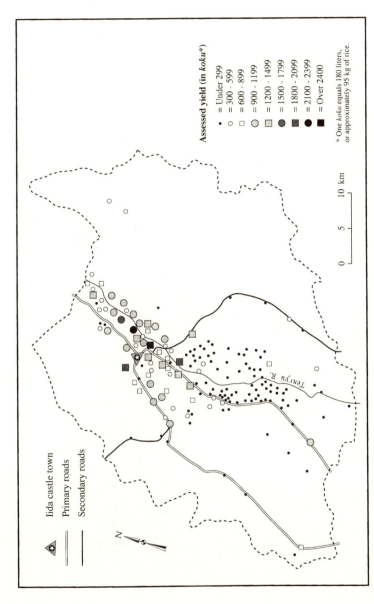

Assessed yield (in *koku**)

•	= Under 299
○	= 300 - 599
□	= 600 - 899
◑	= 900 - 1199
▨	= 1200 - 1499
◕	= 1500 - 1799
▰	= 1800 - 2099
●	= 2100 - 2399
■	= Over 2400

* One *koku* equals 180 liters,
or approximately 95 kg of rice.

Iida castle town

Primary roads

Secondary roads

N

Tenryu R.

0   5   10 km

Map 10. Late Tokugawa Villages in Lower Ina *Gun*. The location and assessed yield (*kokudaka*) of 164 rural settlements, 1834. Important Edo-period trade routes are noted for reference. (Data from Heibonsha 1979.)

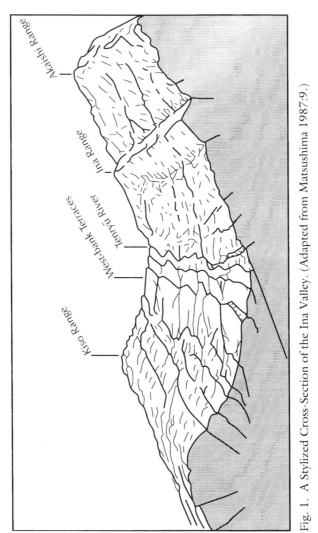

Fig. 1. A Stylized Cross-Section of the Ina Valley. (Adapted from Matsushima 1987:9.)

This concentration of villages in the center of the valley was essentially a product of the seventeenth century. Although the graben offered good soils and level farmland, water control was inadequate to support a large population in the area before 1600. Only with the introduction of advanced irrigation and flood-control technologies could the Tenryū floodplain and terraces be exploited for intensive agriculture.[3] As their potential was developed over the course of the Tokugawa period, these large-scale waterworks appear to have allowed a doubling of agricultural yields in the Ina basin.[4] The concomitant process of demographic centralization can be deduced from a pair of agricultural surveys that bracket the transformation. On the eve of Tokugawa unification, when surveyors in the employ of Toyotomi Hideyoshi ranked the villages of Shimoina in three categories to reflect broad differences in land productivity, "superior" (jō) villages were found to be scattered in small valleys through the mountains (map 11a). Three centuries later, by contrast, when the next systematic agricultural appraisal was undertaken by the Meiji government, all the top-ranked villages in Shimoina, without exception, were clustered within the basin (map 11b).[5]

This populous district was overseen from the castle town of Iida. As was true of most such fortified sites, Iida had originally been located with an eye to its military rather than commercial possibilities. Situated on a promontory of high ground that jutted out over the surrounding farmland between two tributaries of the Tenryū River, Iida only became a significant marketing center by fiat. When Hideyoshi conquered the area in 1593, the vassal he installed on the site enlarged the fortress into a castle compound, constructed a walled town on the terrace behind it, and ordered merchants from nearby settlements on the valley floor to relocate to his newly designated commercial district.

Throughout the Tokugawa period, Iida remained the uncontested political and demographic center of lower Ina. Patterned on a grid after the fashion of the Chinese capitals, the town was dubbed by its inhabitants a diminutive Kyoto of the mountains. Its population at the turn of

3. During the preceding centuries, the valley's population was widely dispersed, forming sizable clusters only along the base of the scarps (where clear springs seeped out through the terrace fill) or at the heads of tributary streams (where surface water could readily be diverted for irrigation). Hirasawa 1953a, 1955.

4. Although official surveys put the productivity increase at only 30 percent between 1593 and 1850 (from a total of 52,000 *koku* to 67,400 *koku*), investigators for the Meiji land-tax reform in the 1870s calculated that actual yields by then were double the latter figure. See chapter 6.

5. On the Hideyoshi and Meiji surveys, see Hirasawa 1954.

the eighteenth century is estimated at 6,500; by 1790, when the rural population of Iida domain numbered roughly 16,000, more than half that number—some 8,500 persons—resided in the castle town and its suburbs. This unusually high urban-to-rural ratio reflected the extent to which Iida by then served as a commercial hub, not only for the domain but for the wider region as well.[6]

LOCAL GEOPOLITICS

The geography of power in Shimoina conformed in a general way to the area's physical terrain and settlement patterns, with the principal fief holders occupying the prime, densely settled lands at the center of the valley. Yet the political landscape even within the basin was strikingly fragmented. Through the gradual rearrangement of Japan's political map in the 1600s, a critical instrument of the Pax Tokugawa, Iida domain—formerly the one respectable power center in the valley— was progressively diminished. By 1750, Shimoina had been reduced to a collection of small estates surrounded by Tokugawa lands: a circumstance that would have important repercussions for the area's commercial development.

A political map of the region at the end of the seventeenth century shows the resulting territorial patchwork (map 12). Shimoina's political core, and its biggest contiguous domain, remained the fiefdom of Iida, but this domain now supported one of the country's smallest daimyo.[7] Surrounding Iida were half a dozen minor parcels, administered by lesser warrior families (*hatamoto*), as well as scattered areas entrusted to another daimyo headquartered in a neighboring province.[8] The bulk of Shimoina's territory, however, was reserved as Tokugawa property. While comprising only a quarter of Shimoina's villages and a third of its

6. In 1672, when the first town census was undertaken, Iida's merchant community numbered 563 households; at the turn of the eighteenth century, the resident commoner population was put at 4,400 persons. Iida's true head count at the time (including retainers, servants, and an adjacent suburb) was probably closer to 6,500, or one in four residents of Iida domain. Furushima 1951:91; Hirasawa 1968:18; Masaki 1978:70–71.

7. By 1672, Iida had been scaled back from an original 120,000 to a mere 20,000 *koku*.

8. The daimyo in question, Matsudaira, was a second son of the Tokugawa branch family that had been installed in Nagoya castle. Although his primary base of power (Takasu domain) lay in neighboring Mino Province, Matsudaira was given scattered lands in the Ina valley in 1672 with a putative net yield of 15,000 *koku*. Accessible sources on southern Shimano's early Tokugawa political history include Shimoina Nōchi Kaikaku Kyōgikai 1950 and Ichimura 1966.

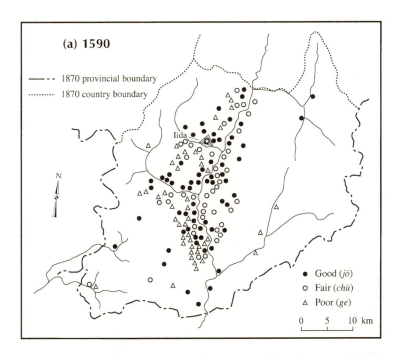

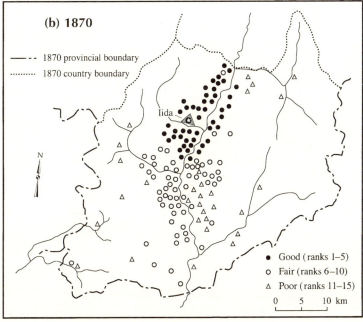

Map 11. Lower Ina *Gun* Village Rankings, ca. 1590 and 1870.
(a) shows classifications employed by surveyors under Toyotomi
Hideyoshi; (b) represents a three-part grouping of village rankings assigned in 1873 on the basis of estimated overall productivity per unit of arable. (Adapted from Miura 1988:28 and
Chiyo-mura 1965:36–37.)

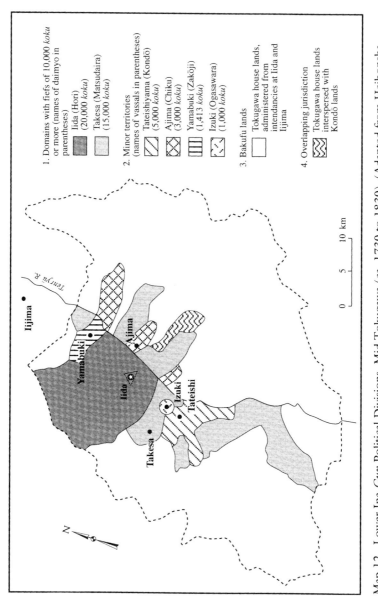

1. Domains with fiefs of 10,000 *koku* or more (names of daimyo in parentheses)

▨ Iida (Hori)
   (20,000 *koku*)

▨ Takesa (Matsudaira)
   (15,000 *koku*)

2. Minor territories (names of vassals in parentheses)

▨ Tateishiyama (Kondō)
   (5,000 *koku*)

▨ Ajima (Chiku)
   (3,000 *koku*)

▤ Yamabuki (Zakōji)
   (1,413 *koku*)

▨ Izuki (Ogasawara)
   (1,000 *koku*)

3. Bakufu lands

□ Tokugawa house lands, administered from intendancies at Iida and Iijima

4. Overlapping jurisdiction

▨ Tokugawa house lands interspersed with Kondō lands

Map 12. Lower Ina *Gun* Political Divisions, Mid Tokugawa (ca. 1730 to 1830). (Adapted from Heibonsha 1979:154.)

arable, these so-called house lands encompassed more than two-thirds of the region's terrain, including most of its high-quality forests.[9]

In short, local history and Bakufu fiat had created a parcelized geopolitical landscape in the Ina valley. Despite the region's potential to support a much larger domain, the spatial scale of local authority was held in check; the resources of a rich agricultural core were divided among several petty fief holders. It was this fragmented terrain that would become the hearth of Shimoina's vigorous but contentious protoindustrial development.

The Lacquerware and Textile Industries

In the early eighteenth century, the area in and around Iida became home to three putting-out complexes, or clusters of related cottage industries, geared to supraregional markets for lacquerware, textiles, and paper goods, respectively. Each of these three crafts had a basis in pre-Tokugawa tribute exactions. But during the watershed Genroku era (1688–1704), newly introduced techniques dramatically improved both the quality and the speed of production in lacquerwork, textile production, and papermaking alike. These changes first permitted Iida's craft workers access to a national market at the turn of the eighteenth century. Unfortunately, despite their acknowledged importance, relatively little is known of the former industries, and only the broad outlines of their development can be ascertained.

LATHE- AND LACQUERWARE

Hardwood craft items constituted one of the earliest regional specialties of mountainous Shinano province.[10] In Iida, lacquerware production dates to the 1670s, when a merchant from Owari settled in the area and invited artisans from Ōmi (Shiga Prefecture), an

9. The house lands (*tenryō*) in Shimoina had a putative yield of between four and ten thousand *koku* and were mostly administered from an intendancy (*jinya*) in Iijima, north of Iida.

10. During the Edo period, Shimoina was home to two of Shinano's nine regional lacquer centers: one in Iida, the other in the Niino basin. Other Shinano lacquerware centers included Takatō, Chino (Suwa), Matsumoto, Iiyama, Ueda, Kiso-Fukushima, and Kiso-Hirasawa. Like Iida, most of these were castle towns, where demand for such luxury wares was concentrated. Tate 1983.

ancient and venerated center of Japan's lacquer craft, to join him. Within a century, the industry had become a substantial contributor to the castle town's economy. By 1763, over 36,000 *kan* (143 metric tons) of lacquered goods were exported from Iida annually, along with a third as many unlacquered bowls, made of buckeye, chestnut, alder, and other local hardwoods.

The lacquerware complex in Shimoina displayed a relatively simple, two-part spatial organization, integrating two geographically distinct stages of production. On the one hand, artisans of the lathe called *kijishi*, or "wood masters," produced the small, turned wooden objects destined for lacquerwork, as well as ordinary unlacquered bowls, ladles, and trays. These lathe-turners were concentrated in Shimoina's southwestern uplands, an area rich in hardwood forests. Originally migratory, moving their base of operations from one mountain valley to the next every decade or so, the *kijishi* formed semipermanent communities where resources permitted, supplementing their woodwork with swidden (*yakihata*) agriculture.[11] The west bank of the Tenryū River along the Mikawa-Shinano border—including southwestern Shimoina—supported a particularly dense concentration of such communities, which first appeared in the 1670s and continued to grow for the next hundred years. Records show their total numbers in the district fluctuating between 100 and over 250 during these years.[12]

While access to hardwood forests was the main factor in *kijishi* location, the higher stages of processing were constrained primarily by commercial rather than ecological resources. In Shimoina, most finishing work took place in the castle town, where supplies and financing converged. A half-dozen lacquer shops in the town's commercial district stocked the expensive lacquers, procured both from tappers in the surrounding mountains and from other provinces.[13] Nearby were some sixty artisanal households (representing as many as two hundred workers), who were engaged in smoothing the turned wares and applying the lacquer.[14]

11. On the social organization of lathe-worker communities, see Fujita 1981:98ff.; Mizoguchi Tsunetoshi 1983:281.

12. This period saw a general shift of *kijishi* activity from southwest to northeast Japan, a migration reflected in miniature within Shimoina. Gravestones carved with the imperial chrysanthemum—a prerogative granted to the *kijishi* in the Heian period—can still be seen scattered throughout the area. Minamishinano-mura 1976:171.

13. In 1763, twenty horse loads of lacquer were brought into Iida from Izumi, south of Osaka in the Kinai region. See table 3.

14. Imai 1953. For further discussions of the early lacquer industry, see Tate 1983 and Shimoina Chiikishi Kenkyūkai 1982:49.

This relatively simple spatial structure of production belies the complexities of capital relations in the industry, however, for lacquer suppliers were not the only financiers involved. Competing with them for control of the lucrative trade were representatives of its two main groups of skilled workers: master lathe-turners (*wan'ya*), who doubled as foremen and financiers for the other *kijishi*, and lacquerware merchants, who distributed materials (both bowls and lacquers) to the artisans who produced the finished wares. The men who filled these three roles were constantly at odds with one another, petitioning the castle-town magistrates at various times for permission to circumvent each other's monopolies. In the end, however, the six lacquer suppliers in Iida emerged with the greatest profits in the trade, securing monopoly privileges on procuring and selling all lacquer in the fief in exchange for paying regular fees to the domain.[15]

THE TEXTILE SECTOR

In contrast to lathe- and lacquerware production, which essentially constituted a single two-stage enterprise, the Ina valley textile industry involved several distinct product lines. Also in contrast to the lacquerware complex, its spatial division of labor was not limited to the valley and its hinterland. Rather, both cotton and silk production were national industries, whose territorial organization encompassed broad areas beyond the region's borders.

The basic story of textile development in the valley, as in much of mountainous interior Japan, involved two contrasting lines of development. On the one hand, imported cotton gradually supplanted both hemp and locally grown cotton for local use. Thus, by the mid 1700s, despite scattered evidence of local cotton production in the early Edo era,[16] raw cotton constituted one of the most important imports into the Ina valley.[17] Beyond supplying their own needs, the women of Shimoina

15. Purchasing and stocking the expensive lacquers tied up significant sums of capital, but the domainal monopoly rewarded such investments handsomely. Some of the largest fortunes in late Tokugawa Iida were amassed by the town's lacquer *ton'ya*. Shimoina Chiikishi Kenkyūkai 1982:49.

16. Shimoina is known to have produced cotton cloth for tribute even before the start of the Edo period, and cotton is believed to have been grown as early as 1589 in central Shimoina's Toraiwa (Takagi) village. Hirasawa 1965a.

17. In 1763, some 2,200 horse loads of raw cotton grown in Mikawa province were imported into the castle town. While 85 percent of this total was merely re-exported to points farther north, the Iida area absorbed 343 horse loads of raw cotton, in addition to 135 loads of the woven cotton cloth and 200 loads of used clothing that were also brought into the area. Calculated from data in tables 3 and 7.

may by this time have been processing a modest volume of raw cotton for market. Willowing (*watauchi*, a process in which the seeds were removed from the raw cotton), proto-spinning (the making of *yoriko*, loose ropes or hanks of ginned cotton), and spinning eventually became common by-employments throughout the region, especially in the southern portion of Iida domain.[18]

Meanwhile, locally produced silk grew into an important interregional export, marketed to towns and cities in the surrounding area.[19] Like the growing trade in raw and semiprocessed cotton, the silk economy testifies to the increasing division of labor across Japan in the middle Tokugawa period. In this case, however, the relevant geographical context must be expanded yet further to include relations with China. Shimoina boasts the oldest tradition of silkworm rearing in Shinano province,[20] but it was the Bakufu's 1685 restrictions on imports of raw silk from Q'ing China, slashing Kyoto's main source of supply to one-third its former level, that opened the door for commercial production.[21] Less than a decade after the international import restrictions were imposed, eight expert female reelers were enticed to Shimoina from neighboring Mino province; shortly thereafter, shipments of raw silk from the Iida area to Kyoto's Nishijin district began.

In spatial organization, the silk industry represented something of a hybrid. At the national level, Shimoina's sericultural complex was part of a far-flung commercial network dealing in silk products; in fact, silk marketing (along with cotton) gave rise to some of the most extensive and elaborate wholesaling relationships of the Tokugawa period, in-

18. Cotton cloth was woven in the area as well, although mostly on a noncommercial basis. Local weaving was greatly boosted in the 1830s by the introduction of upright looms, which increased the productivity of Shimoina's weavers nearly fivefold. It has been alleged that the majority of rural households in Iida domain had such looms by late Edo times. Shin'yōsha 1984.

19. Except where otherwise indicated, the following account is based on Hirasawa 1952.

20. The first record of local sericulture, surviving in the remote northeastern village of Ōkawara, dates to the Kamakura period (1185–1333). Hirasawa 1952.

21. Raw silk imports from China peaked during the Kanbun years (1661–72) at 200,000 catties (300,000 lbs.) per annum; in 1685, the shogun limited them to 70,000 catties. Hoston 1986:111.

22. These debates took place during the 1930s between the so-called farmer-labor faction (*rōnōha*) and lectures faction (*kōzaha*) of Japan's academic Marxists. Hattori Shisō, a member of the latter group, cited developments in the silk industry to stress the revolutionary changes in Japan's domestic commercial relations during the Tokugawa period—changes that he saw as transforming domestic workers into de facto wage laborers, and

vesting the industry with considerable importance in later debates over the nature of Japan's protocapitalism during this period.[22] At the regional level, on the other hand, sericulture was confined to a relatively small area, concentrated in the valley core and in a few outlying villages along the Ina road. Unlike lacquerware or even cotton-thread production, silk-related activities did not lay the foundation for a broad territorial division of labor within Shimoina. Before the opening of Yokohama in 1859, sericulture and filature had become important byemployments only in the castle town and vicinity.

What limited production to this area was a combination of environmental and transport considerations. Silkworms flourished best in the mild climate of the valley floor, and the distance between sericultural and filature operations had to be minimized, for cocoons were highly perishable (it was not unknown for a load of cocoons to hatch moths before delivery). As a result, while Iida was a distribution center for a cocoon trade of some size, the hinterland of that trade appears to have been limited to villages along the major packhorse route leading into the castle town.[23] Once the thread had been reeled, of course, this constraint no longer operated. Durable, valuable, and light in weight, silk floss could profitably be transported over long distances—an essential precondition for the Ina valley's having found a niche in the national silk industry in the first place. Iida merchants were actively engaged in the floss trade, brokering thread that had been reeled outside the immediate area as well as that produced within the domain. The daimyo's efforts to share in the profits of the trade testify not only to its continuing importance through the period, but also to the geographical mobility of silk processing.[24]

Shimoina's role in the national silk industry was essentially confined to these first two stages of production (sericulture and reeling). Commercial weaving and dyeing was largely the prerogative of lowland metropolitan centers, where demand was concentrated; Iida exported only

---

commercial capital into industrial capital. Germaine Hoston (1986:110) summarizes Hattori's argument; for more on the broader context of this debate, see Hoston 1986, ch. 5; K. Takahashi 1978; and Yasuba 1975.

23. For further discussion, see Masaki 1978:58ff.

24. Attempts to regulate the industry were rife. For instance, reelers were not to allow others from outside the fief to buy up cocoons in their names, and it was illegal either to act as an intermediary in such dealings or to lodge would-be cocoon buyers from other domains. Later, in 1823, the *han* further forbade skilled reeling women from leaving the domain to work elsewhere. That such laws were considered necessary suggests that the spatial organization of silk processing exhibited considerable flexibility.

a single horse load of locally produced silk textiles in 1763. To serve the local market, however, a small weaving district did develop. The most popular local cloth was *tsumugi*, a nubby, textured material made throughout the sericultural districts from the broken silk fibers that did not merit transport costs to Kyoto. *Tsumugi* was intentionally woven in such a way as to minimize its sheen so that it could be mistaken at a distance for cotton—an attractive feature for commoners, who were technically forbidden to wear silk under the country's sumptuary laws. The growing market for textiles within the valley supported a small number of specialty dyeing establishments as well, concentrated in the area just north of Iida.[25]

As this brief overview suggests, Shimoina's lesser protoindustrial complexes were organized on strikingly different scales. Lacquer work was essentially a local operation. Except for the procurement of lacquer itself, all stages of the hardwood industry—from lathe-turning to finishing—took place within the Ina valley and its immediate hinterland. Textiles, by contrast, spawned consistently broader commercial networks. In the larger branch of the industry, Shimoina women processed raw cotton grown far to the south of the region's borders, spinning and weaving it for local (often domestic) consumption. In the smaller silk sector, by contrast, local households engaged in primary (sericultural) as well as secondary (filature) processes, partly for extraregional export. In neither textile line, however, did the Ina valley market finished textiles. Even with silk, Iida's role was only to supply semiprocessed materials to higher-order finishing centers in Kyoto.

These contrasts in extraregional integration, however, should not obscure a fundamental commonality in the intraregional structure of these diverse enterprises. Each climaxed in the castle town, where merchants and suppliers congregated. Each also evolved some degree of hierarchical exchange linking those urban financiers with outworkers in surrounding villages. In schematic terms, the landscape of lacquerware and textile production might thus be conceived as a series of overlapping circuits of production and finance—each with its own distinct shape and

25. From early Edo times, the Iida region had supported a market for simple indigo dyeing, using local materials at first but later importing superior indigo leaves from Ise. As the wealthier class of commoners came to wear crested kimonos, the call for more elaborate dyeing increased. Shrine festivals supported the dyeing industry as well, creating a market for elaborate banners, gaily colored *happi* jackets, and draperies. Shimoina-gun 1911:66.

size—which intersected in the commercial nexus of the castle town.[26] This sharply articulated linkage between an urban core and its rural periphery was arguably the most marked spatial feature of Shimoina's lesser protoindustrial complexes.[27]

The Paper Crafts

In a general way, similar patterns may be seen in the region's paper complex, whose dimensions overshadowed both the hardwood and textile crafts. While the area's lathe-turners, lacquer artisans, spinners, and reelers are counted in the tens and hundreds, those who labored to produce the region's hairdress ties, decorative cordage, umbrellas, and craft paper numbered in the thousands.

The establishment of such a large-scale commercial paper complex in southern Shinano was in some ways surprising. As a survey by the German geographer J. J. Rein in the 1880s suggests, most of the best-known and most important Japanese papers of early modern times were produced in densely settled agricultural zones, close to the country's main markets or within easy reach of maritime shipping lines.[28] The rise of the Ina paper crafts accordingly represents something of an anomaly. As the following investigation will show, the process by which this protoindustrial complex took hold in central Shimoina was intimately linked to the area's fragmented political terrain.

Partly because of its ties to domainal patronage, the paper industry is relatively well documented, allowing its territorial organization to be reconstructed in detail. The following discussion follows a geographical as

26. Compare William Cronon's (1991:278) description of Chicago at the turn of the century: "The city's hinterland was actually thousands of overlapping regions, each connected in myriad ways to the thousands of markets and thousands of commodities that constituted Chicago's economic life. Each different commodity had unique sources of supply and demand—and hence a unique set of environmental linkages to the natural world."

27. The importance of cities as coordinating points in the protoindustrial landscape is increasingly recognized; as Jan de Vries (1984:8–9) notes, "Large-scale rural-industrial production for distant markets depended on communications and co-ordination, that is, on cities." For related arguments, see Gutmann 1988:20, 110; Hohenberg and Lees 1985; Kriedte, Medick, and Schlumbohm 1986; Lees 1989; Poni 1985:313.

28. In addition to Shikoku's Iyo and Tosa (which together accounted for 40 percent of traditional paper production at the time), J. J. Rein noted significant sites in the provinces of Yamato, Suruga, Uzen, Mino, Musashi, Kai, Izu, Iwaki, Echigo, and Echizen. Rein 1889:402–3. On paper production in Tosa, see Roberts 1991, ch. 6.

well as temporal progression, situating Shimoina in the national map of Japanese papermaking before narrowing the focus to its intraregional organization. The narrative culminates with an early-nineteenth-century battle over taxation rights: one territorial conflict whose resolution would not redound to Iida's benefit.

THE ORIGINS OF THE INA VALLEY
PAPER INDUSTRIES

In contrast to the clear environmental logic of the Shinano lacquerware industry, natural resources cannot wholly explain the development of the province's paper crafts. Tokugawa paper production depended on three ecological resources: mulberry bark, a steady supply of pure water, and winter sunshine. But many areas of central and western Japan, including the lower Ina valley, offered all three in abundance. To convert a minor tribute item into a major regional export, a series of politico-economic forces had to converge on the valley as well.

If the development of the packhorse trade was one such force, competitive domainal mercantilism was another. The direct instigators behind Shimoina's paper industry were from the valley's warrior elite: a pair of domain lords whose headquarters faced each other across the Tenryū River. The more important of the two was Hori Chikasada, the heir to the domain of Iida, whose family had been moved to southern Shinano in 1672 from an area near Edo. The original Hori fiefdom had supported a flourishing paper industry, boasting not only papermaking but also a lucrative sideline: the production of paper hairdress ties (*motoyui*), which came into vogue among both men and women of the samurai class after 1600.

Shortly after the Hori family was installed in Iida, Chikasada is credited with having taken several steps to create a *motoyui* industry in the Ina valley. He allegedly enticed skilled artisans to relocate to Iida from the neighboring province of Mino, effecting the necessary transfer of technology to the region; he allocated an area near the castle town for the requisite outdoor facilities; and most important, he ordered his lesser retainers to take up *motoyui* production, mobilizing a sizable pool of labor.[29]

At first, Iida's hairdress ties were made with paper imported from Mino. But before long, *motoyui* production served as the catalyst for a

---

29. Imamaki 1959. According to the *Iida Saiyakuki*, a chronicle believed to have been compiled in the 1730s, the first local *motoyui* were produced in 1692. Hirasawa 1969, part 1:30.

local craft-paper (*sarashigami*) industry as well. Papermaking on a small scale had been recorded at scattered villages throughout the region during the medieval period, but the demand generated by Hori's hairdress-tie industry expanded this minor craft into a major regional industry. As such, papermaking, like hairdress-tie manufacture, was clearly boosted by the local daimyo's mercantilist policies. Yet in this case the connection remained indirect. Based exclusively on peasant household labor, this second sector of the local paper complex was less indebted to domainal initiative and less susceptible to domainal control.

As papermaking began to flourish in Iida domain, a lesser feudal authority named Chiku, whose holding lay to the east across the Tenryū River, began encouraging his retainers to produce a related item, paper umbrellas (*bangasa*). The technology transfer necessary to initiate local umbrella production evidently took place without his intervention; according to local lore, an umbrella maker from the Gifu area fell ill while traversing Chiku's territory and, to repay the barrier guards who nursed him back to health, taught them the craft. Thereafter, however, Chiku—like Hori—played a strategic role in mobilizing samurai labor for the nascent industry. As with hairdress ties, umbrella manufacture was in fact established in the Ina valley primarily to address the fiscal imperatives of the local elite: not only the mercantilist requirement of generating export income for the domains but also the need to augment the stipends of hereditary retainers.[30]

Regular access to metropolitan markets, assured by the packhorse network, boosted paper production in the Ina valley well beyond its original basis in domainal patronage. Commoners as well as samurai soon became involved in manufacturing paper as well as hairdress ties and umbrellas for sale outside the region. In addition, another related industry was eventually established, independently of political support, producing colorful paper cordage (*mizuhiki*) used for decorative and ceremonial purposes.[31] Although the scale of production for decorative cordage and umbrellas would remain relatively small until the nineteenth

30. Shinano Kyōikukai Shimoina Bukai 1934:194–96. For members of the warrior class to engage in commodity production may seem unorthodox, but it was hardly unusual. As financial hardship struck domains across the country in the mid to late Tokugawa era, side employment became "common for all but a limited number of high-ranking and well-off samurai." Yamamura 1974:48.

31. Deployed in Shintō rites since the Heian period, *mizuhiki* were more commonly used in the Edo era for secular purposes, usually being tied into elaborate knots and affixed to gifts and ritual displays. Shinano Kyōikukai Shimoina Bukai 1934:128.

century, the hairdress-tie and papermaking industries quickly became major elements of the region's commercial economy. The 1763 survey of the packhorse trade revealed that, in addition to 19 horse loads of umbrellas shipped to interior Shinano, Iida was already exporting some 208 loads of paper and fully 444 loads of hairdress ties annually to ur- ban areas outside the region.

THE ANATOMY OF PRODUCTION

Narrowing our scope to the valley and its environs brings into focus the configuration of the paper industry within Shimoina. Map 13 shows the distribution of the major paper and related craft produc- tion sites in the area. It also demonstrates their relationship to two other types of specialized settlements: the outlying packhorse villages, and eight settlements near the core identified as commercial satellites. The satellites were important rural wholesaling centers, mostly along the perimeter of the castle town—villages of a type unhelpfully referred to in the Japanese literature as "rural places" (*zaigō*), but which in fact coor- dinated most of the area's protoindustrial production and marketing.

A close look at this map shows that, within the commercial heartland, each sector of the paper complex formed a distinct spatial cluster. Um- brella production was confined to a single community east of the Tenryū, decorative cordage to a single village in the west; hairdress ties were made at only two sites on the southern edge of the castle town; and even the most diffuse craft, papermaking, was confined to half a dozen contigu- ous settlements near the banks of the river. Probing the logic behind these locational patterns reveals that a distinct confluence of political, environmental, and commercial imperatives shaped the geography of each sector, putting the configuration of the whole slightly, but signif- icantly, off-center with respect to the region's political map.

The production sites for hairdress ties and umbrellas testify clearly to their origins in domainal patronage. Both industries were centered ad- jacent to the headquarters of their respective patrons, or more precisely, next to the residential quarters of their retainers. The immediacy of this relationship was possible, however, only because neither craft was sig- nificantly affected by ecological considerations. The major material in- put for both was paper itself—a high-value, lightweight good that could bear the cost of transport over considerable distances—and the labor processes in both crafts left them relatively free from environmental con-

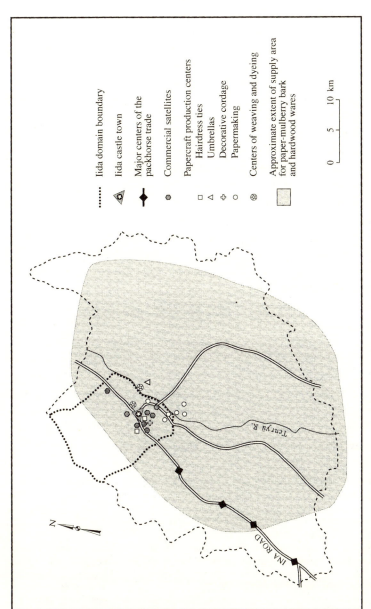

Map 13. The Territorial Division of Labor in the Shimoina Region, Late Tokugawa.

Iida domain boundary

Iida castle town

Major centers of the
packhorse trade

Commercial satellites

Papercraft production centers
  Hairdress ties
  Umbrellas
  Decorative cordage
  Papermaking

Centers of weaving and dyeing

Approximate extent of supply area
for paper-mulberry bark
and hardwood wares

0   5   10 km

Tenryu R.

INA ROAD

N

straints.[32] The only direct physical requirement was the need for a sunny area in which to stretch and dry the paper cords. Any south-facing slope in the valley where frames could be erected offered sufficient sun exposure to accomplish this task. Since the Tenryū tributaries had carved numerous ravines into the terraces on either side of the river, an abundance of suitable sites was available. The choice among them could accordingly be determined by other than environmental considerations. It is no coincidence that most of the industry's frames were to be found on the slope that formed the southern boundary of the castle town; lying immediately adjacent to the largest block of samurai residences, this area was particularly convenient to its core labor force.

Umbrella manufacture was similarly unconfined by physical requirements. Climate played no significant part in umbrella production, which did not require large amounts of water and took place almost exclusively indoors. Furthermore, most of the resources used in the craft—including paper, hardwood hubs, persimmon juice, and rapeseed oil—were required only in relatively small amounts, and all were easily transported.[33] For these reasons, umbrella making, like the hairdress-tie industry, was essentially free to mirror the political terrain in which it was nurtured.

Papermaking and decorative cordage, however, represent a sharp contrast. Both were independent commoner undertakings, whose geography was shaped not by patronage ties but by the distribution of ecological, commercial, and demographic resources. Papermaking in particular was significantly constrained by the physical environment. The paper mulberry might be grown almost anywhere in the region, but its processing required both winter sun and an ample supply of pure water.[34] Sunshine, critical for timely drying of the wet pulp sheets, was simply more readily available in the Ina valley than in the surrounding mountains. The valley floor enjoyed warmer days and less winter precipitation than the surrounding uplands.[35] At the same time, the springs at the base of the terraces on either side of the Tenryū River yielded a highly reli-

32. See ibid.: 105ff., on the elaborate labor processes involved in hairdress-tie manufacture; cf. Pomeroy 1967:33.
33. Ecological conditions were not wholly irrelevant to umbrella production. Different varieties of young bamboo were preferred for spokes and handles, and the Chiku fief's position at the intersection of their respective source areas may have afforded a slight geographical advantage for umbrella production. The point should not be stretched too far, however; umbrella making was successfully undertaken elsewhere in the valley in other periods. Shinano Kyōikukai Shimoina Bukai 1934:155–61.
34. For a detailed discussion of the papermaking process, see Shimohisakata-mura 1973:428ff.; cf. Rein 1889:399ff.
35. On climatic variation across Shimoina, see Anan-machi 1987:159–221.

able year-round supply of pure water. It was these environmental resources that drew Shimoina's paper manufacturing to the cluster of settlements shown in map 13, all of which were located within a relatively
compact zone straddling the Tenryū south of Iida.[36]

Decorative-cordage production, too, was initiated independently by
commoners, and its location was also dictated in part by ecological constraints. The process of manufacturing cordage was identical to that for
hairdress ties, the only difference being that cordage was made from
lower-quality paper scraps. As with hairdress ties, so with decorative
cordage, the requisite outdoor work spaces (for stretching and drying
the damp spills of paper) had to be cleared from a south-facing ravine
slope. In this case, however, the chosen slope was located, not at the
edge of the castle town, but farther south, adjacent to the papermaking
zone. Labor was available here because the local villagers, who lived at
the top rather than the base of the terrace, lacked the necessary springs
to make paper (generally a more profitable undertaking).[37]

In sum, even within Shimoina's compact industrial core, each sector
of the paper complex occupied a highly specific niche. At a regional level,
centripetal forces acted powerfully on the complex as a whole; in addition to the concentration of labor, capital, and political elites in the valley's core, the desirability of proximity to other elements of the area's
integrated protoindustrial economy contributed to this effect.[38] But the
precise location of each step in paper processing was determined by a
conjunction of political, environmental, and commercial forces unique
to that craft. In practice, this meant that those sectors established by
commoner initiative were not confined by political boundaries, despite
important indirect linkages to a local lord's patronage policies. Papermaking in particular followed a locational logic dictated primarily by
ecology and demography, which created suitable sites on both sides of
the Iida border. The resulting mismatch between the geography of production and the geography of domainal power—the variance of bound-

36. Perhaps several hundred peasant households in this area were able to make paper
each winter in their courtyards. Shinano Kyōikukai Shimoina Bukai 1934:200.
37. Interestingly, this village was also involved in paper production in another way:
processing paper scraps into decorative cords at home. Ibid.:130.
38. The effect in question is essentially an external economy of scale, or "agglomeration economy," where the advantages of a shared labor pool, ease of information flow,
and proximity to suppliers and consumers encourage firms to locate near other enterprises
engaged in the same or related work. Taaffe and Gauthier 1973. Agglomeration economies
have been extensively theorized in contemporary industrial and regional geography; for a
review of recent work, see Schoenberger 1989.

aries alluded to above—would have profound implications for the subsequent development of Shimoina's economy.

THE PROTOINDUSTRIAL LANDSCAPE AS CONTESTED TERRAIN

By the turn of the nineteenth century, the production of paper wares had become the commercial lifeblood of the lower Ina valley, remaking the local landscape in its image. Settlements throughout the mountainous arc surrounding the Ina basin had begun growing paper-mulberry bark for the commercial manufacturers located closer to the core. A second tier of villages specialized in supplying craft paper to artisans in the inner ring of settlements around the castle town, where finished products were made. But protoindustrial production for national markets not only reconfigured labor and resource use throughout the area; it also introduced conflict between the political and commercial orders that had conspired in the industry's rise. The ultimate inability of the former to control the latter was illuminated most decisively in the so-called Paper Monopoly Riot (*kami ton'ya sōdō*) of 1809.[39]

In 1807, a merchant from one of the rural centers of the Iida paper industry requested the right to establish a monopoly on all wholesale paper transactions in the domain. Merchants in the castle town were furious, and quickly countered with their own proposal. Two years later, they had secured the domain's approval to form an exclusive paper dealership, or *kami ton'ya*. From now on, all wholesale paper transactions were under the monopoly's jurisdiction; no one was allowed to conduct such transactions without a license issued by the new *ton'ya* organization. This included not only paper dealers but all hairdress-tie and cordage manufacturers in the area—except for those within the castle town, who would be allowed to buy the paper they needed without paying surcharges.

The new ruling created serious hardships for numerous participants in the disrupted paper-craft network. Most disgruntled were residents of Imada village, an important center of paper manufacture on the east bank of the Tenryū River, beyond the limits of Iida domain. Under the new system, Imada workers had no choice but to procure their inputs from, and sell their finished products to, an agent in the castle town, requiring an inconvenient half-day journey across the Tenryū. To aggra-

39. The following account is adapted from Hirasawa 1972a, ch. 2; Imamaki 1959; Shiozawa 1987–88.

vate their frustration, the advances they were accustomed to receiving from their previous broker were no longer forthcoming; requests to borrow advances from the new paper *ton'ya* were denied. These grievances finally prompted Imada residents to stage an uprising against the new monopoly, in which they managed to inflict considerable property damage. Since local authorities had no jurisdiction over the instigators, who resided in Bakufu territory, the shogun's officials were called on to try the offenders.

The ensuing deliberations, while resulting in punishment for those who had destroyed property, turned out very badly for the domain. In its judgment of 1815, the Bakufu determined that Iida officials had overstepped their legal bounds in setting up a monopoly with such wide-ranging prerogatives. Simply put, no daimyo had authority to tax economic activities that took place in surrounding fiefs. Although the new monopoly could continue to assess a surcharge on paper produced within Hori's territory, it had no legal jurisdiction over papermakers from Imada or any other village outside its boundaries. As a result, while producers within Iida domain were ordered to continue paying the surcharges as before, all fees collected from villages beyond its borders were to be restored. In all, forty-five Imada papermakers received restitution in the settlement. By reinforcing the geographical limits of domainal tax-collecting powers, the shogunate had reminded all concerned that this integrated commercial circuit overlay fragmented political terrain.

In such a context, the creation of a domainal monopoly backfired. Iida succeeded only in imposing a burden on its own papermaking villages, putting those inside the fief's borders at a disadvantage relative to their neighbors in other territories. Within a decade, papermakers in the domain were complaining that the additional surcharges were ruining them. The tax-free zone beyond the pale had been given a decisive competitive advantage—one that eventually forced much of the papermaking business out of Iida territory. While a few smaller villages maintained their production levels, all the large traditional papermaking centers in the domain went into decline.[40]

To make matters worse for Iida, the regional struggle over profits soon surfaced in other areas as well. In 1821, for instance, the castle-town merchants tried with mixed success to edge all brokers from the satellite villages out of the lucrative Edo market for hairdress ties. But the compass

40. The number of papermakers in Iida domain dropped sharply in the years following the imposition of the paper tax, from 259 in 1810 to just 179 two decades later. Imamaki 1959, part 2.

of competition had already widened sufficiently to render this a largely anachronistic gesture. By the 1810s, what threatened Iida's cordage merchants was less an access of brokers than a dearth of workers. Shimoina was now faced with rising competition for its skilled young artisans, who were being lured away to newly established production sites in Mikawa and Echizen, outside of Shinano province altogether. In 1819, the castletown merchants responded by forbidding all cordage makers to leave the domain. When this proved unsuccessful, new petitions were submitted to disallow the pawning of balls of paper twine, the proceeds from which were financing young apprentices' illicit escapes.[41]

From the domain's viewpoint, then, an apparently favorable settlement had proven a Pyrrhic victory. Although it had managed to establish a revenue-generating monopoly on wholesale paper sales within the domain, Iida had won the battle only to lose the war. Yet from a national perspective, the fact that local daimyo proved limited in their ability to squeeze revenues out of Japan's emergent protoindustrial networks was not necessarily to be mourned. The net result of such failures, if comparative cases are any indication, may have been to open up space for autonomous commercial development.

Conclusion

Articulating the interplay between state and capital in geographical terms reveals an important discrepancy between the two. While both political and commercial institutions are territorial,[42] they exhibit distinctive spatial structures. The former typically take a fixed, contiguous, and clearly bounded areal form; the latter tend rather to be "linear," integrating production and exchange at scattered points across a widely variegated landscape. In addition, commercial networks are more fluid than their political counterparts; the spatial peculiarity of capital is precisely its mobility. In a competitive state system, the resulting geographical mismatch between the political and commercial orders can work to the benefit of those who wield mobile capital. As Fernand

41. Ibid.
42. I am using the concept of territoriality here in the sense specified by Robert Sack (1986:19): "the attempt by an individual or group to affect, influence, or control people, phenomena, and relationships, by delimiting and asserting control over a geographic area."

Braudel has observed, commerce is "a game played on an infinitely wider plane than . . . that of the state and its particular preoccupations."[43]

Recent work on the development of early industry in western Europe confirms the salience of Braudel's observation. To be sure, the spatial dynamic of protoindustrialization is most often conceived of as located along an urban-rural axis; the migration of craft production from towns to the countryside is consistently stressed as lying at the core of the protoindustrialization process. Yet while it is given less theoretical attention, the significance of variations in sovereignty for promoting and shaping that process—both directly, through industrial patronage, and indirectly, through inducing periodic shifts in production sites—is clearly evident in many empirical studies.[44] Comparing developments in late medieval Tuscany and Sicily, the Italian historian S. R. Epstein stresses the importance of political pluralism for promoting economic diversification and growth in an earlier period as well: "Tuscany suffered from Florence's nearly unfettered domination. While establishing its territorial state, Florence encountered no rivals, had little need to come to terms with competitors, and so was able to substitute a general monopoly for the previous multiplicity of partially competing ones." In Sicily, by contrast, a "plurality of powers allowed for greater competition between formally constituted economic interests, and provided interstices between conflicting demands which could be exploited in a productive fashion."[45]

In Japan as in western Europe, a protoindustrial, protocapitalist economy arose in a pluralistic political landscape. The fossilization of an essentially feudal geography under the Tokugawa regime inspired deliberately competitive mercantilist policies, strikingly similar to those of European principalities; attempts by provincial lords to stimulate local industry for supraregional markets, and to sustain a positive trade balance with neighboring domains, are a staple of the Tokugawa literature.[46] The same dynamic is evident in Shimoina, where local feudal authorities initiated or supported several commercial enterprises, whether by enticing skilled men and women from neighboring regions to immigrate

43. Braudel 1982:554. See de Vries 1984, especially the final chapter, for a discussion of economic competitiveness in the context of premodern European regional and urban development. C. Smith (1976c:353–67) offers a general discussion, including several suggestive passages on Tokugawa Japan.

44. For a general discussion of this point, see Gutmann 1988:7, 73–75.

45. Epstein 1991:46, 49.

46. The most extensive recent analysis of domain mercantilism is Roberts 1991.

to their domains, by mobilizing their retainers' labor, or by donating industrial sites. Like their European counterparts, they also attempted to manipulate the geography of both processing and wholesaling in such a way as to benefit their own accounts.

Yet domainal attempts to manipulate industrial geography were not always successful. The slowing of growth during the eighteenth century in the highly regulated castle towns, with the concurrent rise of industrial and commercial activity in the countryside, suggests the fluidity of the country's commercial networks in response to local price differentials.[47] As in contemporary Europe, the result was a continual tension between the political and the commercial orders—one whose outcome redounded at least at times to the benefit of the latter. As the mutability of Shimoina's paper industries suggests, the same political pluralism that encouraged industrial patronage and mercantilist trade policies could also limit local authorities' powers over the commercial realm; in Japan as in Europe, an uneven political terrain generated intense competition, not only in creating commercial ventures but in retaining them as well. Both imperatives must be taken into account if we are to understand the nature of Tokugawa protoindustrial and protocapitalist development.

Reconstructing Japan's regional political economy in this way allows us to reconceive conflicts between lord and merchant throughout Japan as a clash between two qualitatively different sociospatial orders: on the one hand, a rigid geopolitical hierarchy, and on the other, a more fluid set of commercial circuits organizing production and exchange. Southern Shinano, with its unusually small domains, admittedly represents a skewed sample of the Tokugawa countryside. Numerous daimyo elsewhere controlled much larger holdings than those found in the Ina valley, and one can only expect that a different relationship between lord and merchant emerged in regions where a single domain was large enough to embrace a relatively integrated regional economy. Moreover, we know little as yet about the range of merchants' potential responses to local political conditions, the nature of environmental limits and other determinants of industrial location, or the extent to which actual or threatened "capital flight" may have shaped the development of the Tokugawa economy as a whole. Nonetheless, I would argue that Iida and its neighbors represent in microcosm the juxtaposition of unequal domains that characterized Japan *as a whole*, making Shimoina a useful field in which to explore the nature of those incongruities between po-

47. T. Smith 1973.

litical authority and commercial economy that pertained at all levels of the early modern Japanese economy. The partitioning of the political landscape was certainly carried farther in Shinano than in many other provinces, but the situation of the country at large was structurally analogous to that found on a smaller scale in the Ina valley.

If the political fragmentation that played a role in western Europe's economic growth contributed even in a limited way to the dynamism of the Tokugawa economy, however, it did so under three significantly altered circumstances. Highlighting those differences in turn provides an opportunity to rethink the conclusions that have been drawn from the European literature.

A first critical difference was the absence of warfare. In contrast to every state in early modern Europe, Tokugawa Japan remained free of military engagements for over two centuries. Moreover, for the same two hundred years, the Tokugawa regime succeeded in enforcing its ban on private vendettas between feuding lords. That aggressive mercantilist policies arose even in this distinctly unwarlike context raises the question of whether Europe's bloody battles were in fact indispensable to its economic growth.[48]

Sharp restrictions on foreign trade were a second distinguishing feature of early modern Japan. Tokugawa "seclusion" may be a myth; it is now known that the shogun engaged in active diplomacy and managed exchange throughout the period.[49] But the volume of Japanese commodity trade (especially after the 1680s) fell far short of Europe's, a condition that has often been decried as a setback for Japanese economic development. E. H. Norman, for instance, called attention to "the comparative weakness of the Japanese merchant who lacked such opportunities for the accumulation of capital through trade and plunder as were enjoyed by his counterpart in 16th–17th century Europe."[50] A more nuanced interpretation, I suggest, would recognize the benefits of those policies as well. In addition to stabilizing the domestic political order, curtailing foreign trade created important opportunities for import substitution. Shimoina, for

48. The notion that Europe's recurring wars stimulated technological innovations and mercantilist policies is articulated by scholars across the political spectrum, from Perry Anderson (1974:36ff.) to Nathan Rosenberg and L. E. Birdzell (1986:137). But the historian Joel Mokyr, for one, disagrees, contending "it was peace, not war, that was the innovating force in manufacturing," and concluding that "war and military preparation did not add conspicuously to the material prosperity of Europe" (Mokyr 1990:185).

49. Tashiro 1981; Toby 1984; see also Jansen 1992.

50. Norman 1975:158. For similar claims, see Braudel 1982:592–93; Anderson 1974:494.

one, directly benefited from the slashing of Chinese silk imports in 1685. The resulting surge in demand for the limited domestic supply of cocoons allowed commercial sericulture and silk reeling to take root in this area, laying the groundwork for southern Shinano's participation in what would later become Japan's primary export industry.

But perhaps the most novel spin the Japanese case puts on the postulated importance of political pluralism comes through the role of its central government. The Tokugawa shogun was not a true sovereign but a suzerain; in the memorable words of Eric Jones, the Bakufu was little more than "broker for a tier of regions quite able to sustain and magnify their growth if they were brought together, offered the opportunity of big urban market growth, and otherwise mostly left alone." And mostly left alone they were: "Collectively the daimyo held one another and the Bakufu in check."[51]

What eludes this otherwise apt formulation, however, is the extent to which the Bakufu also kept the daimyo in check. The Tokugawa shogun may well have had little capability to force his positions on domain lords; as Philip Brown has put it, the Bakufu was in many ways a "flamboyant state" notable more for its hortatory proclamations than for effective intervention in the affairs of the nation.[52] Yet one essential tenet of the shogunate's domestic rule was its insistence on mediating interdomainal disputes. To the extent that it was effective in doing so, central power in the Japanese case ironically acted to limit rather than enhance direct (local) governmental control over the economy.

What we find in Tokugawa Japan, then, is the paradox of a central government intervening precisely to enforce decentralization.[53] Through judgments like the one of 1815 defending Imada villagers from taxation by the daimyo of Iida, the Bakufu could effectively prevent domain lords from extending their political reach as far as would have been necessary to control the economic reticula that had been cast across their lands. And what they could not control, they could not destroy. By preventing local authorities from strangling a growing commercial economy, political pluralism in Japan, as in western Europe, appears to have given mercantile networks crucial opportunities to gain ground.

51. Jones 1988:166.
52. Brown 1991.
53. This in turn casts a new light on the debate over Tokugawa "absolutism"; see Berry 1986 and White 1988. It is also possible, however, that such policies generated political hostilities that helped destroy the regime in the long run. I am indebted to Thomas Smith for suggesting this point.

CHAPTER FOUR

Spatial and Social Differentiation

Students of Japanese history have debated for decades whether the agricultural, demographic, and commercial developments of the Tokugawa period enriched or impoverished the average peasant. Attempts to answer this important question have generated heated controversies, notably over the extent of growth and the equity of its distribution. At present, most scholars are agreed that, in aggregate terms, agricultural yields did outstrip population increases, leading to a rise in per capita output (or "intensive growth"). But the debate over how equitably those yields were distributed is less easily resolved.[1]

This chapter proposes a historicogeographical approach to the distribution controversy. Rather than generalize about the well-being of Tokugawa society as a whole, or even of particular strata within it, I submit, we need to reformulate the equity question in spatial terms. Here, too, geography mattered. From one part of the landscape to the next, not only the extent but the very form of social differentiation varied considerably.

Working with data from the Ina valley, I have attempted to sort out those variations on two contrasting scales. The first half of the chapter sets up a schematic spatial grid for the region as a whole, distinguishing

1. As David Howell (1992:269) dryly notes, the term "revisionist" no longer seems apt to describe the positive assessment of Tokugawa economic history, heralded in English more than two decades ago by John Hall (1968; see also Hanley and Yamamura 1971; T. Smith 1973; Yamamura 1973). For reflections on the achievements and limits of scholarship in this vein, see Morris and Vlastos 1980; Totman 1986.

four basic settlement types with broadly similar development paths. Two types of communities that lay off the major trade routes, where commercial opportunities were largely limited to either agriculture or forestry, are contrasted with two types of crossroads settlements or market towns, whose commercial economies were more diverse. Published demographic data for the region are then marshaled to demonstrate that each of these major settlement types exhibited a distinctive trajectory of social development.

The second half of the chapter narrows the focus to the regional core, where the greatest opportunities for accumulation—and the greatest disparities of wealth—were concentrated. Attention centers on two contentious and interlocking relationships in the protoindustrial economy: one between merchants and laborers, and the other between merchants in the castle town and their rural counterparts. These would-be captains of the region's commercial networks fought each other repeatedly to secure monopolies over wholesaling rights, the key to accumulation in the protoindustrial economy. Tracing their conflicts reveals once again the mismatch between commercial and political geography, which is visible not only at the boundaries but at the center as well. Merchants inside the castle town were simply unable to assert control over all the protoindustrial complexes of the Ina valley; just as the areal extent of commercial enterprise exceeded the reach of political institutions, so the center of commercial power was slightly but significantly offset from the seat of political authority.

While the discrepancy at the boundaries (as argued in the previous chapter) enabled the commercial reticula to elude domainal taxation, the comparable offset at the core created a space in which rural elites were able to maneuver their way toward increasing power and privilege. In the end, the satellites grew to rival Iida—partly by the exuberance of the less regulated commercial life beyond the castle-town walls, but equally by forces promoting social differentiation there. It was this dynamic, whereby the profits generated in the satellite villages became concentrated in a few merchants' hands, that paved the way for the entrance of those communities' wealthy commoners into the inner sanctum of Shimoina's merchant elite. At the same time, it also set the stage for increasingly frequent protests by craft workers against the privileges of the few.

Before proceeding with the analysis, two disclaimers are in order. First, the simple spatial grid proposed here represents a deliberate departure from the mathematically generated, multitiered hierarchy of central places deployed by regional systems theorists.[2] Although the latter offers a package of powerful tools with which to categorize settlements and analyze the resulting spatial hierarchy, I am skeptical about the value of applying those tools to the Shimoina landscape, on both empirical and epistemological grounds. From a practical point of view, I have found it imperative to work with a more modest set of categories in order to take advantage of the scattered, uneven data available. More fundamentally, the attempt to delineate a fixed settlement hierarchy presupposes a kind of structural stasis for the Tokugawa period as a whole, a stasis that must be demonstrated rather than supposed. And even where a settlement hierarchy remains relatively stable, central-place networks may represent administrative artifacts overlaying a surprisingly fluid and elusive agrarian landscape.[3] In any case, my project is fundamentally different from that of the regional systems theorist: rather than seeking to analyze the Japanese spatial order at a fixed point in time, my aim is to suggest the developmental trajectories of different types of communities through a century or more of economic evolution. The axes of differentiation chosen here reflect this agenda, being rooted in historical processes with particular relevance for this region.

A second disclaimer concerns the limitations of the demographic analysis. A full-scale investigation of the copious demographic materials surviving for Shimoina is beyond the intent of this study; my aim here is not to probe the intricacies of household structure and decision making, but to trace the general contours of demographic change as part of a broad canvas of historicogeographical transformation. In order to cast as wide a net as possible, I have consulted only published demographic data, mostly drawn from the thirty-odd village histories for the region.[4] A survey of this kind cannot claim to be conclusive; not only are the data incomplete, they also support several competing hypotheses. If I have been emboldened nonetheless to draw attention to consistent geographical correlations, it is in the hopes of pointing the way to a more nuanced assessment of how Tokugawa economic development was lo-

2. On the evolution of the central place model, see chapter 1, note 19 above. The pioneer in applying central place theory to the study of East Asian societies has been G. William Skinner (e.g., 1977).

3. For a brilliant exposition of this argument in the context of southern India, see Ludden 1990; cf. Thomas Keirstead's observations on the "public realm" (*kokugaryō*) of medieval Japan as a series of nodal points in a fundamentally fluid social realm (1992:17).

4. For judicious assessments of the nature and limits of the temple registers from which these numbers are abstracted, see Hayami 1979 and Cornell and Hayami 1986.

Table 10 *Settlement Typology for Late Tokugawa Shimoina, Showing Primary*
Commercial Opportunities in Each Settlement Type

	Inside the Core	Outside the Core
On the main trade route	*Satellite* Protoindustry Commercial cropping Food processing	*Outpost* Transport
Off the main trade route	*Basin village* Industrial crops Food processing	*Mountain village* Forestry

cally experienced—and spatially patterned. For as even a provisional analysis attests, class relations in the Tokugawa countryside were deeply intertwined with regional relations, rendering discussion of social and spatial differentiation truly inseparable.[5]

The Twin Axes of Settlement Differentiation across Shimoina

The later Tokugawa division of labor across Shimoina evolved through the intersection of two distinct axes of development: a linear pattern, promoting commercial enterprise along the major trade routes, and a centralizing dynamic that enhanced population growth and capital accumulation around the castle town. Through the intersection of these two structuring principles, Shimoina's villages became specialized into four fundamental types, whose commercial sectors were oriented primarily toward protoindustry, transport, agriculture, and forestry, respectively (table 10).

5. It should be noted that in addition to the villagers and castle-town residents whose social relations are analyzed in this chapter, Shimoina also supported roughly a hundred households classified as outcastes (*eta* or *burakumin*) and an unknown number of migratory, gypsylike folk locally known as *pon*. On the local *buraku* communities, see Tsukada 1986:143–45, 349; on *pon*, see Furusato no Imeeji Kankōkai 1971:108–9.

The categories appearing in the second row of table 10, distinguishing between villages oriented predominantly toward agriculture and those whose primary commercial opportunities lay in forestry, need no explanation. Rooted in physical geography, these are ubiquitous terms in the discourse about Japanese villages of the period. Satellites and outposts, by contrast, constitute a more unusual dichotomy. The emergence of these settlement types was a historical rather than ecological phenomenon, a product of the Ina valley's development as a trade corridor and protoindustrial region. The following discussion briefly traces the origins of specialized transport centers along the Ina road before returning to the fourfold typology and how it organized social differentiation.

THE RISE OF SPECIALIZED TRANSPORT CENTERS ALONG THE INA ROAD

In the seventeenth and early eighteenth centuries, transport provisioning for Shimoina's growing long-distance trade was an important occupation in villages immediately surrounding the castle town. As shown by the distribution of packhorse drivers at the time of the 1764 settlement, transport-related work remained an important employment in the heart of Iida domain even after three-quarters of a century of development in the lacquer, textile, and paper crafts. But with the continued growth of these protoindustries—and with the attendant opportunities to invest in agriculture and food processing in the core as well—villages in the vicinity of the castle town gradually phased out of packhorse work. As stock raising was pushed to the margins of the increasingly densely settled valley landscape—a process repeated across Japan—packhorse work too passed on to settlements in the mountains, particularly those to the south along the Ina road.[6]

This process spawned two very different types of rural commercial centers, or *zaigōmachi*, in Shimoina (see map 13). The first were the satellites, clustered near Iida, where agricultural processing and artisanal work was concentrated. The second were the outposts or crossroads settlements, at a distance of half a day's journey or more from Iida

6. The displacement of stock from core areas was a logical outcome, for while horses can exert ten times as much effort in an hour as a human being, the energy output of the two *per unit of caloric intake* is almost identical. As William F. Cottrell (1955:viii) has noted, "Where land is plentiful, population sparse, and draught animals available, there may be an economy in substituting draught animals for manpower; but with increased population and competition of land for the production of food and feed, the situation may

to the south, specializing in cargo shipping and handling. Although the latter too functioned as market towns for the surrounding countryside, their contrasting economic foundations gave them very different production complexes, generating a series of distinct population characteristics as well.

The clearest indicator of outpost emergence is the shifting distribution of horse populations out of the regional core (map 14). Dashina village, located on the Enshū road very near Iida, had as many as ninety horses in 1714. Although a number of Dashina residents continued to operate as professional pack drivers until at least 1773 (the only village in the core area for which this was true), the number of horses in the village had fallen to forty-one by the 1840s. Neighboring Naganohara saw a similar decline during this period, as transport functions were surrendered to outlying villages farther to the south.[7]

A comparable progression can be documented in the number of full-time packhorse drivers claimed by each village. Four settlements in the immediate vicinity of Iida (as well as one, Niino, that occupied a smaller irrigated basin to the south), all of which had claimed sizable numbers of professional packhorse drivers at the time of the Meiwa settlement, registered not a single one in a second survey undertaken half a century later. Conversely, in four outpost villages located further south along the Ina road, the number of drivers jumped between two- and fivefold during the same period (table 11).

Typical of the new outpost settlements was Namiai, a mountain village with a paucity of arable land, twenty-eight kilometers south of Iida along the Ina road. The first survey to be undertaken in the village, at the unusually late date of 1780, revealed that the village had no paddy fields at all (the first reference to an irrigated field would not appear until 1833); the holdings of the village's ninety-five households averaged only about half a *koku*. Going by the rule of thumb that one *koku* of grain was roughly equivalent to one person's annual rations, local self-sufficiency in staples would have required that each of Namiai's peasant households have access to lands with a minimum yield of four *koku*, or eight times as much as was actually available. Moreover, the local dry

---

be reversed, the survival of man being more important than the feeding of work animals" (quoted in Wrigley 1988:39 n. 12). The increasing reliance on human power was an important element of what Hayami Akira (1989) has termed Japan's "industrious revolution."

7. Naganohara's equine population, maintained at twelve through the early 1800s, dropped to half that number by the 1850s. Tatsuoka-mura 1968:367–70, 483.

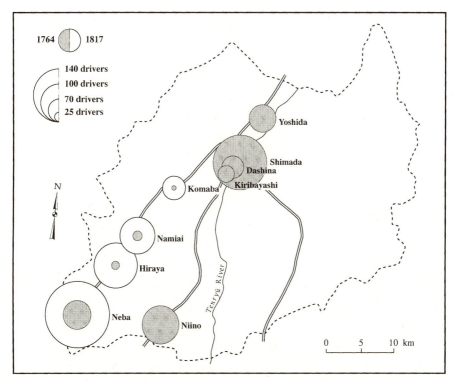

Map 14. Changes in Packhorse Driver Populations in Nine Shimoina Villages, 1764 to 1817. (Data from Furushima 1951:277.)

fields were stony, and the growing season was short. Clearly, nonagricultural employments were critical for the residents of Namiai.[8]

As early as 1722, this roadside village was home to 50 horses and 58 cattle, although it had only two commercial establishments (a timber dealer and a pawnbroker). Four decades later, most villagers depended on forest work or on servicing the packhorse trains that passed through; the village in 1764 claimed twenty-seven professional packhorse drivers. That number rose quickly in the ensuing decades, however, as villages nearer the core gave up packhorse driving for paper- and textile-related work. By 1817, ninety-six of Namiai's residents were able to earn their livelihoods leading pack trains; a census taken ten years later counted 107 horses and 50 cattle in the village, double the total of the earlier survey. While agricultural development remained minimal, scores of residents were now employed in the inns, teahouses, stores, and stables that

8. Namiai-mura 1984:395–415.

Table 11   *The Numerical Distribution of Packhorse Drivers along the Main Commercial Corridors of Southern Shimoina, 1764–1817*

	1764	1817
Core Villages		
Shimada	144	0
Niino	100	0
Yoshida	72	0
Dashina	59	0
Kiribayashi	44	0
Outpost Villages		
Neba	76	175
Hiraya	23	119
Namiai	27	96
Komaba	12	62

SOURCE: Furushima 1951:277.

had sprung up to service the growing through-trade. Villagers of all ages worked to keep the animals supplied with grass and fuel, and weaving straw horseshoes—which had to be replaced daily—became a universal by-employment on winter evenings.[9]

The encouragement of transport facilities was not the only way in which the packhorse trade stimulated the economy of villages along the Ina road. Commercial production of all types appears to have occurred earliest, and to have been most pronounced, in villages along the main trade routes. The mid eighteenth-century cargo surveys document a wide variety of agricultural products being imported into the castle town from "villages along the Ina Road": rice, barley, buckwheat, soybeans, azuki beans, tobacco, rapeseed, cocoons, hemp, and persimmons, as well as mountain products including walnuts, chestnuts, medicinal plants, and wooden craft objects. The volume involved, moreover, makes it clear that these villages were not merely marketing a small surplus while retaining an essentially subsistence economy. In the uplands of Shimojō, close to the southwestern packhorse routes, persimmons alone—one of the first commercial crops grown in the region—accounted for as much as 10 to 15 percent of peasant household income in an average year, and up to 30 percent in good years.[10]

9. Furushima 1951:280ff.; Yamauchi 1980:10–15.
10. Shimojō-mura 1977:800. Dried persimmons were widely used in New Year's decorations.

Commercial development thus varied across the landscape, not only in a concentric core-periphery pattern but along a cross-cutting linear pattern as well. So, too, would the social changes attendant upon that development. The following discussion accordingly locates villages relative to the transport grid as well as to the castle town in order to discern the ways in which social processes reflected growing regional disparities.

POPULATION GROWTH RATES

A first criterion by which to distinguish villages is a comparison of their relative rates of overall population growth. A survey of published population figures, covering most of the 140-odd villages that made up Edo-period Shimoina, suggests that, in general, population growth varied inversely with distance from the road, tending to be highest at important crossroads.

Nine Edo-period communities that were later consolidated into Achi village formed a representative microcosm that dramatizes the pattern found throughout the region (map 15). The two villages at Achi's core (Komaba and Onogawa), which together straddled the crossroads of the Misaka pass and the Ina road, grew handsomely beginning in the mid 1700s, when the packhorse trade rose to prominence in the area. The population of this core area had more than doubled by 1850 over mid eighteenth-century levels, only partly owing to field reclamation. Villages fronting the Ina road but at some distance from the crossroads grew at much more modest rates (30–50 percent over the same timespan), while those located farthest from this important trade route stagnated or actually lost population.[11]

In Shimoina's mountainous hinterland, proximity to transport was particularly crucial. Developments in two distinct zones within that hinterland highlight the contrast. On the one hand, isolated between two steep and largely unpopulated ranges to the east, were Ōkawara and Kashio (together forming present-day Ōshika), as well as the more southerly Minamiyama; on the other hand, straddling the most frequently traveled pass between the Ina and Kiso valleys to Iida's west,

---

11. Achi-mura 1984, 1:571–77. The quasi-urban character of centrally located Komaba in the 1800s is suggested by the variety of licensed tradesmen the settlement supported (including *sake* brewers, pawnbrokers, shopkeepers, blacksmiths, and confectioners), and by the existence of seven licensed waterwheel operators, who primarily pounded rice for households that did not grow their own food. For more on the diversification of Komaba's commercial community, see Itsubo 1987.

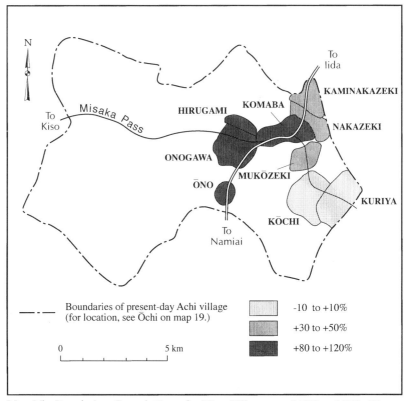

Map 15. Population Growth Rates for Nine Villages, ca. 1700 to 1850. (Data from Achi-mura 1984:571–77.)

was the settlement of Seinaiji. All were mountain villages, whose agricultural land in the 1600s consisted mainly of dry fields and extensive "burned-field" (*yakihata*) or swidden plots; forest work was a mainstay of local subsistence in all four at the opening of the period. But their situations diverged dramatically in the course of the eighteenth century. Ōkawara, Kashio, and Minamiyama, located in some of the most remote interstices of the county relative to the Tokugawa trade routes, were stripped of the craft specialties they had developed for tribute in the early Edo period (silk reeling and papermaking). Instead, these communities now exported token amounts of agricultural products (grains, beans, and tobacco) and a variety of forest goods (timber, hardwood objects, nuts, medicines, and furs). But their most important connection with the commercial economy was labor export, young adults from these villages migrating in large numbers to seek work as woodsmen, field hands, and day laborers elsewhere in the region and the country.[12]

Strategically located Seinaiji, by contrast, in addition to benefiting from one of the highest rates of recorded land-clearing in the county, was able to develop a thriving commercial economy based on two lucrative exports: tobacco and wooden combs. Both were introduced at the turn of the eighteenth century, and quickly grew into major enterprises for the village. Physical advantages fail to account for Seinaiji's singular lead in these industries; tobacco could be and in fact was grown in all of Shimoina's mountain villages,[13] and the hardwoods required for comb manufacture were widely distributed. But only Seinaiji was located on a major trade route, where it was able to take advantage of the growing market for both products. Decorative wooden combs were sent primarily west to the Kiso valley, where they found a ready market among the many pilgrims and travelers on the Nakasendō. The area's tobacco, however, could be sent either eastward, downslope to Iida, or westward, over a nearby pass to Kiso for trimming and packing; much of it was shipped ultimately to Edo, where it was popular among the women of the Yoshiwara for its mild flavor and the ease with which it could be kept alight.[14]

The results of these divergent economic paths for overall population levels in the three areas were marked. Ōkawara and Kashio barely maintained their mid-Tokugawa population levels to the end of the period; their combined total in 1842 was lower than in 1733.[15] During the same years, Minamiyama village saw its population fall steeply, allegedly because of overlogging and the resultant depletion of the area's economic mainstay.[16] Seinaiji, by contrast, more than doubled.[17]

PATRON-CLIENT RELATIONS AND RENTERS

The opportunities that generated population growth created another phenomenon that tended to vary along the same continuum: the timing and extent of changes in the status of nonelite com-

12. Ōshika-mura 1984, 1:580–81.
13. Ikuta-mura 1981:500.
14. Seinaiji-mura 1982, 1:322, 419–21.
15. Miura 1988:27; Ōshika-mura 1984, 2:888.
16. Despite a partial recovery in the later eighteenth and early nineteenth centuries, Minamiyama in 1850 held only 90 percent as many people as it had supported nearly two centuries earlier. Yasuoka-mura 1984, 1:379–80.
17. Seinaiji-mura 1982, 1:313.

moners. *Hikan* was the most prevalent term used in Shimoina to denote the hereditary servile class, who were numerous among the early Tokugawa peasantry.[18] In Shimoina, even into the mid Edo period in many villages, population registers listed such persons as belonging to the elite family (*oyakata*) they served. *Hikan* could be bought, sold, pawned, and inherited along with the land they worked. They were obligated to perform corvée each year in their *oyakata*'s fields, and in many cases they had to buy their way out of these obligations if they wished to achieve independence.[19]

The gradual liberation of this class, and its replacement by freer forms of labor, has long been recognized as one of the hallmarks of Tokugawa agricultural development.[20] Within Shimoina—where one village history claims that not a single settlement was without *hikan* at the beginning of the period[21]—there is a very clear, and again unsurprising, pattern to that process: villages along the road saw the greatest and the earliest weakening of the *oyakata-hikan* relationship during the Tokugawa period.

Already by 1700, such arrangements appear to have died out completely in settlements adjacent to the castle town, and to have begun decaying in southwestern villages near the major trade routes. In general, *hikan* proved more persistent along the more isolated east bank of the Tenryū River. At the end of the first century of Tokugawa rule, only those east-bank villages nearest Iida had begun to see a decline in these servile bonds. Far northeastern Ōshika had seen early protests for independence, but these proved unsuccessful, giving way to both violent disputes and the buying out of *hikan* contracts in the mid eighteenth century. The forestry-based villages of Minamiyama, far from the packhorse routes, were the last holdouts of the *oyakata-hikan* system in Shimoina.[22] Parallel trends may be observed with regard to the long-term servants known as *genin* and *hōkōnin*, who were increasingly replaced by tenants and shorter-term contract labor in a sequence that again followed the roads.[23]

18. Alternate terms for this group include *nago*, *kobun*, *kadonomono*, and *kadoya*.
19. Ōshika-mura 1984, 1:613.
20. This view is most clearly articulated in English in T. Smith 1959.
21. Shimohisakata-mura 1973:395. Shimojō village historians contend, however, that there is no evidence for the *oyakata* system in any of its fourteen Tokugawa hamlets; Shimojō-mura 1977:471.
22. Furushima 1974a:391; Ōshika-mura 1984, 1:612; Imamaki 1959, part 1; Hirasawa 1950:40.
23. Shimohisakata-mura 1973:390–93; Igara-mura 1973:345; Hirasawa 1950:40.

A third social-structural change that accompanied long-distance trade was a rise in the number of transient workers who rented housing in a settlement. Most renters hailed from poor villages, often in the mountains, from which they traveled in family groups to seek employment in more prosperous settlements in the core or along the roads.[24] The majority worked as day laborers, chiefly in agriculture, with a few of the more skilled earning their living as carpenters, blacksmiths, coopers, and the like.

Although their presence often went unrecorded in regular population registers, renters became the subject of several domainal surveys in the Ina valley. These were particularly frequent in the 1750s and 1760s. By this time, Iida *han* had begun to attract a large number of inmigrants, and domain officials ordered landlords to report the place of origin and present abode of any outsider they were harboring. In one village after another, the reports submitted in response to these orders attest to a strong correlation between roadside location and high renter populations.[25] As a proportion of the total settlement population, renters in Shimoina ranged from 10 percent in villages along minor roads to 20 or even 30 percent in the satellites of the castle town.[26]

HOUSEHOLD SIZE AND SEX RATIOS

These changes in population growth and social organization together produced a variety of secondary characteristics distinguishing roadside villages from their more isolated counterparts. One was household size. In agricultural areas across the region, the trend during this period was for average household size to decrease (attributable both to a declining birthrate and to the breakup of large extended households).[27] Representative of this trend was Yamamoto, an upland agricultural village southwest of Iida, where the average household had roughly six members in the mid 1700s but only 4.5 a century later.[28] Shimotonooka, a more prosperous village closer to the core, saw a de-

24. These families were usually very small, consisting either of a single adult, a childless couple, or a couple with one child. Itsubo 1987, part 1.

25. See, e.g., Kamisato-machi 1978:248; Igara-mura 1973:329–37.

26. The 10 percent figure comes from the village of Chikudaira (along the relatively quiet Akiha road on the east bank); Shimohisakata-mura 1973:356–60, 512. The higher figures were recorded in the immediate vicinity of Iida. See Kamisato-machi 1978:238–39; Itsubo 1987, part 1.

27. This trend was visible throughout Tokugawa Japan, at least until the 1840s; Hayami and Uchida 1972:473.

28. Yamamoto-mura 1957:133.

cline from 6.2 in the early 1700s to a low of 3.9 in the late 1700s.[29] Villages immediately along the road, however, exhibited a different pattern. Here, a rising population of hired laborers, who lived with their employers, raised household size over time. This was true of both transport-based and protoindustry-based settlements.

Dashina makes a particularly interesting case study in this regard. As a roadside village near the core, Dashina until the early 1700s was heavily dependent on the packhorse trade. Average household size peaked in Dashina at approximately 8.0 at the same time that the local horse population reached its maximum, in the second decade of the 1700s. By 1720, however, the village elite began phasing out of transport work, shifting investments toward newly lucrative opportunities in commercial agriculture. Thereafter, as the pack trade (and its associated service population) migrated to settlements farther south along the Ina road, Dashina's demographic profile began to follow the pattern of its agricultural neighbors: average household size declined during the succeeding century by nearly half, to 4.5.[30] In the outpost of Namiai, on the other hand, the same logic led to a reverse trajectory. In the 1750s, before the village had a sizeable *chūma* industry, the average Namiai household consisted of 3.8 members; by 1859, increasing use of hired labor in the transport industry had raised that figure to 5.5.[31]

Like transport centers, protoindustrializing areas also witnessed an increase in average household size. Both larger offspring sets and more live-in help appear to have contributed to this trend; in addition, families in centers of craft-related development may well have kept their sons and daughters at home for a longer period than had previously been customary. The castle town, with the densest concentration of both protoindustry and live-in hired labor, consistently had the largest households in the domain, the average there rising in the course of a century from 8.6 (in 1705) to 10.0 (in 1827).[32] But the same trends were evident in nearby villages as well. Shimotonooka, a prosperous agricultural settlement in the core, first witnessed a decline from 6.2 persons per household in the early 1700s to a low of 3.9 in the late 1700s, but saw its household size rise again to 5.3 by the later Edo period after its residents became heavily involved in the paper industry. In this case, the increase cannot

29. This was followed by a slight rise in the later Tokugawa period. Igara-mura 1973:349.
30. Tatsuoka-mura 1968:374.
31. Namiai-mura 1984:417–18.
32. Hirasawa 1968:19.

be accounted for by renters, who declined after the 1760s, or by hereditary servants (*genin*), for only two households in the village included any *genin*. The rise would appear to have been due rather to some combination of rising birthrates, prolonged retention of offspring within their natal households, and an increase in live-in hired help.[33]

A similar pattern is evident in the papermaking villages of the Tenryū's east bank, where a striking difference in household size between papermakers and nonpapermakers developed. In 1794, papermakers averaged 5.9 members per household, compared to nonparticipants in the trade, who averaged only 4.1. In this case, the difference can be partially accounted for by a high rate of live-in hired hands in the protoindustrial families, and a low rate of out-migration of family members in search of other work. But even after making allowances for these differences, one historian contends, the offspring sets of papermaking families must have been larger.[34]

In short, the variation of demographic growth rates across Shimoina would appear to confirm the hypothesis that rising birthrates accompanied (and may indeed have been caused by) the development of large-scale commercialized putting-out industries: the central argument of the controversial protoindustrial thesis.[35] It is beyond the scope of this study to ascertain with any certainty or precision what drove the population growth in these protoindustrializing villages of southern Shinano, a problem that would call for careful scrutiny of long-term records for individual villages. But the general correlation observed here is suggestive enough to recommend such a study.[36]

Finally, in the matter of sex ratios as in the other characteristics discussed above, crossroads settlements, agricultural villages, and relatively

33. Igara-mura 1973:347–51. Another possibility, of course, would be a declining death rate—a factor whose spatial patterns have yet to be explored in any detail.

34. Imamaki 1959, part 1:58.

35. Sources purporting to find a linkage between the spread of domestic manufacturing, lower ages at marriage (particularly for women), and rising fertility in early modern Europe include Levine 1977, Braun 1978, Kriedte 1983, and Kriedte, Medick, and Schlumbohm 1986. Among the more important critiques are Coleman 1983, Gutmann and Leboutte 1984, and Spagnoli 1983. For a balanced overview of the literature on early industry and population growth in eighteenth-century Europe, see Gutmann 1988.

36. Osamu Saitō (1983) has argued forcefully that the development of protoindustries in Japan did *not* lead to earlier marriages and larger families, as postulated for Europe. Nonetheless, some evidence from elsewhere in Japan does suggest the kind of dichotomy found in Shimoina between demographically stable rural villages and more rapidly growing protoindustrial communities; compare T. Smith 1977 with Morris and Smith 1985. See Tatsuoka-mura 1968:544–45 for a confirmation of the same trend within Shimoina.

isolated mountain villages each exhibited distinct trends. The general tendency in agricultural areas was for the ratio of males in the total population to fall, from an early average of 108 males for every 100 females, until the sexes achieved a near numerical parity by the end of the Tokugawa period. This trend was most pronounced, and began earliest, in areas near the core where commercial agriculture developed; it was delayed and muted, but still evident, in remote mountain villages. In settlements directly on the road, by contrast, an influx of adult males—tradesmen and field hands as well as laborers in the transport business—offset this tendency, keeping the male proportion of the population high throughout the period. Even within the castle town, sex ratios were highest in the commercial wards, where more *genin*—protoindustrial laborers, transport workers, and shop assistants—were concentrated.[37]

Dashina, which made a clear transition from transport to commercial agriculture, exemplifies the contrast. This village's sex ratio was 106 in the first two decades of the 1700s, when it was heavily involved in the packhorse trade. By the end of the eighteenth century, with a decline in packhorses and related transport activities, that figure had fallen to 102; in the 1840s, there were fully as many women as men in the village.[38] To take an example from the more isolated east bank of the river, the sex ratio in Toraiwa village, a papermaking and agricultural center directly across the Tenryū from Iida, ranged from 110 to 116 through the 1700s; between 1800 and 1850 it had declined to the 104–106 range, and after 1850 it did not exceed 102.[39] The forestry-oriented hamlets of Minamiyama in the southeast, on the other hand, did not show any marked decline in sex ratios until the 1840s.[40] How these sex-ratio differentials are to be explained is an essential but thorny question. While a conclusive answer may elude even in-depth examination of the population registers, the Shimoina data strongly suggest some kind of flexible demographic response to broader economic conditions, likely involving sex-selective migration and possibly sex-selective infanticide as well.[41]

37. According to a 1710 census, the proportion of *genin* in the town as a whole averaged 15 percent, but reached 38 percent in the commercial wards. Some 56 percent of *genin* were male. Hirasawa 1968:17–18.
38. Tatsuoka-mura 1968:370–71. Throughout this discussion, following the demographic convention, sex ratios are expressed as the number of men per 100 women.
39. Shimohisakata-mura 1973:358.
40. Yasuoka-mura 1984, 1:378–80. Many more examples could be cited in evidence of this contrast. For crossroads villages, see Achi-mura 1984, 1:572; for the commercial corridor, Kawaji-mura 1988:167–68; Achi-mura 1984, 1:573; Yamamoto-mura 1957:130–32; and Shimojō-mura 1977:622.
41. While the practice of infanticide in early modern Japan is not in doubt, its extent

CLASS STRATIFICATION

These demographic changes combined with the vagaries of harvest and the market to produce another characteristic distinction, in this case marking off all villages in or near the commercial corridor from their more remote counterparts: namely, the emergence of a sizable group of households with little or no property. Whether measured in landholdings or pack animals (the latter more pertinent in the transport-centered villages), the growth of a nonpropertied class was a concomitant of commercial development throughout the region. This did not necessarily mean that the largest landholders (or horse owners) saw their wealth increase, either in relative or absolute terms; nor should it be assumed that birth into a propertied household was any guarantee of remaining wealthy for life. Indeed, the Shimoina evidence suggests considerable fluctuation of fortune among even the richest families. Nonetheless, in all of the corridor villages, population growth and rising nonagricultural employment opportunities supported the emergence of a capital-poor class, permanently changing the region's social complexion.

A downward slide in the size of the average landholding was a nationwide tendency in the mid to later Tokugawa period, and it has been widely recognized that this process was accentuated in the country's national and regional cores.[42] Not surprisingly, within Shimoina the emergence of farm families with very small holdings took place earlier and more completely along the roads, where farm size fell owing both to population growth in excess of reclamation and to independent household formation by former *genin* and *hikan*. In Chikudaira, for instance—a crossroads village on the east bank of the Tenryū—a landless class that had not existed in 1699 had emerged by 1767. Half a century later, in 1824, households with less than 3 *koku* worth of land accounted for over half of the village population. Similar trends were evident in three neighboring villages as well.[43]

---

and motivations are highly controversial. The case for sex-selective infanticide as a widespread element of rational family planning is argued in Hanley and Yamamura 1977, ch. 9, and Skinner 1993. (Thomas Smith's research on Nakahara, while locally confirming this scenario, was couched as an isolated case study; see T. Smith 1977 and Eng and Smith 1976.) Mosk 1978, by contrast, ascribes the low fertility of the period primarily to the poor diet, hard work, and consequent low fecundity of peasant women—a view corroborated in important respects by Saitō 1992.

42. For instance, T. Smith 1959; Hanley and Yamamura 1977.
43. Shimohisakata-mura 1973:435.

The several hamlets of present-day Achi village, discussed above in the context of population growth, also illustrate the difference that location along the road could make. Onogawa, the crossroads settlement that grew fastest during the period, also saw the most rapid emergence of a land-poor population. Not coincidentally, the lower class of peasants in Onogawa protested on ten separate occasions against the elite of their village, a record unmatched by their counterparts anywhere else in Shimoina.[44] Tax data from Dashina, the village whose switch from transport work to commercial agriculture has been discussed above, reveal a comparable progression, indicated by a steady downward slide in the modal tax bracket.[45]

Similar patterns appear to have characterized animal property in the outpost villages. Although we do not have earlier data for comparison, ownership of Niino's 107 horses and 50 cattle was decidedly skewed in 1817. More than two-thirds of the households in the village could not lay claim to a single pack animal, while one-sixth owned four horses apiece (the number required to make up a standard pack train). In addition, while lower peasants might own cattle, horses belonged almost exclusively to the more privileged *honbyakushō* class. The many households in the village that owned neither land nor animals of their own operated as woodsmen or *chūma hōkō* (drivers who worked under contract for wealthy animal owners), for there was little land to rent in the area.[46]

Years of crop failure and food crisis contributed decisively to the ongoing process of differentiation in all villages, ratcheting families downward out of middle-peasant status and swelling the ranks of the landless. But again, the effects appear to have been most accentuated along the commercial corridor and in the regional core. In the outpost of Namiai, the Tenmei crisis of the 1760s caused a number of formerly landed peasants (*honbyakushō*) to lose ownership of their homes in the village center, and had a permanent polarizing effect on the settlement's social structure.[47] In eastern Yamamoto, the number of families with no land

44. By 1809, Onokawa's largest landowner held over ten *koku* worth of land, while over half of all village households held title to lands with a putative yield of less than a single *koku*. The growth of a land-poor class is summarized in Achi-mura 1984, 1:542.

45. Whereas the modal tax bracket in Dashina in 1640 was made up of households that paid 30 to 50 *hyō*, a century later it had fallen to 10–20 *hyō*; by 1776, the most populous group paid only 1–5 *hyō*, and after the 1840s, less than a single *hyō* in taxes. Tatsuoka-mura 1968:558.

46. Namiai-mura 1984:430; Furushima 1951:285.

47. Namiai-mura 1984:430.

at all more than doubled in the decade following the Tenmei crisis, during which time the settlement also lost four households out of its previous total of sixty-nine.[48] And in Shimada, at the center of the richest agricultural district in Iida domain, numerous peasants lost their farms during the Tokugawa period's worst crises. Profiting from their neighbors' hardships during both the Tenmei and the Tenpō years, one wealthy family here was able to increase its holdings more than tenfold in two generations.[49]

DISCUSSION

Social development across Edo-period Shimoina displayed a distinctive spatial pattern. From demographic measures such as population growth, household size, and sex ratios to such class indicators as labor forms and land distribution, the basic parameters of social organization and well-being have been found to vary in related ways with respect to distance from both the castle town and the transport grid. The principal findings are summarized in table 12.

Typological comparisons at this level of generality do not yield any direct answers to the question of whether, or for whom, Shimoina's Tokugawa development enhanced the quality of life. They do, however, suggest two important ways in which we would do well to rethink the process of social differentiation in the Edo period.

At the most general level, the range of variation documented here shows how misleading it can be to generalize about demographic processes across an economically diverse region, even one as small as a single county. Since Shimoina's population was clustered in the basin, aggregate statistics (were they available) would likely reflect the demographic characteristics of satellite communities and their agricultural neighbors. But such a composite sketch would blind us to the staggered pace and local inversions of social change across the region—a richer texture that even a simple fourfold typology has the power to reveal. Whatever else this exercise accomplishes, I believe it demonstrates conclusively the importance of positioning demographic data on the most finely specified socioeconomic grid available.

More pointedly, the Shimoina data suggest the limits of a diffusion model for explaining spatial variations in Tokugawa development. To

48. Yamamoto-mura 1957:132.
49. The household in question increased its tax liability from 9 *hyō* per year in 1781 (early Tenmei) to 92 *hyō* by 1842 (Tenpō 13). Tatsuoka-mura 1968:566–68.

Table 12 *General Trends of Social Development in Edo-Period Shimoina, Distinguished by Settlement Type*

	Inside the Core	Outside the Core
On the main trade route	*Satellite*	*Outpost*
	High population growth	High population growth
	Early transition to freer labor forms	Early transition to freer labor forms
	Numerous renters	Some renters
	Rising household size	Rising household size
	Rising sex ratios	Rising sex ratios
	Early and marked emergence of land-poor class	Marked concentration of pack animals
Off the main trade route	*Basin Village*	*Mountain Village*
	Moderate population growth	Population stagnation or loss
	Moderately early transition to freer labor forms	Late retention of unfree labor forms
	No renters	No renters
	Falling household size	Delayed fall in household size
	Early fall in sex ratios	Delayed fall in sex ratios
	Moderate development of land-poor class	Least development of landlessness

be sure, some of the variables tracked here do conform to a model whereby cores (and corridors) represented hearths of "advanced" development, which eventually diffused to their more "backward" hinterlands. Important examples are the movement toward less servile forms of labor and the trend toward increasing concentration of capital assets, both of which appeared first in crossroads settlements before turning up elsewhere in the region. For other variables, however, the difference between the commercial corridor and outlying areas was one of opposition rather than delay. While most communities gained population, for instance, Shimoina's mountain villages stagnated or even declined; while crossroads settlements attracted significant numbers of renters, off-road villages never seem to have done so; and whereas agriculturally oriented communities saw declines in average household size, satellite villages and outposts alike saw average household size increase. In short, the delayed-but-parallel model of social change suggested by the terms "advanced" and "backward" is not an accurate description of Tokugawa development. Projecting a temporal rhetoric onto spatial variation does not adequately represent the relational nature of economic and social processes across an integrated landscape.[50]

To get a clearer sense of how those processes operated, it is appropriate now to narrow our analytical compass to the regional core. As is clear from the settlement comparisons elicited above, both commercial development and social differentiation were most pronounced at the center of the regional economy. The following section accordingly focuses on the nature and consequences of protoindustrial development for the castle town and its satellites in the valley's agricultural heartland.

Differentiation at the Core

In the early years of the Tokugawa period, the most lucrative commercial opportunities in the Ina valley were concentrated in the region's political core. Like any castle town, Iida was home to the area's wealthiest consumers; it was also the designated entrepôt for both local and long-distance trade. This state of affairs was meant to last. In Tokugawa Japan, the fourfold Confucian social hierarchy had been solemnly inscribed on the settled landscape. All inhabitants of the countryside were ordered to occupy themselves full-time with cultivation, while persons of every other estate—warrior, artisan, and merchant alike—were segregated into clearly demarcated urban spaces. Yet occupational ghettos were more easily established than enforced, and over time—in Shimoina as elsewhere in Japan—the rigid sociospatial divisions of the populace began to break down.

By twentieth-century standards, personal mobility and capital liquidity were certainly limited in early modern Japan. Powerful forces bound individuals to their inherited livelihoods and locations, including kinship networks, regional loyalties, guild monopolies, the resistance of established communities to outsiders, and market constraints on many key assets (land, in particular, was legally untransferable). Yet while the propertied classes may not have been able to move their families from one

50. For a more extended critique of the diffusion model of Tokugawa development, see Wigen 1992.

settlement to another at will, or to shift their capital from one form of investment to another with ease, neither was the commercial order entirely immobile. Just as peasant pack drivers arose to challenge the transport monopoly of the official post stations, so rural landlords in time assumed a wide array of commercial functions, gradually undermining the privileges of the domain's designated merchants.[51]

The most important prerogative thus weakened was the castle town's monopoly on wholesaling. It is virtually impossible in the Tokugawa context to separate merchandise distribution, particularly at the wholesale level, from the production process, since putting-out merchants were generally deeply involved in both. Yet even before it became a means of control over production, wholesale marketing was the sphere of commerce in which the merchants of the castle town had the most entrenched privileges. Not surprisingly, it was over those privileges that they waged their fiercest struggles with the rising elite of the suburbs and satellite villages.

EARLY CONFLICTS: CASTLE-TOWN MERCHANTS AND THEIR RURAL COMPETITORS

In Iida, the battle to confine marketing to a designated commercial district began as early as the third decade of the 1600s. The first contests pitted the thirteen original wards within the moated castle town—an area south of the Nosoko River called Tatemachi ("the vertical wards")—against five wards to the north of the river, known as Yokomachi ("the horizontal wards") (map 16). Tatemachi was the heart of Iida's samurai and merchant communities; Yokomachi, by contrast, had been annexed to Iida to serve as a post station, exempted from direct taxes in exchange for providing transport services to domain officials.

The intended division of labor between the two areas lasted less than a generation. Beginning in 1628, and again after fires broke out in Tatemachi in 1644 and 1684, the thirteen southern wards had to beseech the daimyo repeatedly to forbid the northern wards from entering into retailing or wholesaling activities competitive with their own. Temporary restrictions were imposed in 1664, and reinstated in 1703, giving the original wards exclusive rights to deal in the most lucrative commodities: grain, salt, fish, cotton, textiles, metal goods, paper, tobacco, straw wares, and a variety of foodstuffs. Even within Tatemachi,

---

51. On the commercial activities of the wealthy landlord (*gōnō*) class, see T. Smith 1956, Pratt 1990, and Walthall 1990.

the distribution of commercial activity was strictly mandated. Packhorse drivers were ordered to unload and sell their wares in specific neighborhoods, creating a small-scale periodic market that rotated on a twenty-six-day cycle through the thirteen Tatemachi wards.[52]

Even after these provisions for orderly commerce were passed, however, disagreements between the moated city and the adjacent post-town district continued over three particularly important commodities: tea, salt, and dried fish. Yokomachi merchants were permitted to sell certain grades of tea as an extension of the area's continuing mandate to lodge and serve all travelers headed north from the city. The right to deal in salt, on the other hand, was firmly denied.[53]

Meanwhile, new sources of competition had reared their heads in the rural outliers. When Hori took over Iida *han* in 1672, commercial activity was already well under way in the surrounding area, for the first act of the town magistrates upon his accession included a request that he order all rural merchants to relocate inside the town walls. While the new daimyo did take a number of positive steps to foster commerce and industry in the domain, his policies tended to strengthen merchants in nearby rural settlements rather than eliminate them. In addition to encouraging land reclamation and the planting of such cash crops as mulberry, he stimulated cottage industry, liberalized restrictions on *sake* brewing, and lowered the fees exacted from castle-town merchants. These actions proved a great stimulus to commercial development, but only the last enhanced the position of the castle-town merchants specifically. In fact, the eight satellites that would play a dominant role for the rest of the Edo period took shape in these expansive early decades from Genroku to Kyōhō (1688–1736): Minoze (Kamiiida) to the west, Nosoko and Ichida-Haramachi to the north, and Yama, Yawata (Shimada), Atagozaka, and Upper and Lower Chaya to the south (map 17). By the 1710s, these villages posed a sufficient threat to the castle town to spark strongly worded petitions for restricting their commercial prerogatives.

In 1717, a complaint was lodged by the Tatemachi merchants alleging that the castle town was being impoverished and commerce thrown into chaos by the rampant growth of illicit market transactions in the neighboring villages. In response to the merchants' petition, domain officials investigated the extent of commercial development in the villages both north and south of Iida. It found that 198 houses had recently been erected along the roads leading into the castle town; that the res-

52. These regulations took effect in 1673. Furushima 1974a:373; Shiozawa 1987:13.
53. Shiozawa 1987:14.

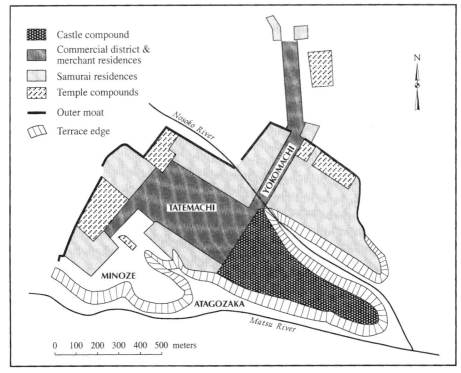

Map 16. The Layout of Iida Castle Town in the Tokugawa Era. (Adapted from Shinano Kyōikukai Shimoina Bukai 1934:111.)

idents of 165 of these were conducting retail or wholesale operations; and that while most of the goods they offered for sale could be justified in the name of service to the packhorse and passenger traffic along the roads (for instance, *sake*, straw sandals, confections, and hot meals), their wares also included a number of long-distance trade goods (particularly textiles) that should by rights have been unloaded in Iida.

The daimyo's response to these findings was decidedly ambivalent. While striking a pose of preserving the townsmen's monopoly rights as before, the domain in fact gave permission for those businesses already in existence in nearby communities to remain in operation, and refrained from issuing a rigid ban on future commercial growth there. Instead, a complicated series of directives was issued concerning what could and could not be sold in each of four classes of settlements. All villages that lay along the road and served travelers were permitted to sell ten items of everyday consumption (including *sake*, sweets, tobacco, smoking pipes, straw goods, and foodstuffs), but were barred from all other commerce. Three villages that straddled the Sanshū and Enshū roads on their approach to the town from the south (Upper Chaya, Lower Chaya, and Atagozaka) were permitted to sell twelve additional items (including copper, raw cotton, textiles, paper, soy sauce, grains, candles, oil, tea, salt, and fish, although the last three were permitted at a retail level only). Eight additional exemptions were made for Kamikawaji, a way-station farther to the south. But three particularly prosperous settlements adjacent to Iida—Minoze, Ichida-Haramachi, and Yawata—were given the broadest latitude of all. These communities emerged with virtually the same rights as any of the original thirteen wards; only wholesaling in tea, salt, and fish was forbidden to them, jealously guarded as the prerogative of the castle town alone.[54]

In short, despite repeated attempts to curb them, rural merchants gradually won a legal basis on which to compete with their Iida counterparts in the wholesale trade. As with the upstarts in the transport business, these illicit commercial operators had managed to become sufficiently entrenched for the petitions of their designated competitors to go largely unrequited. It was from this springboard that rural merchants would launch their later expansion into protoindustrial enterprise.

SOURCES OF ACCUMULATION IN THE SATELLITE TOWNS

By 1750, the villages ringing the castle town on the fertile floor of the valley supported a great diversity of opportunities for wealth generation, including commercial cropping, food processing, moneylending, and trading as well as protoindustry. But the foundations of the satellites' fortunes lay in agriculture.

Exemplifying this pattern was Shimada, originally a farming village located on a broad, low terrace directly south of the castle town. With fertile soil amply watered by an early irrigation project, Shimada residents boasted that theirs was the community into which all the domain's rural families longed to see their offspring marry.[55] In 1717, when only the richest tenth of the population in most nearby villages owned sufficient land to pay twenty bales (hyō) or more of rice in taxes, nearly half

54. Itsubo 1988:27–29; Shiozawa 1987–88, part 3:17–18. The restrictions on Upper and Lower Chaya were later eased, and a number of residents of both communities subsequently emerged as wealthy paper merchants, oil wholesalers, and confectioners.

55. Kega-mura 1987:i.

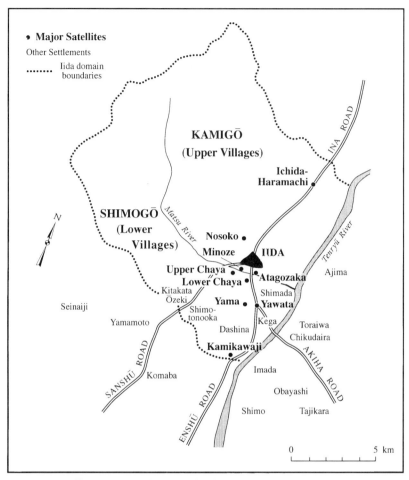

Map 17. Satellite Towns and Selected Villages Mentioned in the Text.

of Shimada's residents were in this tax bracket or higher, and two of its wealthiest landlords paid over 100 *hyō* apiece.[56]

As the latter figure suggests, Shimada may have had the largest average landholdings in Iida domain, but its agricultural wealth was distributed very unevenly. It was in this village that one family, profiting from its neighbors' distress during the Tenmei and the Tenpō eras, was able to increase its holdings more than tenfold in sixty years.[57] Yet well before the mid-Tokugawa crises, agriculture brought significant profits to

56. Kamisato-machi 1978:236; Matsuo-mura 1982:163.
57. See n. 49 above.

the area's larger landholders. In addition to such natural advantages as a mild climate, abundant water, and fertile soil, Shimada landlords enjoyed an unusually favorable tax break as well. Lying in a flood-prone area that had not been adequately diked until the early Tokugawa years, much of the village's land consisted of "new fields" (*shinden*), reclaimed only after the initial surveys of the Tokugawa period had determined tax liability. Since new fields were less heavily taxed (and in some cases hidden from the tax collectors altogether), landlords in this village were able to retain an exceptionally high percentage of the harvest from their fields. While the normative distribution was one-third to the tenant, one-third to the landlord, and one-third to the domain, landlords in Shimada were routinely able to keep four to five times what they paid in taxes. In the most extreme case, the landlord's take from one field was later alleged to be as much as twenty times that of the daimyo.[58]

While Shimada represented an extreme in this regard, it was not alone. Unreported reclamation, double-cropping, and other productivity gains increased the net return for many landowners in the villages surrounding Iida to double or triple what they paid in taxes. Through exploiting this agricultural wealth and investing it in other ventures—ventures made possible by their proximity to both the castle town and the cargo routes—Shimada and its neighbors in the valley floor developed robust commercial economies.

One index of that robustness was occupational diversity. The post station of Ichida, a representative satellite north of Iida, supported craftsmen in over a score of trades.[59] Other satellites specialized in particular services to the castle-town population. The village of Yama (Kanae), for instance, had an unusual asset in the form of a steeply sloped irrigation canal that cascaded down a terrace edge. Yama residents set up no fewer than fourteen waterwheels along this canal, where they pounded more than 9,000 bales (*hyō*) of rice annually for the castle town and vicinity.[60] The neighboring hamlets that came to be known as Upper and Lower Chaya, on the other hand, were the center of a regional confection industry, begun in the early eighteenth century.[61] All of these trades, along with the ubiquitous *sake* brewing and moneylending, offered opportu-

58. Kamisato-machi 1978:462–64; Matsuo-mura 1982:166; Tatsuoka-mura 1968:570. Tsutsui Taizō (1970) suggests that even in less favorable locations, landlords often received 20 to 30 percent more of the harvest than domain lords.

59. These included (in addition to those mentioned for the smaller outposts) stonemason, rush-mat maker, cooper, basket weaver, loom maker, tailor, paperhanger, woodcarver, cobbler, dyer, and pedlar. Takamori-machi 1972:567–600.

60. Kanae-mura 1969:146.

61. Iida Shimoina Kashi Kumiai 1982.

nities for rural families to accrue mercantile profits. But it was in the protoindustries that the greatest opportunities for accumulation would eventually be found. As the key to those opportunities lay in the characteristic relations of production in the Tokugawa export crafts, a closer look at the deployment of labor and capital in the protoindustrial economy is in order.

DIVISIONS OF LABOR IN THE
PRODUCTION OF COMMODITIES
FOR INTERREGIONAL EXPORT

If one feature may be identified as common among the Tokugawa protoindustries, it is surely their labor intensiveness. An abundance of peasants in search of slack-season income, combined with intense interregional competition for a limited metropolitan market, ensured that wages remained low—low enough, at least, to preclude the kind of capital-consuming innovations that might have significantly increased labor productivity.[62] Nonetheless, to speak of labor in the singular is misleading. All protoindustrial work was not the same, and all protoindustrial workers were not equal. In the Shimoina crafts, labor demands were highly specific by season and skill, and labor supply was sharply differentiated by gender, age, training, and organization. As such, it is important to abstract the main categories of labor from the combinations and hybrids of those forms that made up each sector's work force.

Shimoina's protoindustrial producers can be conceptually grouped into three distinct categories. Most common were rural households that undertook domestic manufacture on a part-time basis during the agricultural slack season. Labor in this case was structured primarily on the basis of sex, age, and family position, with different members of the household typically performing different tasks or stages in processing. Broadly speaking, women and children appear to have assumed responsibility for jobs that were seen as secondary, tedious, or requiring fine motor skills, while adult men, in addition to performing what were usually seen as the core industrial operations, also conducted the family's business transactions with brokers. The manufacture of paper was characterized by precisely these arrangements. Strategically located

62. The importance of labor supply, and especially of the seasonal complementarity of labor demands between agriculture and early industry, is increasingly recognized in the European and American literature as well. See, e.g., Gullickson 1986:197; Poni 1985:308; Earle 1987:174.

peasant families (rather than artisanal guilds) retained control over the process, and household-level production remained the rule in the papermaking zone.

At its simplest, the resulting arrangement would be an atomized one, characterized by disconnected, although often neighboring, units of production. For numerous reasons, however—including variability in household composition over time—certain families might be unable to carry out all phases of production. Although this could be solved by hiring outside help, it sometimes gave rise to interdependencies between producing households. The latter would in turn lead to a more complex picture at the hamlet level. Materials might eventually be passed back and forth between several households before the finished product was returned to the merchant or his agent. Umbrella manufacture in Ajima village, for instance, generated complex arrangements that might accurately (if awkwardly) be termed "multihousehold domestic labor networks."[63]

In cases where this incipient division of labor was taken further, a second major class of protoindustrial laborers emerged: domestic workers who specialized in a single phase of production. In the Shimoina paper industries, such workers were typically women or youths engaged in the less remunerative, purportedly unskilled ends of the cordage business. Their work required no special facilities or formal training; it also offered no prospects of advancement. When demand for Iida's hairdress ties was at its peak, more than a thousand women in and around the castle town are said to have been engaged in the repetitive tasks associated with their production, either twisting the rough initial paper ropes or winding balls from the prepared cord. Such work was in fact sufficiently female dominated to give rise to an unusual local term for prostitution: *hayakan-base*, or "hasty cord-making."[64]

This largely female protoproletariat co-evolved with the third and last major group of protoindustrial workers, the professional artisans. Far fewer in number, and exclusively male, these workers performed the skilled procedures that were considered the heart of each craft. All appear to have operated out of hierarchically organized workshops, typically with one master supervising half a dozen journeymen and apprentices. The masters of these artisanal shops were clearly the aristocracy of Shimoina's protoindustrial work force. Their income was maintained in

63. The most detailed picture of local umbrella making dates from the early 1930s; see Shinano Kyōikukai Shimoina Bukai 1934:167.

64. Hirasawa 1969, part 2:19.

part by limiting entry into the craft to a handful of novices each decade; in at least one case, this was reinforced by a restrictive guildlike association (*kō*) established among the masters themselves. In Shimoina, it was the hairdress tie (*motoyui*) and decorative cordage (*mizuhiki*) industries that developed the most powerful artisanal guilds. Members of both guilds were sufficiently powerful to negotiate with area merchants over the prices of both inputs and products. Documents also attest to their leadership in late-Edo protests over the price of rice—a commodity for whose starch there was no substitute in their work.[65]

THE COMMERCIAL NEXUS

Each sector of the Shimoina protoindustrial complex made use of a different constellation of these basic labor types. But regardless of the wide variation in labor forms across the region, the commercial infrastructures that underwrote the Shimoina protoindustries took essentially the same form in every case. All were organized into one or more putting-out networks, uniting financiers with producers in a series of hierarchical relationships. While these arrangements inevitably bred tensions between labor and capital, they also set the stage for conflicts between rival mercantile networks.

In time, the majority of Shimoina's protoindustrial funds came to be supplied by the class of provincial capitalists known as *ton'ya*.[66] These wholesaler-suppliers included not only townsmen but village landlords as well (the so-called *gōnō* or wealthy peasants), for "as capitalist and organizer no one could compete with the wealthy peasant, who knew the needs of the local market, the characteristic defects or excellences of local products, and even the reliability and quality of workmanship of individual workers."[67] Whether based in urban or rural areas, *ton'ya* typically employed intermediary agents or brokers, called *nakagai*, through whom materials were distributed to common laborers or artisans for processing at piecework rates. Relations between merchants, brokers, and producers in Shimoina were typical of those in domestic and small-workshop manufacture throughout the preindustrial world. They tended to be local, personal, enduring, and markedly unequal, characterized by

65. Ibid., part 3:3.
66. For general treatments of *ton'ya* organization and the role of mercantile capital in Tokugawa Japan, see Toyoda 1969:84–91; Hauser 1974; Hirschmeier and Yui 1965:28ff. Iida was also served by ten to thirteen licensed moneylenders (*shichiya*), who consolidated small savings accounts and made funds available to the merchant community. Hirasawa 1972b:101. For a suggestive study of similar credit arrangements in the Nagoya area, see Toby 1991.
67. T. Smith 1956:170.

episodic if not chronic indebtedness on the part of the workers and monopsony privileges on the part of the merchants. Nonetheless, within this general hierarchy of power there was room for considerable variation. What William Reddy notes for eighteenth-century France was equally true of early modern Japan: putting-out operations could be either "tight" or "loose," depending on the specific relations between laborers and these various representatives of capital.[68]

Typically, laborers and artisans were bound to a given *ton'ya* by complex debt arrangements. In the lacquer industry, for instance, the foreman of the lathe-turners (known as a *wan'ya* or bowl merchant) forwarded rice and miso to the woodworking communities in the mountains on credit, securing in exchange the right to collect their finished bowls and other wooden wares. These he would sell to a rural lacquerware merchant (*shikki ton'ya*) in one of the satellites, who in turn advanced both bowls and lacquers (obtained from a lacquer dealer [*urushi ton'ya*] in the castle town) to nearby lacquer artisans. In each case, extending raw materials on credit ensured the *ton'ya* exclusive access to the finished products, which would often be paid for at least in part with more raw materials.[69]

Similar relations of production obtained in the paper crafts. Many papermakers appear to have operated independently during the early decades of the eighteenth century, selling their finished products directly to packhorse drivers or other pedlars, but by the later 1700s, paper *ton'ya* located in or near the castle town had gained control of the bulk of the business. These wealthy wholesalers would advance both mulberry bark and money to rural families, financing papermaking operations in the villages in exchange for exclusive rights to market the finished paper.[70] Similar merchant-suppliers emerged in the production of hairdress ties and decorative cordage as well.[71] Even mulberry bark came to be controlled by specialized merchants (*kōzō ton'ya*), who forwarded cash loans to debt-strapped cultivators (often at tax-collection time) in exchange for later repayment in the form of mulberry bark.

Just as a hierarchy developed among protoindustrial laborers, so a pecking order formed among Shimoina's merchantry. Since each stage

68. Reddy 1984:23.
69. Masaki 1978:77–79; Imai 1953.
70. Paper in turn came increasingly to be collected by specialized agents of the castle-town *ton'ya*, who themselves clustered around the town. Masaki 1978:77; Imamaki 1959, part 1:59.
71. Shinano Kyōikukai Shimoina Bukai 1934:110. Interestingly, cordage merchants were located not in the communities where these products were made, but in the centers where the raw paper was processed.

of processing gave rise to its own *ton'ya*, partially processed materials were continually changing hands between merchants; and since small differences in the terms of those exchanges accrued over time into significant profits or losses, *ton'ya* in related lines of work were continually jockeying for position. One such rivalry involved a tug-of-war between lathe foremen, lacquerware merchants, and lacquer suppliers, in which the latter emerged with a corner on the hardwood craft complex. But similar struggles rent the paper complex as well. Here, it was the dealers in craft paper who managed to secure the most profitable position. Dispensing mulberry bark to papermakers and supplying paper to dealers in the related crafts gave them leverage over all branches of the trade—not only the local producers of hairdress ties, decorative cordage, and umbrellas, but the extraregional consumers served by the packhorse trade as well.

RIVAL CENTERS

As we have seen, the higher stages of processing in all of Shimoina's export-oriented crafts took place in Iida and a handful of surrounding settlements. There was considerable spatial overlap among the lacquer, paper, and textile complexes, as well as integration with other by-employments. For instance, an occupational breakdown for Minoze (Kamiiida)—a suburb of Iida that abutted the western border of the castle town—indicates that hairdress-tie and cordage laborers, lacquer artisans, and tobacco cutters worked side by side in the same neighborhood. Similarly, the all-important *motoyui* industry was dispersed through several settlements, surrounding the castle town on all sides. Despite a degree of overlap, however, each industry had a traditional center, giving a distinctive economic character to each of the various satellites. The undisputed core of the lacquerware complex was prosperous Shimada to the south; for silk reeling (and the associated dyeing and weaving industries), what is now Kamisato to the north; and for the various paper crafts, a smattering of villages east of Iida, at the base of the terraces on both sides of the Tenryū River (see map 13).

If the production process tended to cluster in a few hamlets, the merchants who controlled those processes were even more selectively located. While intermediate brokers might be scattered throughout the producing areas, the more powerful wholesaler-suppliers were tightly concentrated at the region's core. Yet in the paper industry, largest and most lucrative of Shimoina's protoindustrial complexes, their locus included not one but two commercial centers: Iida itself, and a compact trio of prosperous settlements to the south, including Shimada. Each area offered a different basis for control of the paper trades.

Iida's merchants, in the eighteenth as in the seventeenth century, enjoyed a number of important advantages over all others in the region. Chief among these were official protection, exclusive wholesaling rights for select commodities, and proximity to the domain lord and his retainers. But Iida was perched inconveniently on high ground, overlooking the valley from a projecting terrace several hundred feet above the Tenryū floodplain. As noted above, its original merchant community had had to be forcibly relocated within the castle walls from more natural commercial sites nearby.[72] The other center of *ton'ya* activity, by contrast, was essentially a constellation of three settlements: Shimada village, the largest and most prosperous agricultural settlement in the region; its similarly situated neighbor, Kega village; and the shrine-entrance town (*monzenmachi*) of Yawata, a semiautonomous commercial district serving Kega and Shimada. Straddling the juncture of two long-distance trade routes leading into the valley from the south and east, and located at the center of the papermaking zone, the Shimada-Kega-Yawata triangle generated intense competition with Iida, eventually giving rise to some of the most influential paper dealers in the region.[73]

The conflict between this area's merchants and their castle-town rivals finally came to a head in the Paper Monopoly Riot of 1809. Briefly discussed in the preceding chapter in the context of taxation rights, this incident can also be read as a symbolic culmination of the tensions engendered by Shimoina's commercial development. Beginning with a rural merchant's proposal to establish a wholesale paper monopoly, and climaxing in an uprising against the paper industry elite led by artisans from across the river, it illuminates better than any other the crosscutting conflicts created by growing social and regional disparities.

THE PAPER RIOT REVISITED

The incidents leading to the uprising of 1809 were initiated by a paper merchant from Kega village, one Hayashi Shinsaku. In 1807, Shinsaku petitioned the Iida daimyo for permission to establish

72. Hirasawa 1972b.
73. For instance, of eighteen cordage merchants in Iida domain located outside the castle-town walls in 1809, eleven resided in the settlements just south of the castle town. Hirasawa 1969, part 2:15.

the first licensed paper monopoly in the domain. In exchange for exclusive control of the wholesale paper trade, he offered to serve as a tax farmer for the *han*, collecting the *goyōkami* (a surcharge on all paper sales within the domain) that had been the *han*'s prerogative since the turn of the eighteenth century.[74] Shinsaku also proposed a new surcharge (amounting to roughly a 2 percent sales tax) on dealers in hairdress ties and decorative cordage, who would be exempted from paying the *goyōkami* on the paper required to produce their cordage. While not averse to these proposals, the daimyo's officials solicited the reactions of paper-industry merchants and brokers in and around the castle town before approving Shinsaku's petition. Those from the surrounding villages proved amenable, but the castle-town merchants were adamantly opposed. Charging that the proposed assessment was too high, and would severely inconvenience cordage makers as well as merchants from other domains, Iida's merchant community further alleged that extending this type of privilege to an upstart merchant from a peasant community would undermine the castle town's prosperity. Many Iida merchants were galled that their rural competitors were already acting as brokers for paper made elsewhere. Further opposition was voiced by artisans in the cordage enterprises, both in the castle town and particularly across the Tenryū River in Imada (Tatsue)—a paper-craft center intimately connected with Iida but formally outside Iida *han*'s jurisdiction.

In the face of such protest, the *han* deferred a decision on Shinsaku's proposal until autumn of the following year. Negotiations continued between Shinsaku, the town merchants, and the cordage makers, but the three parties were at an impasse. As it began to appear that Shinsaku's request would be withdrawn, representatives of Tatemachi and Yokomachi, the two commercial districts of the castle town, put forward a joint petition to form a monopoly of their own. In a victory for the castle town, this second petition was approved in the fall of 1809.

As noted in the previous chapter, the new ruling created serious hardships for numerous participants in the paper-craft network. Many former brokers from the villages of Kega and Shimada lost their commercial privileges; not a single merchant from Shimada's commercial district of Yawata was licensed in the new system. The out-workers in the paper crafts were saddled with new burdens as well. Most disgruntled were the papermakers of Imada village, where paper sales were crucial for generating the revenue required to pay the village's monetized taxes. More-

---

74. On a similar merchant-run (or "liturgical") tax collection system in China, see S. Mann 1987, ch. 2.

over, the rural merchant who had traditionally handled Imada's business was unable to continue doing so. Under the new system, Imada workers had no choice but to procure their inputs from, and sell their finished products to, an agent in the castle town. Leaders from four east-bank villages requested that they be allowed to deal through a broker in one of the nearby satellite communities rather than have to haul their goods all the way to Iida and back (a round-trip that took more than a day), but their request was denied.

It was these grievances that finally prompted an uprising against the new monopoly. Late on the night of 12/5, 1809, the papermakers of Imada stealthily rowed across the Tenryū River into Iida territory. Gathering on the west bank, they lit their torches and candles, set up a great cry, and began marching toward the houses of several powerful merchants who were profiting from the new system. The commotion soon roused their counterparts from the castle town and surrounding villages, who joined them in a quest for revenge. As the procession neared Kega and Shimada, hundreds of paper-craft workers, from both within and without Iida domain, joined the angry crowd. Before the night was over, four houses were attacked, including Shinsaku's and three belonging to brokers licensed by the new monopoly. All were broken into, their furnishings smashed, and their food stores destroyed before the crowd dispersed.

The investigation that followed uncovered more than a lone instance of property destruction. Testimony before the Bakufu's magistrates alleged routine infractions of the new *ton'ya* arrangements as well. In an attempt to circumvent the surcharges, papermakers and artisans in the related industries had begun dealing with unlicensed brokers: in one case, an oil merchant from Yawata, but more often the ubiquitous packhorse drivers. In a serious blow to rural merchants, these practices were ordered halted. Yet, as noted in chapter 2, this proved a Pyrrhic victory for Iida; within a decade, papermakers inside the domain were complaining that the additional surcharges were ruining them. Only large peasant households with cheap domestic labor were able to continue to compete with producers from outside the domain, who did not share the burden of the new taxes; many formerly employed in the industry were said to have run away or been forced to become pedlars.[75]

The paper producers thus displaced were not on the whole a wealthy lot. Although papermaking had begun as an occupation of the rural elite,

75. The main source of the foregoing account is Imamaki 1959. For other details, see the different reading given this incident in chapter 3.

wealthy families had been gradually phasing out of production, transferring their capital into other investments. Those who could were accumulating land from poorer peasants, many of whom proved unable to meet their tax obligations in the aftermath of the Tenmei crisis. In the process, paper production had become a winter by-employment of middling and poorer peasants, an evolution that has been traced by comparing paper taxes with land taxes over time in Dashina and Kiribayashi villages. Whereas the two series were roughly parallel from the 1760s through the 1780s, suggesting that large landholders were making proportionately large amounts of paper, this relationship was broken in the 1790s, immediately following the Tenmei crop failures. At this point, large landlords dropped out of papermaking, and the average holding of the remaining producers dropped accordingly.

As this shift occurred, papermakers with small holdings became increasingly dependent on wealthy brokers to finance their operations. And while the producers in this period became steadily poorer, those who financed them grew dramatically richer. One of the three such brokers whose houses were destroyed in the 1809 riot, the descendant of a middling peasant lineage in Dashina village, became a designated lender (*goyōtashi*) to Iida *han* the following year. In exchange, he was entitled to take a surname (Kinoshita), use a family crest, and assume the honorary rank of *shōya*, putting him on an equal footing with six of the top ten holders of power in the castle town. Kinoshita went on to amass considerable landholdings in the 1810s and 1820s. By 1830, he was the largest landlord in his village.[76]

Kinoshita was not the only rural magnate to lend money to the daimyo during these years. Indeed, the geography of merchant loans to the increasingly hard-pressed Iida *han* lord constitutes a vivid index of the distribution of economic wealth within the fief. It clearly betrays the growing superiority of the satellites, particularly those in the so-called Lower Villages (Shimogō), the southern part of Iida domain, where the paper industries were headquartered (see map 17). As early as 1750, the Hori had tapped three merchants in this area, in addition to three in the castle town, for forced loans. By 1828, fifty-nine merchants from the Lower Villages extended such loans, more than the total from the castle town itself. More remarkably, the combined contributions to the *han*'s finances from these Lower Village merchants, 3,280 *ryō*, amounted to more than two and a half times that of their counterparts in the north-

76. Imamaki 1959, part 1:50.

ern or Upper Villages (1,270 *ryō*), and came within 1 percent of equaling the amount collected from the castle town proper (3,305 *ryō*).

The list of villages in which these local notables resided reads like a roster of the major centers of crossroads trade and protoindustrial production in the *han*. But head and shoulders above the rest stood Shimada, the domain's wealthiest agricultural village and hearth of its lacquer and cordage industries. A single wealthy Shimada magnate in that year supplied the *han* with 500 *ryō*, while his neighbors contributed another 300 *ryō*—together enough to support the daimyo and his retinue in Edo for nearly half a year. By contrast, the highest individual loans to come from the castle town and the Upper Villages did not exceed 200 *ryō*. Not surprisingly, the same disparity would show up in private finance as well. By the 1840s, the Lower Villages' elite had joined those of the castle town and the papermaking centers across the river in underwriting the mortgages of Upper Village peasants. Even the hazardous business of lending to the *han* appears to have exacerbated the social differentiation that made such fortunes possible. When the daimyo made payments on his debts, he often did so in grain, which merchant-lenders would then hold back from the market to sell when prices were high. Such actions surely contributed to the dramatic increase in rice riots in the area; during the last century of Tokugawa rule, the peasants of Iida domain organized eight major protest incidents (more than any of their counterparts in contentious Shinano province), mostly provoked by high rice prices. Needless to say, a manipulated grain market may also have been one of the forces pushing lower peasants into the landless class.[77]

That merchants from the satellite towns came to be tapped for major loans by the *han*—and thereby came to share in the privileges of the castle-town elite—symbolizes the culmination of the two intersecting trends of social and spatial differentiation traced in this chapter. Not only the scale of commercial and protoindustrial development in the villages, but also the simultaneous process of class differentiation within those settlements, allowed rural merchants in this mountain valley to amass considerable fortunes. By the early nineteenth century, when Iida's own

---

77. Of the eight Iida-area protests in the later Tokugawa years, five were violent riots or "smashings" (*uchikowashi*), and an equal number are categorized as motivated by "poor harvest or food shortage" (Fujimori 1960:242–43). On the distribution of lenders to the domain lord, see Matsuo-mura 1982:164; on Iida *han* finances, Tsutsui 1956–57; on Upper Village mortgage financiers, Kamisato-machi 1978:460; on loan repayments and grain hoarding, Imamaki 1959, part 2:51.

merchant community no longer commanded sufficient wealth to finance the *han*'s growing debts, the daimyo was able to turn instead to a sizable pool of wealth engendered through two centuries of protoindustrial development in the surrounding villages—breaching the barrier between Iida and its immediate hinterland in the process by extending the prerogatives of the moated city to a select few in the surrounding countryside.

With this act, the protoindustrial economic regime of Shimoina may be said to have reached its symbolic apogee. Stimulated by the independent packhorse drivers who had begun coursing the Ina road nearly three centuries earlier, the lacquer, silk, and paper trades had served to integrate the landscape of Shimoina into a highly unified economic region, with multiple interlocking and overlapping circuits of production and exchange. But integration of necessity brought differentiation, both territorial and social. In the process of elaborating a broad-based, trade-oriented commercial economy, the lords and merchants of Iida domain had unintentionally created a landscape of sharp disparities. To be sure, the earnings of southern Shinano merchants paled beside the fortunes being amassed by their richest contemporaries in Japan's major cities. Yet their insignificance in the national context should not blind us to their local importance. As Carville Earle has noted of antebellum America, "although middlemen earnings were small, . . . they constituted the foundation for the spatial structure of regional development."[78]

It was thus an already highly differentiated, tightly integrated regional economy that the international silk trade would rapidly transform in the coming decades. A silk production complex had been developing for two hundred years as part of Shimoina's broader Tokugawa economic matrix, and it was well past the embryonic stage on the eve of the opening of Yokohama. Protected by stiff international trade barriers since the 1680s, and nurtured at a critical stage by an activist domain lord in search of export earnings, sericulture had become deeply rooted in Shimoina's regional economy. But no observer in the early 1850s could have imagined the extent to which it would come to dominate that economy during the succeeding seventy years.

78. Earle 1987:174.

PART TWO

The Region Inverted, 1860–1920

CHAPTER FIVE

Mobilizing for Silk

The First Quarter-Century

If the last decades of Tokugawa hegemony were years of escalating social tensions in Japan, they were also years of stepped-up economic activity. After more than a century of relative stability, long-term forces for growth had been set loose within the rural economy beginning in the early 1800s. Population was on the rise, as was agricultural productivity, boosted both by cheap fertilizer from the Hokkaido herring fisheries and by the diffusion of improved cultivation techniques. But the most dramatic stimulus would come in the 1850s, with the country's forced entry into the global economy.

By the mid nineteenth century, European and American ambitions in the Pacific had begun to threaten Japan's two-centuries-old tradition of carefully screened contact and minimal trade with the outside world. Already by 1800, China had emerged as the center of a profitable commerce in tea and silk. Over the next fifty years, this trade would become the impetus for Western economic and military penetration of East Asia, culminating in the treaty port system, which would eventually extend to China's neighbors as well. In the wake of the Opium War of 1842–43, first Britain and then other European countries demanded the "opening" of China, on terms highly favorable to the European powers.

Anxious not to be bested by the British or the Russians, U.S. policy makers also began jockeying for position in the western Pacific. Animated by a vision of the potential for trade between China and California, influential Americans quickly concluded that they must secure access to Japan, if only for a coaling station and provisioning stop for that trade.

139

By 1853, only a decade after the conclusion of the Opium War, a small but formidable squadron of American warships had arrived at the entrance to Edo Bay, ready to transgress the barrier laid down by Tokugawa Ieyasu in the early seventeenth century and demand the right to enter Japan. Rebuffed the first time, Commodore Matthew Perry returned with a larger fleet in February of the following year, when he managed to negotiate an agreement that gave him most of what he sought: the return of shipwrecked sailors, the appointment of a U.S. consul, a vaguely defined right to purchase goods at designated ports, and a most-favored-nation clause. Despite public outcry in Japan, commercial treaties followed quickly over the next three years, and in July 1859, the first two ports—Nagasaki and Yokohama—were officially opened to trade.

The resulting rupture of the carefully bounded world of Tokugawa trade faced the Bakufu with a crisis, threatening not only its legitimacy as a government but the economic stability of the country as well. On the one hand, in order to bring its coinage into line with prevailing world values for gold, silver, and copper, the Bakufu was forced to initiate a complete overhaul of its currency, resulting in rapid inflation. At the same time, preoccupation with defense led to heavy expenditures on weapons and military technology, creating a drain on the national treasury, while the popularity of Western textiles caused a further outflow of currency. In order to pay for these desirables, the Japanese had to find something of local manufacture that the foreign traders were willing to buy. The obvious initial candidates were the mainstays of the China trade: tea and silk.

As it happened, circumstances converged at the moment of Japan's debut in world trade to create an unprecedented void in the world's raw silk supply. A silkworm blight had broken out in southern France in 1852, sweeping north through France and east into Italy by the end of the decade to decimate Europe's prime sericultural regions. A system of rigorous silkworm egg inspections effectively eliminated the problem in France only in the 1870s; Italy's silk production did not fully recover for another decade. Yet China was unable to respond to the increased demand, its economy being thrown into disarray by the Taiping rebellion of 1850–64. With these disruptions in Europe and China, Japan's traders for a brief period profited from a voracious European market for raw silk and silkworm eggs.

It was this opportunity that would propel Shimoina into the new global order. In 1859, as soon as foreign buyers arrived in Yokohama,

Iida's commercial community attempted to export a variety of the region's specialty wares. A local merchant secured a land grant in the castle town in order to store and ship export goods, and promptly dispatched three horses laden with samples to the port. But the inscrutable Westerners could not be tempted with lacquered bowls, dried persimmons, rice wine, paper umbrellas, or even the handsome hairdress ties that were the region's pride, and this first trading party had returned home demoralized, their load lightened only by the weight of the *sake* with which they had consoled themselves.[1] Before long, however, local merchants discovered what the foreign traders wanted. Raw silk from Iida found its way into the second shipment dispatched to France, and the oldest extant bill of lading for Shinano thread exports, dated 11/20, 1859 (just five months after Yokohama opened for business), is preserved in the Shimoina village of Kamisato. The European dealers were clearly pleased with the local wares; price lists for 1860, 1862, 1863, and 1865 show Iida's first and second grades of thread fetching as much as any on the market.[2]

Encouraged by these early results, the valley's farmers and financiers moved quickly to expand production. But increasing output on short notice was not a simple matter. A shortage of silkworm eggs, for which local producers now had to compete with French and Italian buyers, was the most conspicuous bottleneck. But other resources in the region were also heavily committed. Since transport had not yet been improved, the valley had to continue to feed itself, meaning that staple crops could not be compromised; nor were markets for Shimoina's traditional export goods declining at the time. A growing sericultural and filature sector had accordingly to be accommodated within traditional regional economic structures, remaining but one enterprise among many in a pluralistic space-economy. As a result, the first problem in the drive to expand silk output in Shimoina was one of mobilizing resources in a heavily committed rural landscape, and diverting them out of long-established pathways to augment what had heretofore been a minor local handicraft.

What made this possible was a sustained influx of cash. During the first years after the opening of Yokohama in 1859, Japan had the good fortune to maintain a positive trade balance, owing mainly to the favorable prices for silk during Europe's silkworm blight. When the trade surplus evaporated in 1863, it was replaced by a substantial foreign capital

1. J. Yamamoto 1958.
2. Hirasawa 1952:104–7.

inflow, primarily in the form of long-term loans by foreigners to the Japanese government.[3] This brief but intensive infusion of foreign capital fed both investment and inflation throughout the country. Currency debasement added further fuel to the fire. In the last decade before the collapse of the Bakufu, hundreds of thousands of virtually worthless coins were put into circulation by the southwestern domains to pay for imported arms; after 1876, the new regime succumbed to the same temptation, allowing provincial banks to print unbacked paper notes to help finance its battle against the last remnants of armed resistance. It was in this chaotic, inflationary context that Shimoina's farmers and filaturists scrambled to increase the county's sericultural output.

The aim of the present chapter is to uncover how resources were appropriated to make that expansion possible. The years in question, from 1859 to the early 1880s, bracket a series of profound political changes at the national level: the downfall of the Bakufu, the "restoration" of the emperor in 1868, and the creation of a strong, centralized state. These events in turn involved a complete remapping of the local political terrain, including the overthrow of the local barons, the consolidation of villages, and Shimoina's incorporation into the newly formed Nagano Prefecture (identical in its final boundaries with Shinano province). These political realignments were of signal importance for the region, and will be addressed at length in chapter 6; for the present, however, the focus is on the steady rise in production of cocoons and silk thread, which continued apace throughout these upheavals until arrested by the Matsukata deflation of the mid 1880s.

Since no major component of Shimoina's protoindustrial economy contracted significantly to make way for new enterprises, the essential question is how—and where—the people of the region found land, labor, and capital to accommodate the growing silk industry during this turbulent quarter-century. The details of that accommodation are explored here in two stages, focusing first on the landscape of production, and subsequently on the new geography of finance. In each case, three issues are addressed: the nature of resource demands imposed by sericultural and filature operations; the strategies by which those demands were met; and the implications of those strategies for the spatial configuration of the growing industry. Mapping the first quarter-century of the silk economy's growth in this way reveals a pair of complementary but opposing tendencies. While the geography of silk production dif-

3. Key 1971.

fused steadily outward, involving more and more households and villages across the region, the geography of finance pulled gradually in the opposite direction. Particularly in the more capital-intensive filature industry, local producers were forced to rely increasingly on centralized sources of funding. This in turn opened the door to outside control, preparing the way for the remaking of the region as a periphery in succeeding decades.

Making Space for Silk

CONTINUITIES IN PRODUCTION

Early sericultural expansion in the Ina valley was not achieved by displacing existing protoindustries. From the last years of the Tokugawa regime through the fifteenth year of the Meiji era, virtually every sector of Shimoina's traditional economy was holding steady if not actually expanding output. National demand for the region's paper products, lacquerware, processed cotton, and tobacco held strong, and the valley remained dependent on a basically unchanged transportation system.

The paper crafts continued to reign as Shimoina's foremost enterprise through the first decade of the Meiji era, with area villages producing over 250 metric tons of paper-mulberry bark per year during the 1870s.[4] The majority of paper-industry income in the region continued to be generated by sales of hairdress ties (*motoyui*). A decree of 1871 ended sumptuary restrictions based on the four-tier Confucian class system, banning the samurai from wearing the traditional topknots (*chonmage*) for which these decorative ties were designed. In practice, however, the relaxation of class codes also allowed commoners to emulate samurai styles, and *motoyui* in fact retained their market value (in part by becoming popular among women). Until the Matsukata deflation of 1884, sales rose in most years for which statistics remain (table 13).

Statistics showing that the number of wholesalers (*ton'ya*) in the area's paper industry fell by half during these years (from 106 just before the

4. In the core village of Kawaji, over a hundred householders continued to be employed as papermakers (Kawaji-mura 1988:299–300); in Shimohisakata village, as late as 1880, income from paper production remained more than two and a half times as high as that from silk (Shimohisakata-mura 1973:684).

Table 13 *Sales of Shimoina Hairdress Ties, 1876–1883*

	Production (*maru*)	Value (yen)
1876	612,216	—
1877	657,880	73,171
1882	1,055,245	129,305
1883	1,087,655	137,708

SOURCES: The figure for 1876 represents the sum of individual village totals within Shimoina County as recorded in the first Nagano prefectural gazetteer. This important survey was compiled between 1876 and 1879; most entries for Shimoina villages read "as of Meiji 9" (1876). A two-volume edition published sixty years later as *Nagano-ken chōson shi* (NCS) was the version consulted for this work. The balance of the data are drawn from prefectural statistics compiled by Masaki Keiji (1978:414).

Restoration to 50 in 1880) are sometimes adduced as evidence that the paper industry was in decline.[5] In fact, however, the contraction of *ton'ya* was symptomatic rather of consolidation of control over the trade in the hands of fewer merchants. Between 1871 and 1875, two dealers from Shimohisakata village jointly captured an astonishing 243 *ryō* in profits by brokering the region's paper products to Tokyo wholesalers; one of the two alone managed to corner 80 percent of the trade, an unprecedented feat of market concentration.[6] But there is no evidence of a decline in the total volume of production. As late as 1882, hairdress ties generated more income than any other item of craft manufacture in Chikuma Prefecture, of which Shimoina was then part.[7]

Cotton processing was another traditional industry that throve during this period. Riding the wave of rural prosperity in the later 1870s, the market for cotton textiles in the countryside expanded; locally finished cloth could now find a ready market outside the producer's home.[8] While the majority of the imported cotton continued to be spun and woven for domestic use, 122 professional weavers in the valley were weaving approximately 21,000 yards of cloth for sale annually by 1882. In particular, production of cotton footwear (*tabi*) became a specialty of the villages north of Iida during this period.[9]

5. E.g., Nagano-ken 1971:246.

6. Shimohisakata-mura 1973:685.

7. In fact, the Ina valley produced nearly half of Japan's *motoyui* at the time. Furushima 1960:20; Masaki 1978:410–12.

8. For instance, in 1878, Takagi village's weavers sold approximately 500 *tan* (5,000 yards) of cotton cloth to Iida and Matsumoto. Masaki 1978:416.

9. Chikuma Prefecture ranked third in Japan in this industry and led the nation in production of the heavy soles for these traditional socks. Masaki 1978:416. For additional evidence of expansion in the cotton trades, see Takamori-machi 1972:397; Mukaiyama 1984:260ff.

Lacquerware similarly continued to flourish in early Meiji Iida. In response to an order that all itinerant woodworkers (*kijishi*) declare a permanent residence, many converged on the castle town; by 1877, there were forty lacquer craftsmen in Iida, and over two hundred *kijishi* in the vicinity. Early in the next decade, Iida was producing over 11,000 lacquered bowls annually, and the six lacquer wholesalers, twelve brokers, and twenty-nine retailers in the business handled over 150,000 hardwood items.[10] The local tobacco industry also grew during these years, particularly in the outpost villages south and west of Iida.[11]

The cargo transport business likewise appears to have expanded modestly during this period, while the infrastructure for transport remained largely unchanged. Most cargo continued to be carried on horseback, and even the tripartite organizational division of the Tokugawa period (between relay operators in the post towns, long-distance packhorse drivers, and informal village carriers) was maintained.[12] It is reasonable to conclude that the Nakasendō's transport business fairly collapsed when official subsidies were withdrawn at the end of the Tokugawa period, since in the Meiji era packhorse fares for all cargo passing through the tortuous Kiso valley were 30 percent higher than those for the Ina road.[13] While this surely sent additional business to the latter route after the Restoration, such changes are difficult to gauge from the fragmentary data available.[14]

10. Masaki 1978:421; Shimoina Chiikishi Kenkyūkai 1982:50. In 1882, Shimoina accounted for roughly half of Nagano Prefecture's lacquer production, generating nearly 2,000 yen in sales. Nagano-ken 1883.

11. On the growth of tobacco output, and the deepening division of labor in this industry, see Shimoina Kyōikukai 1962:23–25.

12. The overhaul and reorganization of the cargo-handling business is a complex story in its own right; its outcome, in brief, was a massive displacement of the private packhorse drivers by their (reorganized) post-station competitors. By 1882, only one-quarter of Shimoina's cargo was being shipped under the auspices of the struggling *chūma* organization. For details on the reorganization of the post towns, see Nagano-ken 1971:181; Ōsawa 1966:21. On the subsequent struggle for control of the trade, see H. Yamamoto 1972:211; Masaki 1978:392–95; Nagano-ken 1971:182.

13. Nagano-ken 1971:182.

14. Figures purporting to indicate the total annual volume of overland haulage are not available until the 1880s, and even then they vary from one year to the next by as much as 500 percent (e.g., from just over three hundred thousand *kan* in 1883 to nearly two million *kan* in 1884). Moreover, even the highest figure—given improbably for the recession year of 1884—is the equivalent of only 68,432 horse loads, or fewer than the 73,000 recorded in the Meiwa survey over one hundred years earlier (calculated at 28 *kan* per horse load). Statistics for 1882 are recorded in *Meiji jūgonen kangyō nenpō*; for 1883, in *Meiji jūrokunen Nagano-ken tōkeisho*; and for 1884 and 1885, in *Meiji jūhachinen Nagano-ken kangyō nenpō* (all Nagano-ken). Estimating transport capacity from the valley's rising horse population (which tripled between the middle of the Tokugawa era and 1883) is also somewhat problematic, since most of the expansion appears to have been

The one significant new element in Shimoina's cargo transport scene was an extension of shipping services on the Tenryū River. The new regime relaxed the former legal restrictions on use of the river, encouraging local ferry operators to expand their operations. By 1882, cargo could be sent upriver to Iida from the coast at less than half the cost of transporting the same load by horseback, and rivermen were ferrying an estimated 150,000 *kan* (560 metric tons) of goods a year in and out of the county.[15] Shortly thereafter, a Mikawa construction firm was hired to blast out the rock base of the Yagura waterfall, the last major physical obstacle to traffic in the middle stretch of the river.[16] Nonetheless, substitution of more efficient river transport for packhorses came toward the end of this first quarter-century, and even then supplied only a fraction (roughly 5 percent) of Shimoina's transport needs. In the interim, the county's overland transportation system, like its protoindustries, appears to have continued to operate at the same levels as before, and with basically the same technology as in the late Tokugawa period. In short, silk had to be accommodated into a regional economy whose traditional sectors were commanding sustained, if not increasing, levels of resources across the board.

The growth of silk output under these constraints was made possible by two conditions governing the early development of the industry. One was the very small scale from which it started. Although growth was rapid, silk's total value remained only a fraction of the region's gross product until the turn of the century. But a second and equally important condition was the industry's ability to attract capital from outside the producing regions. Funds were secured primarily from two sources: the dramatic (if erratic) profits earned in selling silk to Western traders, and a torrent of debased paper currency. Taken together, these conditions made it possible for Shimoina's farmers and merchants to multiply the area's total silk output many times over without fundamentally altering the fabric of the larger regional economy. The trick was to mobilize land and labor from within the region by using the lever of capital from outside the region. To understand how this process worked, and what it meant for the economic geog-

---

owing to increased demand for draft animals in agriculture. On numbers and uses of horses in this region in early Meiji times, see Ichikawa 1961 and Shimojō-mura 1977:1183.

15. Nagano-ken 1971:182. This adds 25 percent to the total handled by the overland carriers that year, and explains the sharp drop-off in traffic on the Enshū or Lower road, which closely paralleled the river.

16. Kanaya 1987.

raphy of Shimoina, requires looking at land, labor, and capital in more detail.

LAND RESOURCES

At first glance, finding suitable land for the silk mulberry (*kuwa*) might not appear to have posed major problems in the Ina valley during the early years. The total acreage needed was simply not substantial enough to displace most other crops: as late as 1880, mulberry plots occupied only 7 percent of Nagano Prefecture's arable. When the first comprehensive survey was undertaken in 1881, Shimoina's silk mulberry acreage was estimated at 750 hectares (*chō*), or 7 percent of the county's total cultivated area; moreover, most of that total consisted of unirrigated upland fields.[17] Yet a closer look at the spatial distribution of *kuwa* fields suggests that a complicated set of local trade-offs was required even to boost mulberry acreage to this modest level.

Such data as we have concerning the geography of silk-mulberry production by village during the first quarter-century of trade are depicted in map 18 (for census units, see map 19). While individual village figures are almost certainly questionable, the geographical pattern is clear: most of Shimoina's early *kuwa* expansion took place in the densely populated villages of the Ina basin, with one outlier in the northeast. The southern perimeter of Shimoina was completely left out of the picture. Environmental conditions do not account for this pattern. Because the silk mulberry is a hardy plant, tolerant of most soils and climates, it proved able to survive on marginal plots of almost any description anywhere in Shimoina, provided they were tolerably well drained. To understand why the silk mulberry made no headway in the southern part of the county, one must look rather to differences in history, land use, and market access.

The most suggestive correlation is that of historical precedent. Both of the areas that recorded significant mulberry acreage in the early 1880s—the valley core and the northeastern village of Ōshika—were significant sericultural centers during the Edo period. Even within the basin, the most extensive development of mulberry and silkworm culture was already taking place to the north and east of Iida: precisely the area that had nurtured the area's traditional textile crafts. This pattern underscores the dependency of the early silk trade on highly local

17. For statistics pertaining to sericulture across Nagano Prefecture in this period, see Nagano-ken 1971:144; Nagano-ken 1985:259.

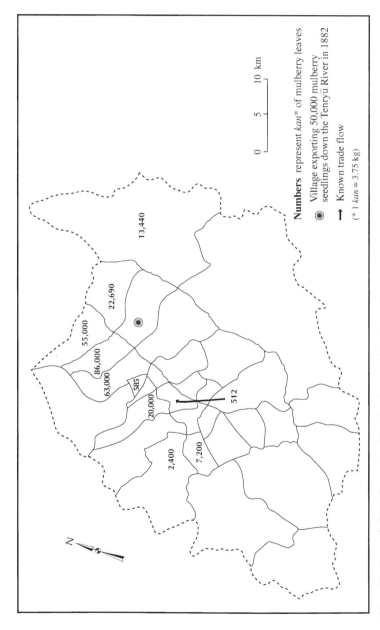

Map 18. Recorded Mulberry Leaf Production and Trade, Early Meiji. (Data from NCS and Toyooka–mura 1975:1050–51.)

Numbers represent *kan** of mulberry leaves

◉ Village exporting 50,000 mulberry
   seedlings down the Tenryū River in 1882

↑ Known trade flow

(* 1 *kan* = 3.75 kg)

0   5   10 km

13,440

22,690

55,000

86,000

63,000

585

20,000

512

2,400

7,200

N

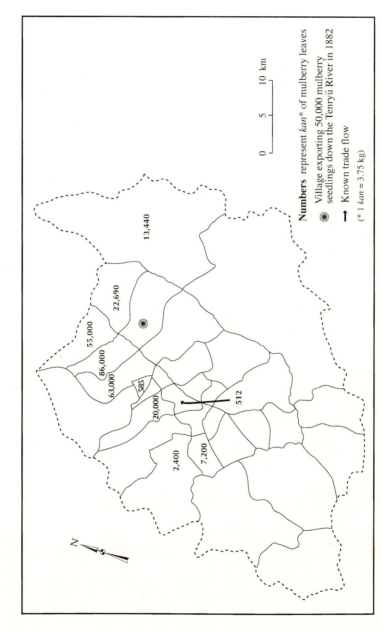

Map 18. Recorded Mulberry Leaf Production and Trade, Early Meiji. (Data from NCS and Toyooka–mura 1975:1050–51.)

raphy of Shimoina, requires looking at land, labor, and capital in more detail.

LAND RESOURCES

At first glance, finding suitable land for the silk mulberry (*kuwa*) might not appear to have posed major problems in the Ina valley during the early years. The total acreage needed was simply not substantial enough to displace most other crops: as late as 1880, mulberry plots occupied only 7 percent of Nagano Prefecture's arable. When the first comprehensive survey was undertaken in 1881, Shimoina's silk mulberry acreage was estimated at 750 hectares (*chō*), or 7 percent of the county's total cultivated area; moreover, most of that total consisted of unirrigated upland fields.[17] Yet a closer look at the spatial distribution of *kuwa* fields suggests that a complicated set of local trade-offs was required even to boost mulberry acreage to this modest level.

Such data as we have concerning the geography of silk-mulberry production by village during the first quarter-century of trade are depicted in map 18 (for census units, see map 19). While individual village figures are almost certainly questionable, the geographical pattern is clear: most of Shimoina's early *kuwa* expansion took place in the densely populated villages of the Ina basin, with one outlier in the northeast. The southern perimeter of Shimoina was completely left out of the picture. Environmental conditions do not account for this pattern. Because the silk mulberry is a hardy plant, tolerant of most soils and climates, it proved able to survive on marginal plots of almost any description anywhere in Shimoina, provided they were tolerably well drained. To understand why the silk mulberry made no headway in the southern part of the county, one must look rather to differences in history, land use, and market access.

The most suggestive correlation is that of historical precedent. Both of the areas that recorded significant mulberry acreage in the early 1880s—the valley core and the northeastern village of Ōshika—were significant sericultural centers during the Edo period. Even within the basin, the most extensive development of mulberry and silkworm culture was already taking place to the north and east of Iida: precisely the area that had nurtured the area's traditional textile crafts. This pattern underscores the dependency of the early silk trade on highly local

17. For statistics pertaining to sericulture across Nagano Prefecture in this period, see Nagano-ken 1971:144; Nagano-ken 1985:259.

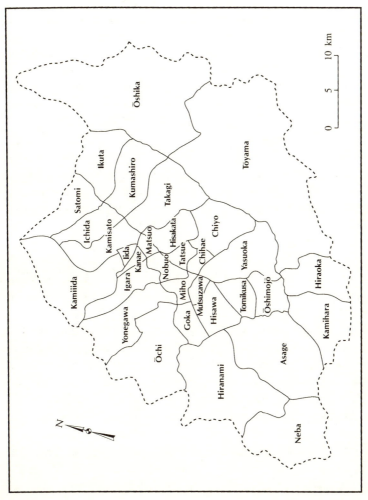

Map 19. Units Employed in *Nagano-ken chōson shi*, Early Meiji.

knowledge. Not until the last years of the nineteenth century would the state undertake major interventions to overcome the spatial inertia of production patterns inherited from the Tokugawa period.

Transport costs were a second essential consideration, affecting inputs as well as marketing of the finished product. While *kuwa* could be planted on any well-drained land, its productivity over time varied greatly according to the amount of nutrients that could be regularly supplied to the soil in which it was rooted. The main fertilizers applied to Nagano Prefecture's mulberry orchards in 1880 were *sake* lees, oil-seed cake, nightsoil, manure, rice bran, soybean cake, and silkworm droppings.[18] Without exception, such organic nutrients would have been more plentiful and less expensive along the main north-south transport corridor through the valley than in the relatively isolated southeast.

But there were further distinctions within the well-traveled north-south corridor as well. In particular, proximity to Yokohama gave a new competitive edge to the county's northern villages—an area traditionally known as "the interior" (*oku*) because of its remoteness from the Pacific coast. With the focus of the silk trade swinging from Kyoto to Yokohama, interior location was suddenly an advantage, and may have hastened sericultural development in the villages north of Iida, particularly along the east bank of the Tenryū. With its closely stacked, well-drained terraces, the east bank was suited to mulberry in any case, and this area soon showed high overall production levels closely resembling those of the traditional west-bank satellites.[19]

The same logic produced opposite results in southeastern Shimoina. Unfamiliar with the intricate rhythms of silkworm rearing, and separated by formidable ridges not only from the county's traditional sericultural centers but from the Yokohama market as well, villagers in this remote area chose instead to grow paper mulberry. The bark of the latter proved more durable—and more portable—than the leaf-mass of its sister tree, and could be packed out profitably to papermaking villages near Iida. Moreover, demand for paper-mulberry bark was strong at the time, since the traditional supply from the valley proper was beginning to contract as the silk mulberry (*kuwa*) displaced the paper mulberry (*kōzō*) in and around the core.[20]

18. Nagano-ken 1971:144.
19. In total output, northerly Kawano village (at 1,398 *kan*) bested the centrally located Matsuo as early as 1874, and Kumashiro was producing over 2,000 *kan* of cocoons annually by 1881. Toyooka-mura 1975:1050.
20. Igara-mura 1973:1121.

Finally, a third constraint on the distribution of the silk mulberry in these early years was its compatibility (or incompatibility) with other crops. This factor came into play in the southwestern outpost settlements, where although *kuwa* was introduced on a small scale, it did not show up in land-use surveys because it was relegated to untaxed wasteland. What kept silk mulberry from becoming a significant upland crop in this area was a conflict with existing agricultural commitments. Tobacco, a highly profitable cash crop during these years, proved physically incompatible with sericulture: if tobacco pollen alighted on mulberry leaves, it was found to poison the silkworms.[21] As a result, farmers in the Ina road outposts were only willing to plant mulberry on steep slopes where tobacco or grain could not be grown. Leaves were collected either from wild trees in the village commons or from casually tended orchards planted on steep hillsides. *Kuwa* might also be planted at the end of so-called swidden or burned-field (*yakihata*) agricultural cycles, but no permanent fields were converted to the silk mulberry in these villages.[22]

The pattern of mulberry expansion in central Shimoina presents a marked contrast. In the hilly country south of the basin, farmers began planting *kuwa* in permanent upland fields, displacing such tree crops as persimmon and paper mulberry.[23] In the valley floor, where land was more intensively used, mulberry bushes were first squeezed onto the narrow ridges between irrigated fields, in unused riparian floodplains, or even in dooryard gardens. To support further expansion, however, permanent fields, including some irrigated paddies, were also converted to mulberry in the valley core. As early as 1869, when the sharp rise in the price of silk had quintupled the cash value of mulberry leaves over late Tokugawa levels,[24] a number of landlords in the Iida area began experimentally converting rice fields to *kuwa*—a practice that was still officially banned by prefectural officials. The most famous was one Hara Rokuzaemon of Kamisato village, who promptly converted four *tan* of inferior-grade paddy to *kuwa*, becoming the first person in Shinano to

---

21. Shimoina Kyōikukai 1962:25; Toyooka-mura 1975:1064. In Ōchi and Seinaiji, where tobacco was most villagers' main livelihood, mulberry was not yet planted at all.

22. Shimoina Kyōikukai 1962:25. The post-swidden mulberry plantation was particularly common in the southern mountains around the Niino basin (Asage), where farmers maintained extensive acreage in *yakihata* fields throughout this period. The gazetteer compiled in the early Meiji period (NCS) reports 169 hectares of swidden fields for Asage village (and none elsewhere in Shimoina).

23. Igara-mura 1973:1121.

24. Nagano-ken 1971:5. Part, but not all, of this price rise was attributable to the general inflation of the times.

take established wet fields out of rice production for this purpose. Hara also kept records comparing gross income per *tan* from the two crops. His findings were compelling: the cocoons produced from one *tan* worth of mulberry leaves brought more than three times as much cash as the same acreage planted to rice (95.50 yen versus 30.25 yen). While Hara's figures did not take fertilizer, labor, or other differential costs into account, sericulture proved sufficiently profitable for *kuwa* plantations to spread rapidly in this area after land use was liberalized in 1871.[25]

In short, by the end of the first decade of the Meiji era, the silk mulberry had entered a variety of niches in Shimoina, creating at least three distinct production zones. In the mountain villages of the southwest, farmers planted mulberry only on marginal upland slopes unfit for grain or tobacco. Closer to the core, in the hilly country skirting the castle town to the south, *kuwa* became a major upland cash crop, displacing persimmon orchards and paper-mulberry plantations in permanent dry fields. Lastly, in the central agricultural areas of the valley, *kuwa* began supplanting grain crops, even in paddy fields, from early in the Meiji era.

LABOR FOR SERICULTURE

If local knowledge, market access, and existing crop requirements constrained farmers' choice of land on which to plant mulberry, accommodating the labor demands of sericulture involved a comparably intricate calculus. The biggest problem was one of timing. Sericultural work was strictly dependent on the weather, traditionally peaking sharply in April and May. This meant it competed not with protoindustrial by-employments, which primarily took advantage of slack-season labor during the winter, but with agricultural work.[26] A successful silkworm crop required constant attention from at least one adult woman during the very weeks of field preparation and planting; it also meant sparing occasional labor during the same weeks to tend the mulberry fields.[27] Increasing Shimoina's cocoon output under these conditions was accomplished through three strategies: staggering the silkworm-rearing season to better fit the agricultural calendar, enlarging the labor force, and enhancing on-farm labor productivity.

25. Hirasawa 1952:126; Tenryūsha 1984:136.

26. On the seasonal tempo of sericultural operations and their role in the rural household economy, see T. Smith 1986:291 and Vlastos 1986, ch. 5.

27. John Lossing Buck's study of eighteen localities in the Yangtze region found that labor requirements for mulberry were higher than for any other crop: 76 days per crop acre were required for rice, and 126 for tea, but 196 days were required for mulberry. Li 1981:148.

The first of these innovations was of limited use during the first quarter-century of silk expansion. In the main, the labor cycle in early sericulture remained stubbornly inflexible in its timing requirements. For the first (spring) crop of cocoons, which accounted for over 70 percent of Shimoina's production during this period, every stage of the sericultural process except orchard maintenance fell within a single feverish six-week period in April and May. During this period, the silkworms required unflagging attention. Since their body weight increased some 10,000 times during their brief life span, their appetites were voracious, and exacting hygienic standards had to be maintained to keep them free of disease.[28] Successful efforts were made in early Meiji times to stretch the sericultural season over more of the year. In 1872, for instance, farmers in Matsumoto devised a method of postponing the hatching of the worms by storing them in cool caves until later in the year, allowing a second crop of cocoons to be harvested in summer; hybrid silkworm varieties were later developed to permit the raising of a third generation in the fall.[29] But success in this case only meant that a second or third round of equal intensity was added in later months. No methods were discovered for reducing or smoothing out the labor requirements associated with each batch. Whether a household chose to raise one, two, or three generations of larvae per year, supplying fresh leaves every few hours and keeping the room warm and clean required nearly constant attention from one or more adults during the later weeks of the silkworms' lives.

Where the development of summer and fall silkworm varieties did make a difference was in extending the sericultural frontier into the mountainous Shimoina hinterland. In the temperate areas near the Tenryū River, spring remained the best season for sericulture; while May was a busy time for field preparation and planting, the agricultural labor peak here fell in June and July, overlapping with the summer rather than the spring silkworm cycle. Moreover, summer and fall silkworms were never as dependable as their spring counterparts; a local saying in southern Shinano claimed that "fall cocoons and miso soup never come out right" (*akigo to misoshiru wa atatta koto nashi*). As a result, the later-maturing silkworm varieties made little headway in central Shimoina during these initial decades.[30] But in the mountainous hinterland, the

28. For a vivid description of the associated labor and its temporal rhythms, see Li 1981:19–20.

29. Smethurst 1986:205–7.

30. In 1883, when summer and fall cocoons accounted for 29 percent of the total crop in the prefecture, they made up only 22 percent of the total in Shimoina *gun*. Nagano-ken 1971:145–46; Nagano-ken 1985:270–1.

situation was reversed. Since these high-altitude villages depended heavily on winter crops that were harvested in the spring, summer and fall varieties of silkworms were essential for sericulture to take hold in these areas. Labor simply could not be freed for rearing spring silkworms here until the 1910s, when improved transport allowed cheap grain imports.[31]

If juggling the timing of labor demands was one way to make workers available for sericulture, another was to increase the labor supply, whether through rising fertility or by encouraging in-migration. Both proved important for the Ina valley in the long run. Shimoina experienced a steady rise in population beginning in the mid nineteenth century, partly because of rising birth rates but largely owing to immigration from the surrounding counties.[32]

Since sericultural work fell primarily to women and girls, it is notable that the northern area of Shimoina—which by the end of the transition years was maintaining the highest overall sericultural yields—had both larger than average households (map 20) and a slightly elevated proportion of females in residence (map 21). The most important pattern revealed in map 20 is the existence of two disjunct areas with smaller than average households: the Iida vicinity, and the far southeastern periphery (a forestry zone that was experiencing considerable out-migration at the time). In the remainder of the county, household size ranged from average (4.8) to slightly large (6.4), with the largest households clustered in the relatively isolated (and agriculturally conservative) Niino basin to the south. The spatial variation of sex ratios demonstrated in map 21 likewise shows important correlations with economic activity. Most important, agricultural villages throughout the valley floor had sharply lower sex ratios (i.e., a higher percentage of women) than did mountain hamlets in the southern and eastern periphery. (The former castle town of Iida stood out as having an unusually high proportion of men within the core zone of generally lower sex ratios, an anomaly that

31. Toyooka-mura 1975:1078.

32. The combined village totals given in NCS, which was compiled between 1874 and 1879, indicate that Shimoina was home to 19,829 households with a total of 95,216 persons; figures recorded in the first prefectural census of 1883 were 21,062 and 105,574, respectively (Nagano-ken 1883). Taking into account the variable dates of the individual village surveys giving rise to the former figures, and the likelihood that the later census was more thoroughly executed, nothing definitive can be averred concerning the 11 percent difference between these two numbers. According to both sources, the resident population was in any case slightly greater than the permanently registered population (by 1 to 3 percent in most villages), indicating a tendency toward additional in-migration. An excess of births over deaths during 1883 of 1,118 (2,681 births versus 1,563 deaths [Nagano-ken 1883]) suggests a natural growth rate of 1.15 percent per annum.

was most likely owing to its concentration of male shop assistants and tradesmen.) As a comparison of the two maps soon shows, the sole area where large households coincided with low sex ratios was north of Iida, along the banks of the Tenryū River.

One important determinant of this pattern was migration. Not surprisingly, information in an early Meiji gazetteer concerning temporary migration for work (*dekasegi*) indicates that, at the end of the first decade of the new regime, these same villages were experiencing a minor net gain of female workers. By 1883, this tendency was marked for the county as a whole, particularly its rural areas. Women constituted only 52 percent of temporary migrants to Iida, but a striking 81 percent in rural Shimoina County (table 14).[33]

A third means of mobilizing labor for silkworm rearing was to enhance the labor productivity of agriculture: specifically, to liberate existing household workers from other springtime tasks by finding substitutes for their customary labor. In Shimoina, as throughout Japan, this was accomplished by the introduction of a complex of labor-saving practices known collectively as the Meiji agricultural system, or *Meiji nōhō*.

The central features of the *Meiji nōhō* were deep plowing (with draft animal power) and the application of commercial fertilizers. Thousands of horses were still kept by Ina valley peasants at the end of the Tokugawa period for use as pack animals and for their manure, but the only agricultural work they had traditionally performed was trampling organic mulch into the paddy mud before transplanting. The introduction of deep plowing methods in the early Meiji era called for much more extensive use of horses and oxen, in upland fields as well as in paddy.[34] A history of the northern Ina valley specifically notes that the use of horses in field preparation there, which began roughly in 1877, was critical in freeing household labor for sericulture.[35]

Commercial fertilizers played a similar role. Organic wastes produced within the valley, whether residential, agricultural, or industrial

33. Oral histories testify that the Ina valley was a favored destination for both men and women from Kiso, who found food cheaper and opportunities for work greater in Shimoina than at home; see, e.g., Chiba 1963. Presumably the same attraction operated in the mountains of interior Mikawa and Tōtōmi, from which many female workers are known to have entered Shimoina seeking jobs in the filatures in later decades (Sawada 1981). What is certain is that in-migration of temporary female labor was much more pronounced in Shimoina than in the prefecture as a whole.

34. For a more extended discussion, see Francks 1984:56ff.

35. Horiguchi 1972:157. By 1879, Shiomina's horse population had reached 9,829; twelve years later it stood at 10,118. A survey undertaken at that time found that 81 percent of the county's horses were kept primarily for agricultural use. Ichikawa 1961.

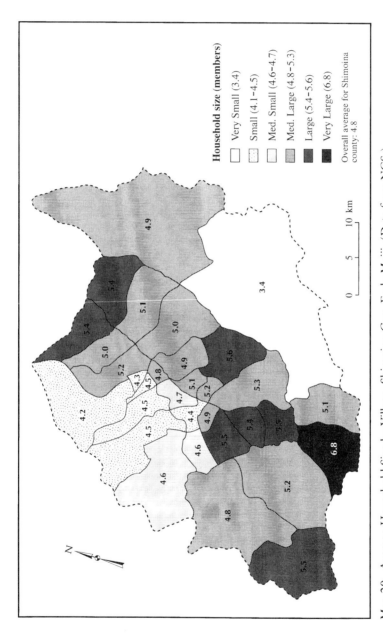

Map 20. Average Household Size by Village, Shimoina *Gun*, Early Meiji. (Data from NCS.)

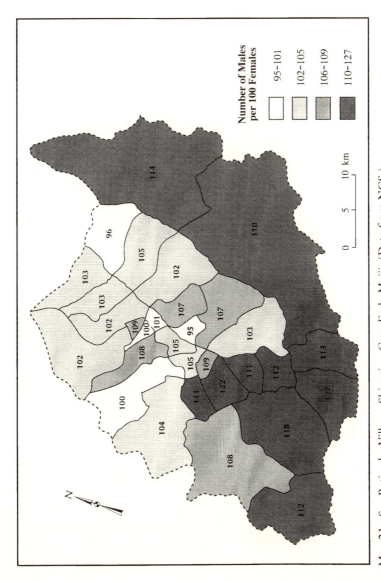

Map 21. Sex Ratios by Village, Shimoina *Gun*, Early Meiji. (Data from NCS.)

Table 14 *Temporary Migration Patterns for Shimoina, Iida, and Nagano Prefecture as of January 1, 1883*

	Out-Migrants	In-Migrants	Net Migration	Percentage of Total Net In-Migrants
Rural Shimoina				
Men	1,026	1,064	+38	17
Women	770	932	+162	81
TOTAL	1,796	1,996	+200	100
Iida				
Men	201	516	+315	48
Women	137	474	+337	52
TOTAL	338	990	+652	100
Nagano Prefecture				
Men	13,810	12,503	−1,307	—
Women	9,351	10,347	+996	100
TOTAL	23,161	22,850	−311	100

SOURCE: Nagano-ken, ed., 1883.

by-products, had been carefully husbanded and returned to the fields since at least the early Tokugawa period. As the silk economy expanded, however, the scope of commerce in soil nutrients increased; by 1882, business was sufficiently brisk to support eight fertilizer dealers in the former castle town.[36] In addition to a rising output of organic by-products, several lime kilns were booming in the Ina valley at this time as well, and additional fertilizers were imported into Shimoina from the northern Ina valley.[37] The ability to purchase fertilizers was crucial for sericultural expansion in the valley, since grasses and leaves had traditionally been harvested from the commons in late spring. Reducing this labor-intensive activity created substantial labor savings during the period of peak sericultural activity.[38]

A final source of springtime labor savings was the ending of transport corvée. Several Shimoina villages had been enlisted to help the strug-

36. Iwashima 1967:75.

37. A roster of items sent south from Takatō to Iida in 1875, for instance, mentions seventy-nine horse loads of pressed soy cakes, twenty-eight loads of oil-seed cakes, another twenty-eight of rice bran, and eleven of ashes. Takatō-machi Kyōiku Iinkai 1966:183.

38. On the procurement of green manure from the commons, see Wigen 1985, ch. 2.

gling post towns of the Nakasendō as early as 1715, and the particularly heavy traffic of the later Tokugawa period had increased this loathsome labor tax appreciably. (The most demanding year by far was 1861, when the imperial princess Kazunomiya processed from Kyoto to Edo to marry the shogun. Some 10,000 porters and 2,000 horses assisted her retinue of 25,000 as she passed through the Kiso valley on her way to Edo. From Shimoina alone, 155 villages dispatched 6,000 men, 650 horses, and a prodigious quantity of provisions to assist the company, which required a full day and night to pass each station along the road, and ten days to wend its way through Shinano.)[39] Although officially abolished in 1869, transport corvée in the southern Japanese Alps effectively continued for half a dozen years more, but by the second decade of the Meiji era, this work had shifted entirely to professionals. This, too, freed rural household members for alternative springtime employment, since the bulk of official traffic had traditionally been during the spring.

In sum, the labor requirements for an expanding sericultural sector were met in three principal ways: by developing late-maturing silkworms to accommodate a variety of agricultural calendars, by expanding the total pool of workers (through both natural increase and immigration), and by freeing farm workers from springtime tasks (primarily through commercial fertilizers and labor-saving cultivation techniques). With these innovations, sericulture was able to spread into rural hamlets across much of the region and the prefecture; as early as 1882, two-thirds of all Nagano households were raising silkworms, and cocoons, floss, and silkworm eggs together accounted for a remarkable 39 percent of gross prefectural product.[40]

CAPITAL FOR SERICULTURE

The ability of rural households to buy commercial fertilizers brings us to the question of capital. Farm families embarking on silkworm culture might go far without spending a single yen; the main resource required was hard work. Yet even on a modest, cottage-industry scale, initiating sericulture required access to a certain amount of cash.

The capital outlay required to take up sericulture may have been small, but it was not negligible. Mulberry could readily be propagated by cutting and rooting branches from existing stock, but where that was not

---

39. Hirasawa 1958. Shimazaki Tōson's famous novel *Before the Dawn* includes a vivid description of the logistical nightmare created by Kazunomiya's progress through the Kiso valley; see Shimazaki 1987, 135–40, 149–55.

40. Nagano-ken 1971:144–45.

possible, seedlings had to be purchased.[41] Silkworm eggs had also to be purchased, at least for a family's first sericulture venture and often thereafter.[42] Baskets for the growing larvae had to be made or bought and fitted into shelving racks. Finally, labor had either to be mobilized through kinship networks, hired in at going wages, or liberated within the household. While the women who worked in Nagano's prewar sericultural rooms speak vividly of having simply slept less and worked more during silkworm season, even this arrangement often necessitated some form of labor-substitution expense, whether to buy commercial fertilizers or to rent a neighbor's horses.

An additional requirement for undertaking sericultural operations was indoor space suitable for raising silkworms. Again, many families simply gave up part of their living space each spring to make way for the voracious larvae; local women tell stories of having slept (and, in one memorable case, of having given birth) between racks of munching caterpillars.[43] But there were limits to the scale of operations that could be conducted in this manner. As early as 1872, wealthy Iida-area families began adding new rooms and even second or third stories to their dwellings to house silkworms, initiating a practice that would spread widely throughout Ina in the later Meiji and Taishō periods (many of the largest farmhouses in the valley today having been built during these years).[44]

More concentrated than sericulture itself was the distribution of related commercial enterprises. After two decades of silk expansion, all of the wholesalers, brokers, and retailers connected with Shimoina's silk industry—with the single (albeit important) exception of silkworm egg retailers—were located exclusively in the former castle town and its nearest satellites (Kamisato and Matsuo) (table 15). In the long run, as will be discussed below, sericultural diffusion across Shimoina was financed less by the local landed elite than by brokers with Yokohama connections, who advanced egg cards to thousands of rural households in exchange for rights to the finished cocoons. Yet the region's landowning

41. Mulberry seedlings cost 1.4 yen per hundred at wholesale rates in 1882, the equivalent of five to six days' wages for an agricultural laborer. Tatsuoka-mura 1968:1000–1001.

42. A card bearing 1,000 silkworm eggs cost 1.25 yen in Shimoina in 1884 (Nagano-ken 1985:605). Rearing the silkworms that would hatch from one card required the labor of two persons.

43. Shimoina-gun Rengō Fujinkai 1983:19–20.

44. Demand for home improvements of this kind may have been one of the forces sustaining the high local wage rates for men in the building trades, documented in Chiba 1963 and Nagano-ken 1985:634.

Table 15   *The Location of Wholesalers, Brokers, and Retailers Connected with Sericulture in Shimoina, ca. 1880*

	Matsuo	Kamisato	Iida	Other
Mulberry brokers	2			
Seedling retailers	1			
Silkworm-egg retailers	7	3		5
Raw-silk brokers	9	7		
Raw-silk wholesalers	3	2		
Waste-thread brokers		20		
Unspecified			20	
TOTAL	22	32	20	5

SOURCES: Matsuo-mura 1982:604; Kamisato-machi 1978:1099; Iwashima 1967.

class played a crucial supplementary role. Well-to-do landlords from the center of the Ina valley became important agents of sericultural expansion in their villages, mainly by augmenting their own households' production, but also by extending loans to neighbors and relatives. It is surely not coincidental that sericulture took root initially in areas of the county with established capital resources. In fact, the geography of early Meiji cocoon production in many ways mirrored the differential distribution of wealth accumulated during the Tokugawa period. Satellite villages in the vicinity of the castle town (centered in the newly consolidated villages of Matsuo, Kanae, Kamisato, Zakōji, and Takagi) quickly emerged as the area's leading producers, with the landlord class in those villages accounting for the bulk of local cocoon output. While production levels in the heart of Shimoina's sericulture district averaged only three to five *kan* (10 to 20 kg) per household per year, a few wealthy families sold as much as 70, 80, or even 300 *kan* (up to 1,100 kg) of cocoons annually during the first Meiji decade.[45]

THE GEOGRAPHY OF SERICULTURAL PRODUCTION

Map 22 shows the balance that had been struck between the various forces shaping Shimoina sericulture at the end of the first Meiji decade. By this time, most villages in and around the populous agricultural heartland of the valley were producing cocoons. The landlord class was particularly active in the new industry: planting mulberry

45. On sericultural conditions in the core area during these early years, see Kawajimura 1988:292ff.; Kanae-machi 1969:559–60; Hirasawa 1952:126ff.; Masaki 1978: 405–6.

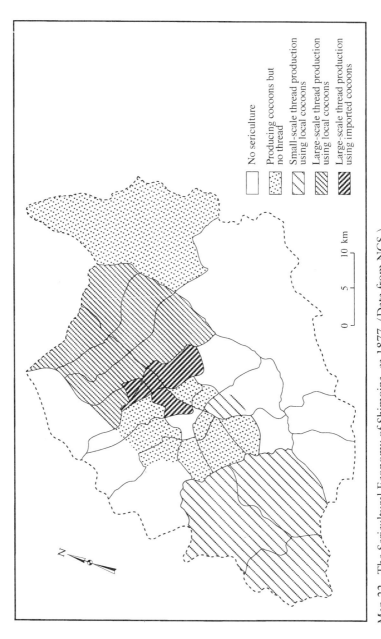

No sericulture

Producing cocoons but
no thread

Small-scale thread production
using local cocoons

Large-scale thread production
using local cocoons

Large-scale thread production
using imported cocoons

0   5   10 km

N

Map 22. The Sericultural Economy of Shimoina, ca 1877. (Data from NCS.)

(mostly on marginal strips of land but, in a few cases, in prime, irrigated fields), nourishing the plants with commercial fertilizers, employing paid help to tend the silkworms, and even adding rooms to their houses in order to expand their capacity. But sericulture was not limited to village notables. Already by 1876, more than half of all farm households in centrally located villages were involved in cocoon production.[46]

Away from the fertile valley terraces, sericulture was slower to take hold. No mulberry was grown as yet in the isolated southeastern quadrant of the county, but small quantities of silkworms were being raised both in the far northeast and in the transport-oriented outposts south of Iida along the Ina road. The former represented a carryover from the Edo period, while the outpost villages were relative newcomers to sericulture. In both zones, however, production levels per household were low; wealthy households in outlying areas fell far short of attaining the high outputs of their counterparts in the core. Innovations diffused later to these peripheral areas as well, with a lag of six years or more. Some mountain villages relied on wild strains of mulberry, and primitive rearing technologies, throughout the first decade of the Meiji era, only gradually converting after 1880 to improved strains and deliberately fashioned rearing rooms. Finally, production in the mountain villages was heavily biased toward the less reliable summer and fall cocoons.[47]

Cutting across these major zones, however, was a second axis of differentiation, distinguishing two contrasting types of spatial organization. In both the marginal outlying areas and the northern core, territorial specialization in the growing sericultural enterprise was minimal; locally harvested leaves were fed to local silkworms. North of Iida, as a rule, most mulberry leaves appear to have been processed within the villages where they were grown.[48] The southern terraces, by contrast, developed a considerably deeper spatial division of labor, with some hamlets and even whole villages specializing in mulberry production while others focused to a greater extent on producing cocoons.

46. Kanae-machi 1969:559–60; Matsuo-mura 1982:604; Kamisato-machi 1978: 1099.
47. In Ōkawara (present-day Ōshika), for instance, all but 5 of the village's 228 households raised silkworms in 1872, but the highest household cocoon yield was a mere 13.8 *kan*. Average output here was less than 4 *kan*, and in Ōchi and Yamamoto, less than 2 *kan*. On early sericulture in the county's periphery, see Miura 1964:15; Yamamoto-mura 1957:462; Ōshika-mura 1984:226; Achi-mura 1984, 1:720.
48. Cargo records indicate that the village of Kumashiro shipped 50,000 mulberry seedlings down the Tenryū River in 1884, bound for Shizuoka Prefecture (Tatsuoka-mura 1968:1000–1001). See map 18. Yet there is no evidence of local trade in leaves north of Iida, and village production ratios of leaf mass to cocoons hewed closely to realistic figures (ranging from lows of 12:1 to modest highs of 22:1; see following note). NCS.

The emergence of this kind of territorial specialization in the former Shimogō, or southern area of Iida domain, is worth noting. Mulberry leaves were a bulky, perishable cargo, incapable of bearing transport costs beyond a narrow range; they were also required in enormous quantities on a daily and ultimately hourly basis at the peak of the sericultural season. Sustaining an annual cocoon output in the range of 80 *kan* (300 kg)—a level attained by numerous wealthy households in the satellite villages—required conveying five to six metric tons of leaf mass from the fields to the feeding rooms, and that during the busiest time of the year.[49] Under these circumstances, there were strong incentives to locate mulberry fields as close as possible to the centers where the larvae were reared. Yet scattered evidence suggests that, in the area south of Iida, mulberry leaves were traded over a distance of up to ten kilometers. Just as they had done with paper-mulberry bark in preceding years, farmers in the hilly Nanbu uplands evidently packed silk-mulberry leaves north to sericulturists in settlements nearer the core.[50] The proliferation of this trade in the Iida area is further verified by 1879 tax records for Matsuo, which list two brokers specializing in *kuwa* leaves.[51]

In short, the landscape of sericultural production in Shimoina during the first twenty-five years after Perry arrived was richly differentiated, reflecting the uneven dispersal of a multitude of relevant resources and constraints across the region. On the one hand, a concentration of previously acquired experience and capital ensured that the traditional agricultural core would see the fastest initial growth, just as inherited spatial inequalities of mercantile connections and expertise acted to keep most commercial services for the enterprise within the Iida area. Yet at the same time the search for both land and labor at the margins and in-

49. The ratio of mulberry leaf mass to cocoon output in this period ranged from 15:1 to 20:1 or higher. Two local figures are available for this period, indicating leaf-to-cocoon ratios of 18:1 for one producer in Kamisato village in 1871 (Hirasawa 1952:127) and 15:1 for Takagi village in 1884 (Takagi-mura 1979, 2:360). Lillian Li has compiled a table showing yield ratios in China from the eleventh to the twentieth centuries; most fall between 11:1 and 25:1 (Li 1981:35).

50. NCS contains two valuable direct references to *kuwa* trade across village lines, attesting that both Yasuoka and Ōchi villages exported mulberry leaves. In addition, *kuwa* trade may reasonably be inferred from the wide discrepancies in reported ratios of leaf output to cocoon production across the region. Obvious leaf exporters included Yasuoka village, which produced an unspecified amount of *kuwa* but reported no cocoons, and Igara village, whose six hamlets produced both commodities, but in ratios ranging from 23:1 to 240:1 (Tenryūsha 1984:136). The hamlet producing 240 times as much mulberry leaf as cocoons was adjacent to Iida, an obvious importer, with a mulberry/cocoon ratio of 4:1.

51. Matsuo-mura 1982:604.

terstices of a heavily committed rural landscape was already drawing sericultural activity outward from this initial core. If silk was to be produced within the framework of a still-flourishing agricultural and protoindustrial economy, it would have to do so extensively as well as intensively, taking advantage of widely dispersed scraps of land, and employing labor on the farms where the bulk of the county's people lived. It was above all the diffusionist thrust of these land and labor requirements that would push the sericultural frontier steadily outward across the region.

The Filature Sector: A New Geography of Finance

The geography of reeling during these years followed a diffusionist pattern similar to that of sericultural operations. Proximity to cocoon-producing regions was the most important consideration in locating a filature; even large firms found it more profitable to operate small factories in the midst of the mulberry fields than to consolidate operations in a central location, to which the fragile cocoons would have to be transported over rough mountain roads. Thus, as sericulture spread into most of Shimoina's hamlets during the second decade of the Meiji era to take advantage of rural domestic labor, reeling operations soon followed. The volume of raw silk reeled in outlying villages was quite low, but by the eve of the Matsukata deflation, small mills had begun to appear in scattered locations outside the valley core.[52]

In contrast to the diffusionist forces that shaped the landscape of production, however, the geography of finance obeyed a centralizing imperative. Particularly in silk reeling—the most profitable sector of the industry—concentration was the rule. At first, most of Shimoina's silk was reeled by hand in domestic workshops, using simple equipment. But as European producers recovered from the silkworm blight of the early 1860s and reentered the market, Japanese exporters faced stiffer competitive standards for thread quality. Meeting those standards required mechanizing the reeling process, which in turn called for substantial capital investments.

52. Yagi 1980:45; Tenryūsha 1984:157. Indeed, a spatially diffused pattern of development has been identified as a hallmark of Japan's textile-based industrialization as a whole (in contrast with the much more concentrated geography of heavy industry); Kandatsu 1985:142.

That such mechanization was initiated repeatedly by local capitalists is compelling testimony to the vibrancy of Shimoina's late Tokugawa and early Meiji economy. By 1878, the bulk of Ina valley cocoons were being processed by wage-workers operating water-powered equipment in modest village filatures—all within the region's borders—most of which were owned by local landlords and merchants. But even those who managed to meet the rising overhead costs were buffeted by wildly fluctuating markets and unpredictable supplies, pushing local filature operations out of business almost as quickly as they formed. In the long run, sustaining large-scale factories equipped with superior reeling technology was possible only by concentrating capital resources from throughout the region—and beyond—in the hands of a limited number of industrialists.

The degree of such concentration was increased, in this as in all sectors, by the bankruptcies and financial restructuring following the Matsukata deflation of 1884–85. But the roots of the subsequent shift toward outside control of Shimoina's economy, and of the accompanying class differentiation within the region, must be sought in the earlier transition period. To recover the first signs of this sea change in the region's economic orientation requires tracing out the paths by which outside capital insinuated its way into Shimoina's silk economy before 1883.

OUTSIDE CAPITAL IN THE EARLY SILK INDUSTRY: THE ERA OF SPECULATION

The dawning of international trade coincided with the twilight years of the Tokugawa regime, not surprisingly a time of minimal state intervention in the economy. During this chaotic transitional period, no state-coordinated investment in Japan's silk-producing capacity took place; on the contrary, existing regulations on commerce broke down. In the resulting regulatory vacuum, while thousands of individual farmers scrambled to increase their own silk output, the efforts of large operators were directed solely toward speculative buying and selling. Two new sources of funds lubricated those operations: foreign exchange, now entering the country through Yokohama, and a flood of debased local currencies.

The entry of foreign exchange was tightly controlled. Japan's unequal treaties of 1858 with five Western powers (the United States, Britain, France, the Netherlands, and Russia) had left one key privilege in the hands of the Japanese: control over all domestic circulation of export-oriented trade goods. To their chagrin, foreign merchants were restricted to their compounds in the treaty ports, effectively prevented from combing the countryside to bargain directly with producers. Moreover, to extract the best possible terms from the foreign buyers, the Bakufu concentrated control over the domestic side of the trade in the hands of a select Japanese merchant elite. In 1867, eight merchant houses, beginning with Mitsui, were selected to form an exchange association with exclusive rights to negotiate with foreign merchants over all export goods passing through Edo to Yokohama.

Members of this association would forward loans, amounting to four-fifths of their export earnings, to regional purchasing agents (*urikomi ton'ya*). These agents would in turn lend cash to subcontractors (*ninushi*) or brokers (*nakagai*) in the countryside (often *gōnō* or *gōshō*, local rural notables or wealthy merchants) to enable them to purchase cocoons and floss on consignment from local producers. As capital flowed outward through this network from Yokohama to the provinces, silk flowed toward the port along the same routes, ultimately for export to Europe. This pyramid, dominated by the eight Japanese merchant houses at its apex, was the only sanctioned route by which Western capital could penetrate the silk districts of late Tokugawa Japan. Foreign merchants were prevented not only from dealing directly with producers but even from contracting independently with Japanese agents in the silk regions.[53]

In the first years after the opening of the ports, capital thus entered Shimoina through the agency of brokers with Yokohama connections, who scrambled to corner the limited existing stocks of cocoons and thread for the booming world market. These dealers were not only the primary vehicle by which the foreigners' money passed into the producing regions; they were also major beneficiaries of the new trade, frequently profiting at the producers' expense. Throughout Japan, opportunistic and shady practices were endemic to the broker's profession: "the declaration of freedom of commerce, together with the opportunities created in the port cities, opened the doors to a new brand of marginal, corner-cutting money makers who would crisscross the countryside buying up export goods and making their quick kills with currency speculation and armament trading . . . wreak[ing] havoc with the well-established wholesaler trade patterns."[54]

53. Unno 1985:174–79. On the contrasting arrangements prevailing among Chinese silk brokers, see Li 1981:91–92.
54. Hirschmeier and Yui 1975:92. Glowing tales of such men's successes are not lack-

Shimoina gave rise to one such quick-kill artist, Tanaka Heihachi. Born in 1834 in the central Ina valley, Heihachi was sent at age eleven to work for a merchant in Iida. To his family's dismay, a cycle of gambles and misadventures, each larger than the last, made up most of Heihachi's young life. An apocryphal story from Tanaka's youth highlights his precocious entrepreneurial bent. Two years into his service in Iida, Heihachi was entrusted with a small sum for an errand. Instead of following orders, he absconded with the money to Nagoya, intent on founding his own fish-vending business. The adventure ended in disaster; his father eventually had to send him back to Nagoya with money to repay debts he had incurred in the process of losing his initial two *ryō*. Before parting with his father's money, however, the intrepid Heihachi once again chose to gamble with capital that was not his own, succeeding this time in making a profit for himself before turning over the sum he owed.[55]

When trade with the West was initiated in 1859, the 25-year-old fortune-seeker promptly left for Yokohama, where he befriended a trader who commissioned him to travel through southern Shinano to purchase cocoons and thread. Ten years later—after a dizzying sequence of windfalls and mishaps, in which he built and lost several fortunes—Iida's Tanaka Heihachi had become famous as Tenka no Itohei, the "silk dealer without peer under heaven." He had also emerged as one of Yokohama's wealthiest merchants. Within a decade, his Tanaka Ginkō would become the official bank for the Nagano prefectural government, playing a decisive role in financing a new generation of silk dealers in Nagano as well. Yet even in his banker's guise, Tanaka Heihachi specialized in the speculative brokerage business rather than in building productive capacity.[56]

The new breed of corner-cutting money-maker epitomized by Itohei won little but contempt from the traditional merchant community. In the Ina valley, the established elite's reactions to the new international trade were strongly colored by a tradition of nativist thought (*kokugaku*), which had spread widely through the area by the 1850s. The influential scholar Hirata Atsutane (1776–1843), who had turned the rhetoric of nativism away from poetics and toward political action, had gained an impressive posthumous following among the rural elite of southern Shi-

---

ing in the literature. Wakao Ippei of Yamanashi Prefecture, singled out by Richard Smethurst (1986:142–50) as a prototypical rural entrepreneur of the age, is commended for amassing his initial fortune during these years by scouring the countryside for "Yamanashi crystal," which he could sell for stupendous profits in Yokohama.

55. T. Takahashi 1917; see also Kobayashi 1967.
56. Nagano-ken 1971:171–74.

nano. Indeed, such was their enthusiasm that the valley's converts by themselves raised funds to publish ten volumes of Hirata's life work, the *Koshiden* (Commentaries on Ancient History), which ran to thirty-odd volumes.[57] Having imbibed antiforeign sentiment at nativist lectures and poetry meetings since the 1830s, the merchant-landlords of the Iida area readily denounced the silk trade as a betrayal of Japan. But they saved their strongest invective for the new breed of unscrupulous dealers. Local silk brokers with connections to the Yokohama trade were perceived to be bankrupting their country, and their own province, for purely personal gain. The nativist Matsuo Taseko, later the Iida area's leading activist in the attempt to overthrow the Bakufu, composed a poem "to be read lamenting the export of thread to other countries," in which she charged:

It is disgusting, the disorder in thread everywhere. After the ships from foreign lands came for the jeweled silkworm cocoons to the country of the gods and the emperor, before anyone was aware of it, the abomination of trading expanded until for those who make mountains of silver, only empty Chinese smiles remained in place of their hearts.[58]

A visiting fellow poet and nativist from Kōshū, a neighboring sericultural center, wrote another poem lamenting the sad state of affairs whereby, in the very regions that produced the coveted cocoons, elders walked about in thin garments pierced by the wind. And in a similar vein, an anonymous activist posted the following notice on one of the main bridges into Iida on a summer night in 1863, addressed to a local silk dealer: "Ōharaya Kenzō, take heed. To sell raw silk to the foreigners is the ultimate insolence. If you cease all such activity forthwith, you will be pardoned, but if you do not, divine punishment will surely be your reward. Signed, Kamiyo Michinosuke [roughly, He who walks in the path of righteousness]."[59]

The market disorder that the nativists decried was heightened by an influx of debased money, which opportunistic dealers used to finance their local purchases. *Nibunkin*, literally meaning "two parts gold," was

---

57. Of 3,745 Hirata disciples in Japan in the later 1850s, Shinano accounted for 627—more than any other province—and a remarkable 61 percent of the Shinano total hailed from the Ina valley. In *Before the Dawn*, Shimazaki Tōson, a native of Kiso, notes the fervor of Iida's Hirata followers (see, e.g., Shimazaki 1987:126, 189). For an astute analysis of the political implications of Hirata's philosophy, see Harootunian 1988; for more on the Ina valley nativist movement, see Masaki 1978:103ff.

58. Walthall 1989:5.

59. Hirasawa 1952:113.

a name given to the severely debased currencies minted by many domains (as well as by the Edo government) in the last years of the Tokugawa era. Most such coins originated in the southwestern domains of
Satsuma, Chōshū, and Tosa, which were forced to go heavily into debt
arming themselves to battle the Bakufu. After the opening of the ports,
when brokers from throughout Japan flooded into the Ina valley to buy
up thread as well as paper, lacquerware, and other goods for export,
they brought *nibunkin* with them, often obtained at a discount in the
course of previous deals in the southwest. A single merchant from Ōmi,
after receiving such discounted currency from Satsuma *han* in exchange
for a delivery to that domain's warehouses, brought 10,000 *ryō* worth
of the debased money into the Iida area and convinced local peasants to
accept it at face value for their raw silk. Through similar practices, unscrupulous dealers in paper, mulberry bark, and hairdress ties as well
had put over 100,000 *ryō* worth of virtual counterfeit in circulation in
the Iida area by the spring of 1869, when the new government outlawed its use.

That summer, riots broke out simultaneously throughout Shinano.
On the first three days of the seventh month of 1869, some thirteen
thousand infuriated peasants stormed the town of Iida, attacked the
property of several merchants who had perpetrated use of the debased
cash, and demanded that the government exchange the old currency
for new at face value. The disturbance soon spread to northern
Nagano, where a mob of thirty thousand rose up against the new prefectural government. Similar protests were touched off throughout the
country, but the most sustained violence occurred in Shinano, which
was later estimated to have had more illicit currency in circulation than
any other province in the nation. In Shimoina, the counterfeit riots
grew into the most massive eruption of popular protest of the nineteenth century.

While the central government made a limited amount of gold available to each domain for redeeming the illegal currencies at less than face
value, the officials of Iida *han* saw that they would have to do more to
appease the local populace. As a first step, they collected 20,000 *ryō* in
fines from merchants found guilty of currency abuses. Using these funds
as collateral, the *han* then issued 30,000 *ryō* of its own bills, which it ex-
changed for the *nibunkin* currency on a one-for-one basis. Private parties in the newly formed Ina Prefecture, which administered former
Bakufu lands in the valley from 1868 to 1870, attempted to head off the
crisis in their territories in a similar fashion. Wealthy landlords and mer-

chants keen on restoring order absorbed the loss of the conversion and proceeded to establish a joint investment fund to recoup their losses.[60]

These interventions signaled a decisive break in commercial practice in southern Shinano. The virtually unregulated buy-and-sell frenzy engendered during the last years of the Tokugawa regime, while creating the climate in which operators like Tanaka Heihachi could acquire fabulous fortunes, had also brought on unacceptable levels of civil unrest. Jolted in part by the uprisings, local leaders helped usher in a new era of more orderly capital accumulation: one where investment in the means of production would be rewarded at least as much as speculative dealing, and where economic activity in the critical export sectors would be firmly presided over by the new Meiji state.

TECHNOLOGICAL CHANGE AND RISING
CAPITAL REQUIREMENTS IN THE FILATURES

To promote investment in the silk industry, state intervention was fast becoming essential. A primary force driving local filaturists toward dependence on the resources of the national treasury was the sheer rate of technological change, continually raising the investment ante.

The first steps in reeling silk—namely, immersing cocoons in a basin of hot water to loosen the ends of the fibers, which were then picked out by hand for threading onto the reel—were little affected by technological advances, and remained the core of the labor process in the filatures throughout the period under consideration. Innovations centered rather on the motive power turning the reel. In the traditional Japanese hand-reeling system called *tebiki*, used throughout Shinano before 1859, the operator's right hand turned a reeling frame, while her left hand gathered, twisted, and fed the fibers onto it. An improved treadle machine (*zaguriki*), in which the reeling frame was turned by cogwheels and a belt connected to a foot pedal, was developed in Kōzuke province in 1855 and introduced into eastern Shinano in 1860, the second year of export trade. Mechanized reeling, by contrast, entailed using an outside source of energy to turn the reel: usually a waterwheel during the early decades, to be replaced by steam at the turn of the century. Often, several modified *zaguri* machines would simply be linked

60. Sources on the Ina valley debased-currency riots include Shimohisakata-mura 1973:640–54; Miyashita 1972, 1980; Hirasawa 1978:236–42. For an account in English of several such uprisings in Shinano, see Bix 1986:194–214.

together and operated by water power.[61] The resulting operation did not necessarily save labor;[62] neither did it enable better yields of silk to be obtained from the cocoons.[63] The sole advantage of the new method was the quality of the silk it produced. Yet the "results of this simple innovation were striking: it permitted the reel to be turned at a faster and more constant speed, producing a silk filament of brighter luster, more uniform size, greater strength, and (because of these qualities) a higher market value."[64] As a result, filaturists across the country were soon clamoring to secure the new machinery.

The problem, for most, was financing. Shortly after treadle equipment was introduced to Shinano in 1860, mechanics in the Suwa basin began manufacturing *zaguriki* for export throughout Shinano and to neighboring provinces as well. But these treadle machines were relatively costly, selling for 0.25 to 0.30 *ryō* apiece. To make the further leap to water-powered machinery required an order of magnitude more capital.[65]

As the capacity of the filatures expanded, however, operating expenses emerged as still more problematic than fixed overhead costs. Although it might take months to secure payment on a sale, massive cash outlays were required on short notice to buy cocoons.[66] There was no flexibility; funds simply had to be available as soon as the silkworms pupated. Within ten days of spinning its cocoon, the imago would emerge, breaking through the fibers and destroying the utility of the silk. The only satisfactory method of forestalling this was to dry the cocoons at high enough temperatures to kill the pupae, a process that required ovens too expensive

61. Most of the "mechanized" filatures in southern Shinano in the early Meiji period utilized wooden reels modified from Italian or traditional Japanese models. Only a few plants were equipped with imported (French or Italian) machinery. Nagano-ken 1971:9; cf. Reischauer 1986:212.

62. A high degree of skill and training continued to be required to operate the new water-powered reelers, prompting an extensive debate in the literature over whether the early Meiji technological advances constituted true mechanization or merely another stage of "manufacture." See, e.g., Yagi 1980:69–70; Kandatsu 1985:134.

63. Li 1981:29–30.

64. T. Smith 1955:55.

65. In the first decade of the Meiji era, the average fixed capital invested in Shimoina's filature plants ranged from five to ten yen per basin. By 1883, with the transition to water-powered reelers, that figure had jumped tenfold to roughly a hundred yen per basin. Hirasawa 1953c:25–26. On the production of treadle-powered reeling equipment in the Suwa area, see Oguchi 1960:191–92.

66. By 1877, each *kin* (0.6 kg) of floss produced represented roughly a hundred yen in short-term capital outlays. Labor costs absorbed only 6 to 7 percent of this total; cocoons alone absorbed fully 70 to 80 percent of all nonfixed capital expenditures in the early Meiji filatures. Yagi 1980:42.

for most sericulturists. Accordingly, until the early twentieth century, rural households were forced either to sell or to reel most of their cocoons during a brief ten-day period after harvest—and brokers were forced to find enough cash to buy them during the same short period.[67] The result was increasing dependence on financiers from outside the region.

PRIVATE FINANCIERS AND THE RISE OF MECHANIZATION

The first major investor in the Shinano silk industry was a Kyoto-based merchant house called Ono, a family at the pinnacle of the Tokugawa commoner elite. The Ono had augmented their fortune in the last years of the old regime as one of the eight designated members of the powerful exchange association designated by the Bakufu to control the domestic side of the silk export trade. In the first years of the Meiji era, the head of the household founded a combine called the Ono-gumi, which became one of Japan's largest dealers in raw silk. When the price of hand-reeled silk began its precipitous decline in 1870, the Ono-gumi responded by founding one of the first privately operated mechanized reeling mills in Japan. By 1874, Ono was running eight filatures in Shinano, including one in Shimoina's Takagi village. With branch operations in Ueda, Matsumoto, and Nagano city, the Ono-gumi was soon the primary bankroller for the entire Shinano silk industry.[68]

Shimoina's first step toward mechanization was taken in 1873. In that year, several dozen daughters of the valley's former samurai families departed Iida for Tomioka, a modern, government-run demonstration mill in neighboring Gunma Prefecture. In fact, of the 556 female workers from around Japan who gathered to work in the Tomioka factory in 1873, one in eight came from Shimoina *gun*, making Shimoina's the largest contingent from any county outside the one where the plant was located.[69] Their mission was a dual one: to persuade commoners that

67. It was possible to kill the pupa by subjecting the cocoon to steam or a salt solution, but silk reeled from cocoons that had been treated in this manner lacked luster. For related discussions, see Li 1981:26; Smethurst 1986:215; for a vivid description of an early twentieth-century cocoon-drying operation, see Saga 1987:41.

68. The combine's first factory, employing imported machines of Italian manufacture, was erected in 1870 at Tsukiji in Tokyo. But when operations in Tsukiji were hampered by a shortage of local cocoons, the equipment was transferred to Nagano Prefecture, where sericulture was already flourishing. Horie 1965:187.

69. The famous Tomioka filature, a state-owned demonstration mill equipped with the latest imported machinery, had begun operations in 1872, one year before the first mechanized reeling mill opened in the Ina valley. On Shimoina's labor contribution to

work in the mills was a safe and respectable job for young women, and to return to their natal villages with the expertise to train a new generation of operatives. It was thus appropriate that, in the same year these local daughters were sent off to Tomioka, a small filature owned by Ono—equipped with Italian-style machinery and modeled on the Tomioka factory—began operating locally in Takagi village.

The manager of this new plant, a young village headman named Hasegawa Hanshichi, was another important agent in Shimoina's early silk development. Hasegawa had traveled to Tokyo to investigate the Tsukiji factory in the early 1870s, touring the new silk mills on the Kantō plain on his way home. He subsequently applied successfully to the Ono-gumi for backing to open Shimoina's first mechanized filature in his home village on the east bank of the Tenryū, a traditional papermaking center directly across from Iida. The new filature used a water-powered cogwheel and belt to drive the reeling frames, onto which twenty women, standing at as many basins, fed continuous filaments. A wood fire kept a large pot of water at a constant boil, and steam was piped to each reeling basin to heat the water in which the cocoons were immersed. The operation produced 45 *kan* (169 kg) of thread during its first year in operation.[70]

The following November, however, the Ono company suddenly collapsed, suspending production at its filatures and dealing a heavy blow to Shinano's nascent silk industry.[71] Most of Ono's assets were transferred to the central government, but Hasegawa managed to purchase the Shimoina plant outright. Although temporarily forced to suspend operations when the Ono-gumi went under, Hasegawa soon secured the backing of an Iida financier and resumed production. Within a year, he had rebuilt his factory along the lines of Tomioka, expanding it to accommodate fifty reelers. Three more factories in Shimoina were equipped with Italian-type machinery that year (1875), to be joined by an additional three in 1876.[72]

That spring, a silkworm pestilence broke out once again in Europe, and by the end of July, the price of Japanese floss had begun to climb

---

Tomioka, see Tenryūsha 1984:139–40, Nagano-ken 1971:151–52; for an analysis of the short-lived Tomioka experiment, see McCallion 1989.

70. Hirasawa 1952:135.

71. In addition to bankrolling every mechanized filature in the prefecture, Ono had served as the *kawasegata*, or international currency exchange agent, for Nagano, and was the sole major financial institution in the region at the time of its collapse. Nagano-ken 1971:149.

72. Hirasawa 1952:133–34; Nagano-ken 1971:149.

once again. Whereas the value of Japan's raw silk exports in 1875 had totaled only 6.4 million yen, the country netted 16.2 million yen, more than two and a half times as much, from silk sales in 1876. At the same time, the Japanese economy embarked on another inflationary cycle, triggered by a huge increase in printed money (condoned by the government in part to cover the expense of subduing the Satsuma rebellion of January–September 1877). In these expansionary circumstances, more and more individuals were enticed into the silk business, and the following two years saw a flurry of investment in mechanized mills. Between 1873 and 1879, twenty-seven filatures were established in Shimoina, over a third of these (ten) in 1877 alone. By 1878, the corner had been turned: the bulk of Shimoina's silk was mechanically reeled. That year, only five years after its first mechanized filature had opened, Shimoina boasted fourteen such factories, employing 665 women to produce 1,207 *kan* (over 4.5 metric tons) of floss. Half a decade later, fifty filatures employing ten or more reelers had been established in the county (map 23).[73]

PARASTATALS AND NATIONAL BANKS

This surge of investment, while managed by the local elite, was underwritten by capital infusions from new parastatal organs. Private capital—even on the scale of the Ono combine—proved too weak at this stage to survive the vagaries of an unpredictable foreign market, leading Shinano's producers to turn increasingly to the state for the stable financing they required.[74] The first quasi-governmental institution to play an important role in filature finance in the Ina valley was the so-called Shōsha, or Commercial Association, of Kamiina, established by landlords and merchants in the northern Ina valley to contend with the counterfeit currency crisis of 1869. Having staved off the immediate threat of riots by purchasing debased currency with their own capital, the members of the Shōsha proceeded to set themselves up as a Western-style corporation and invest in a variety of agricultural and industrial endeavors in the valley, including silkworm-egg and silk-floss production.[75] But even the Shōsha was soon superseded by larger and better-funded institutions.

73. Hirasawa 1952:134–62.
74. This turning to the state for financing is particularly important given that the silk industry has been singled out among Japanese industries for having been substantially mechanized before 1880 through the initiative of private enterprise; T. Smith 1955:66.
75. Iwashima 1967:227; Miyashita 1972.

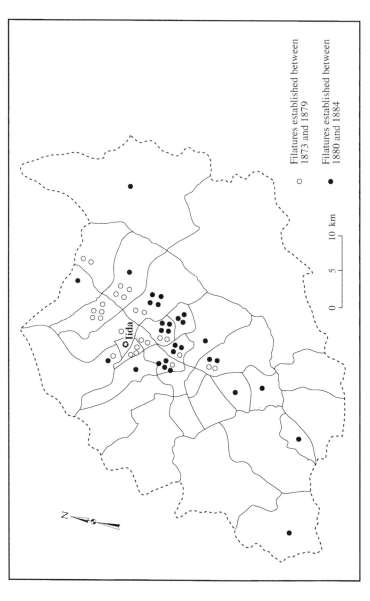

Map 23. The Spread of Reeling Works in Shimoina, to 1884. (Data from Hirasawa 1952:183–85 and Tenryumsha 1984:142.)

Filatures established between 1873 and 1879

Filatures established between 1880 and 1884

When Ina was incorporated into Chikuma Prefecture in 1871, the new governor established a similar organization to boost commercial enterprise. This investment fund, called the Kaisansha (Association for Enterprise Initiatives), raised its capital not just from the local elite but from all prefectural residents, including nonlandowners.[76] Its avowed purpose was to act as a savings scheme. Farmers were informed that the organization was designed to reimburse its members in case of a recession or bad harvest, ensuring them of at least some income in an economic crisis. When it opened its doors in 1875, however, the Kaisansha was chartered primarily to lend start-up capital to new enterprises. In effect, it functioned to concentrate funds from throughout the prefecture in the hands of landlords and merchants. Borrowers were to clear and cultivate new lands for commercial crops, undertake sericulture or stock raising, build irrigation ponds or *sake* breweries, open collieries, and, not least, establish filatures to reel improved silk. The Kaisansha provided seed money for scores of small village filatures established throughout Shimoina, as well as for mulberry plantations and silkworm-rearing operations, before dissolving in 1888.[77]

By this time, the most important source of finance for the silk industry had shifted to the newly burgeoning system of national banks. An act authorizing these institutions had been passed in 1872, but few banks

76. Over 90 percent of the Kaisansha's initial reserves were collected directly from Chikuma residents under the auspices of local officials (*daikuchō*, or district headmen). In 1873, all landowners were assessed a mandatory "contribution" on the basis of their holdings (one *shō* of rice plus one *shō* of dry-field grains per *tan* of land owned); nonlandowners were also required to contribute grain or cash. The total raised in this way throughout Chikuma Prefecture exceeded 14,000 yen. An interest-free ten-year loan in the amount of 38,770 yen contributed by the Finance Ministry in Tokyo more than doubled the organization's lending power. For references, see the following note.

77. During its decade and a half of operation, the Kaisansha extended 269 loans to 267 individuals, for a total value of 14,181 yen. The amounts involved tended to be modest; the average was less than 53 yen, and the highest just 500. Most loans were for one year only, at 12 percent annual interest, and were extended only after land or other property had been secured as collateral. One of the larger loans, in the amount of 200 yen, provided the bulk of financing for a 44-basin steam-powered filature in Zakōji village, which produced 450 *kan* (1,688 kg) of thread a year. Some 88 percent of all loans were made between 1882 and 1885, the four years of recession following the Matsukata deflation. Its management having fallen on hard times during the recession, and its raison d'être being largely usurped by newly established local banks after 1885, the Kaisansha repaid its loan to the central government by 1888 and put itself out of business. Its remaining assets—in Shimoina, 7,398 yen, or roughly 1,000 yen more than had been collected in the county fifteen years earlier—were distributed to the original contributors through village governing bodies over the next eight years. Tenryūsha 1984:150–51; Masaki 1978:306–8.

had been formed until 1876, when the government permitted them to begin printing paper money—secured against the gold-backed bonds issued to the former warrior class (daimyo and lesser samurai) in compensation for their forfeited domains and stipends. This put the system on a profitable footing for the first time, and the next three years saw a veritable explosion in national bank formation.[78] The 117th such institution to be chartered nationwide was founded in Iida in 1879. Nicknamed the *tonosama ginkō*, or "lords' bank," the 117th National Bank was in fact clearly run by businessmen from the former castle town and nearby satellites; the seven original members of the board of directors included one landlord, one lacquer dealer, one *sake* brewer, and two pawnshop owners. The institution these local notables oversaw quickly supplanted the earlier parastatal organizations as the nucleus of Iida's commercial life.[79]

In particular, the newly formed national banks played a crucial mediating role between silk industrialists, Tokyo financiers, and the state, answering a keenly felt need on the part of local filature operators. In 1879, Hasegawa Hanshichi, the area's leading silk industrialist, had sent a petition addressed to Prime Minister Itō Hirobumi through the governor of Nagano Prefecture, asking for a central government loan to support mechanized reeling in the valley. His request was denied. Seeking a broader base of support, Hasegawa spearheaded efforts in 1880 to create a prefecturewide producers' association called the Yūgisha. The heads of forty firms, including three from Shimoina County, joined the organization, whose aims included quality inspections, technological improvements, and securing funding for the industry. One of the first successes of the Yūgisha was in fact to make government loans available to local silk producers through the five national banks in the prefecture, at last opening up a channel whereby central government funds (as well as private finances) became directly accessible to the capital-short silk industry.[80] While this was clearly a strategic victory for local filaturists, it

78. Four banks were founded in Shinano in 1877 alone, all in former castle towns (Matsumoto, Ueda, Iiyama, and Matsushiro), and all based in part on former samurai assets. Ikeda 1965:2–3.

79. While 93 percent of the original stockholders were former samurai (Iida *han* retainers having been compensated at a rate of four yen per *koku* of their previous stipends), contributions from local merchants and wealthy farmers were also important. The 117th National Bank printed forty thousand yen in paper money in a single issue in 1881, one year before Japan's newly created central bank arrogated that function to itself. Kobayashi 1953.

80. Tenryūsha 1984:149. Beginning in 1880, the government also used reserves in the Yokohama Specie Bank—established to assist in the transmission of funds between Japan and foreign countries—as a source of low-interest loans to exporters; Reischauer 1986:224.

also signaled a further aggrandizement of central control over the most dynamic sector of the regional economy.

In short, during its first quarter-century of expansion, the silk industry was served by three generations of financiers. The speculative brokers who dominated silk markets in the 1860s gave way to regional parastatal corporations, which oversaw a period of more constructive and managed investment in the 1870s before yielding in turn to the national banks and producers' cartels that emerged as the main players in the financing of silk production by the early 1880s. This pattern of widening capital recruitment had important geographical effects. While it allowed mechanized reeling plants to spread steadily across the landscape, it also meant that the economic infrastructure of the industry was gradually consolidated. Most telling, the locus of control over the industry could no longer be said to lie solely within the region. Local savings may have continued to provide the bulk of lending funds for the "117th," but they were increasingly supplemented by national funds—giving Tokyo interests both entrée into and a stake in the area's silk industry. By 1880, outside capital had begun to insinuate itself into Shimoina's circuits of production and exchange, laying the groundwork for the more visible changes that would transform its landscape in the decades to come.

CHAPTER SIX

Crisis and Consolidation
The Shifting Locus of Power

While the first twenty-five years of the silk trade brought subtle changes to Shimoina's economic landscape, the first quarter-century of modern state building redrew the region's political map more boldly. Between 1870 and 1895, the Meiji regime set about building a unified and centrally directed state, effectively giving the Ina valley a new set of geopolitical coordinates in the process. A variety of local people challenged and tried to redefine their relationship with the center, initiating diverse movements to press for tax relief, lobby for a national legislature, organize community assistance programs, fight for a railroad, or even overthrow the government. But the nature of their successes as well as of their failures was such as to enhance Shimoina's dependence on the country's increasingly powerful metropolis.

Such an outcome could not have been comfortably predicted in 1868. Although imperial loyalists had handily overthrown the Bakufu and assembled their forces in Edo (soon renamed Tokyo, or the Eastern Capital), they were now faced with the daunting challenge of taking effective possession of the Japanese countryside. Through trial and error, the Restoration leaders managed to consolidate control in fairly short order. During its first decade, the regime concentrated on replacing the parcelized imperium of the Tokugawa state with a standardized, centrally directed system of government. This in turn required arrogating tax rights to a national treasury, ordering a comprehensive land survey (the first in nearly three hundred years), creating new political units in the countryside, and appointing a bureaucracy to administer the whole.

A decade later, in a second wave of institution building, a federal bank was founded, giving Tokyo control over currency and fiscal policy for the first time. Finally, beginning in the 1880s, construction of national telegraph and rail networks was begun.

Seen from the center, these were the most rational of measures, impelled by an urgent need to pull the country together and modernize its economy. But the view from the provinces was different. The new map of local government abruptly empowered some cities and regions over others; fiscal retrenchment brought severe hardship to working people and threw many small merchants out of business; and the decision to fund a national railroad system generated intense competition among alternative routes. In the short run, the very steps designed to impose order on the nation brought chaos and conflict to its constituent parts. From the countryside, the methodical process of national centralization often looked more like a hazardous, high-stakes poker game.

As it happened, Shimoina was not dealt a winning hand. Although local leaders had dominated the short-lived Chikuma Prefecture, in 1876 Chikuma was dissolved, and the valley was subsumed into the much larger Nagano Prefecture. Overnight, Iida was demoted to the status of a minor county headquarters, far removed from the seat of power where local issues were increasingly decided. No prefectural help would be forthcoming for the Shimoina elite in their first struggle with the central government: a movement to reduce the unusually high land values assigned to the valley in the recent national survey. Moreover, as local landlords preoccupied themselves with the largely unsuccessful effort to reduce their taxes, the valley began to be bypassed by its neighbors in other ways. A new rail line was built to the east, usurping the valley's one-time transport role, while the nearby Suwa basin rapidly took the lead in the nation's silk industry. By the early 1880s, when the first wave of Meiji expansion came to an end, a variety of indicators signaled that Iida had already fallen significantly behind the other towns in Nagano Prefecture.

The fiscal retrenchment that followed would disadvantage the region still further. Beginning in 1882, Finance Minister Matsukata Masayoshi instituted a series of harsh deflationary measures designed to end a decade of spiraling prices. Markets for Shimoina's luxury goods were savaged in the ensuing recession, leaving a trail of bankruptcies and discontent across the county. The most acute forms of suffering were largely redressed during the next round of economic expansion. But for the region as a whole, the Matsukata years had lasting consequences. On the one hand, the ranks of local landlords and industrialists were permanently thinned; at the same time, the emerging regional leadership was increasingly brought to heel within a national political hierarchy. When in the depths of the recession a local group was found to be plotting against the government, the discovery only hastened the extension of new instruments of central authority into the area, strengthening Tokyo's influence over the county. By the time the machinery of electoral politics was put in place in the 1890s, Shimoina had been subsumed into the political order as a clearly subordinate player. Its representatives mounted a valiant campaign to bring a national railroad to the valley, but this lobbying effort, too, was defeated. By the turn of the century, the Ina valley trade corridor had effectively been bypassed, leaving local politicians to squabble over the lesser spoils of road improvement funds.

The present chapter traces these successive dislocations in Shimoina's wider world, as well as local leaders' successive attempts to define a new role for the region, during the pivotal decades before the Sino-Japanese War of 1894–95. The progression is chronological, beginning with the evolution of local government in the 1870s, proceeding to the Matsukata deflation of the 1880s, and concluding with a major railroad battle of the early 1890s. Collectively, these episodes underscore the capricious and chaotic character of centralization as it was experienced on the ground, revealing the variable implications of that process for particular places. But they also highlight the complex realignments of spatial and social identities—suggesting that, in the Meiji state, regional rivalry emerged as a sanctioned alternative to more threatening, class-based forms of social conflict.

Redrawing the Political and Economic Map of Central Japan

CHIKUMA PREFECTURE AND THE TAX ASSESSMENT CONTROVERSY

One of the most enduring feats of the first Meiji decade was the creation of a top-down system of local government. After formally dissolving the long-defunct provinces as well as the feudal domains, the oligarchs were faced with the task of mapping out an administrative hierarchy of their own devising. The new political structures crystallized slowly, but once solidified they became potent instruments of central rule.

Political consolidation had begun before the Counterfeit Riots of 1869 broke out. With the imperialist forces' victory over the shogun in 1867, all former Bakufu lands had been promptly seized in the name of the emperor. The shogunate's holdings in Shinano—a collection of small territories totaling only one-tenth of the province's land, but dispersed across its entire length—were designated as Ina Prefecture, to be administered from the town of Iijima (in the northern Ina valley). Former domainal holdings were also declared imperial property, but the daimyo were left in charge of administration, and continued to be responsible for their own armies and finances. In short, while all land in the realm nominally belonged to the new authorities, in practice a patchwork of divided polities remained. But the currency debacle of 1869 touched off a series of riots that swept through the mountains for over a year, revealing the inadequacy of this decentralized governmental structure. Tens of thousands of Shinano residents attacked government offices as well as merchants' property, and the Iijima officials charged with keeping the peace in the former Bakufu lands had to borrow soldiers from neighboring daimyo to quell the disturbances in their far-flung territories.

Shaken by this experience, the new ruling clique in Tokyo determined to consolidate all local administration in the countryside in 1871, for the first time creating a unified treasury and a large, centralized military force capable of policing the rebellious citizenry of the new nation. To clear the way, the last vestiges of daimyo privilege were summarily eradicated; in Iida as in scores of similar towns, the castle compound was dismantled, and the former moat was filled in for use as a commercial thoroughfare.[1] Meanwhile, to replace the former political patchwork, a comprehensive new map was drawn up, dividing the country into seventy-two (later forty-three) prefectures (*ken*) and three municipalities (*fu*), and standardizing the lower levels of the administrative hierarchy as well.[2] The large and unwieldy Shinano province, having never been unified except in name, was split in two; the northern half was des-

---

1. These architectural conversions were merely the symbolic expression of a long attack on feudal prerogatives, from the dissolution of the samurai class to the abolition of the former domains. For more on this process, see Jansen 1986 and Umegaki 1986.

2. The early prefectural system was designed to create a series of government units roughly equal in size (initially averaging 100,000 *koku* apiece). The first draft showed 302 prefectures, but by the end of 1871, that number had been whittled down to 72. Twenty

ignated Nagano Prefecture, while the southern half became the core of a new entity called Chikuma Prefecture.[3]

The new arrangement was amenable to Shimoina residents. Matsumoto City, site of Chikuma prefectural headquarters, was an easy three days' walk from Iida, and the two towns had long-standing trade ties through the packhorse network; in addition, the Ina valley supported one of the new prefecture's wealthier rural communities. As a result, the Ina valley elite were able to become influential members of the bureaucracy that ruled Chikuma's affairs.[4] That influence had tangible results. Chikuma Prefecture followed the Ina valley's lead in forming the joint savings scheme known as the Kaisansha;[5] more tangibly, its officials gave top priority to improving Iida's transport network, slating the Ina road for early upgrading.[6] In addition, the Chikuma prefectural office sided with Ina valley landlords on the issue of local land assessments, setting the stage for the area's first confrontation with Tokyo.

The Meiji tax reform was an immensely ambitious project, requiring that all fields be resurveyed, that a land market be established, and that the yield-based dues of the Tokugawa period be converted to a fixed tax, assessed as a percentage of newly calculated land values.[7] The government's haste in pushing through this massive series of measurements resulted in numerous regional discrepancies, calling for a series of adjustments over the following years. But nowhere did the process of assigning land values engender more social upheaval than in Shimoina, where the issue formed the focal point of political activism for more than a decade.[8]

Differences of interpretation as to what constituted taxable land created a furor in the Ina valley. As soon as the reform was ordered in 1872,

---

years later, a stable order of 43 prefectures and 3 autonomous municipalities had emerged. Steiner 1965:24; Fraser 1986.

3. Also included in Chikuma Prefecture were portions of Hida and Takayama provinces to the west.

4. Sources on the early political consolidation of southern Shinano province include Shimoina Chiikishi Kenkyūkai 1982:16–18; Tsukada 1974:233.

5. The Kaisansha is discussed in chapter 5, nn. 76–77.

6. Igara-mura 1973:1105–62.

7. For an account of the Meiji land-tax reform and its social consequences, emphasizing the benefits to landlords, see Waswo 1977; for a controversial dissenting view, see Smethurst 1986:57–73.

8. Movements against the land-tax reforms were widespread. While the revised system generally worked to the advantage of large farmers and especially landlords, such protest movements were almost always led by the local notable (*gōnō*) class of farmers; see Vlastos 1989:372–82. The account given here of how these measures were carried out in the Ina valley is based on Tsutsui 1972 and Furushima 1956.

Chikuma Prefecture had handed down a series of guidelines for compliance, requesting that landowners report the extent of their own holdings to village officials. The latter were instructed to compile cadastral maps with annotations specifying the size, location, quality, and ownership of every plot; these were to be submitted to prefectural officials by the end of the year. Land values would be calculated as a fixed percentage of net yield, after taking standard deductions for input costs.

Relying on these procedures, Chikuma Prefecture reported the highest rate of paddy increase in Japan—more than double (214 percent) the level of the Tokugawa surveys—and the former Iida domain reported some of the highest local increases in the country. Yet the land-tax bureau was not satisfied and ordered a resurvey. This time, to the indignation of local landowners, interfield ridges, rock piles, and even channel bars in the Tenryū River were counted as taxable.[9] More difficult in the long run, however, were differences of procedure for deriving land values from estimated yields. Whereas the central government allowed a standard deduction of only 15 percent for seed and fertilizer costs, the landlord-dominated Chikuma prefectural administration—which had ordered local surveying to begin before national guidelines were published—had suggested much higher deductions, in the range of 30–65 percent for paddy and 40–75 percent for dry fields. Ina valley landowners had worked on the assumption that they would be allowed the higher deduction rate. When the tax bureau enforced the smaller standard deductions, the land values based on those estimates came out much higher in Shimoina than elsewhere in the prefecture. All told, the new assessments roughly doubled the tax burden in the Iida area from what the villages had originally calculated.[10]

The public outcry was not long in coming. Beginning in the fall of 1876 in the wealthiest core villages (Matsuo and Kamisato), a movement for reform quickly spread throughout the county. For while the Meiji tax revision benefited landlords across the country at the expense of tenants, local elites were nonetheless incensed over what they saw as unfair discrimination against the region. Over the next six years, this cause galvanized central Shimoina's wealthy rural families. It also, how-

---

9. The inclusion of interfield ridges, riparian floodplains, and similar strips of "wasteland" as well undoubtedly reflected an accurate perception on the part of government officials of the potential value of such lands as sites for the planting of silk mulberry. Nonetheless, three years later, the first concession the government made to the local reform movement was to remove these areas from the roster of taxable land. Tsutsui 1972.

10. The average price per *tan* for all lands under the jurisdiction of the Iida branch office was 34.22 yen, while that for Matsumoto was 22.47, and for northern Shinano, 25.28. Tsutsui 1972:51.

ever, gave them a bruising education on where they stood in the new geography of power in Meiji Japan.

THE CREATION OF NAGANO PREFECTURE
AND THE STRUGGLE FOR LAND-TAX REFORM

No sooner had the fight begun than Chikuma Prefecture was dissolved. On June 19, 1876, in a fire some local residents attributed to northern Nagano arsonists, the prefectural offices in Matsumoto burned to the ground. Two months later, Chikuma's territory was divided between its neighbors, and Shimoina was incorporated into Nagano *ken*, which now formed the third largest prefecture in Japan (map 24). All of the former Shinano province now fell within the boundaries of Nagano, whose seat of government lay far to the north, across the central Japanese Alps, in Nagano City—separated from Iida by more than 125 kilometers of rough road (a four- to five-day journey on foot). It was—and remains—the longest distance between county seat and prefectural capital anywhere in Honshu.

With the new center of prefectural administration lying beyond the pale of Shimoina's traditional trading sphere, in an economically and even culturally distinct part of Japan, Iida had suddenly become peripheral to the concerns of its immediate administrators. The impact was felt immediately. Funds for transport improvements—a major component of the prefectural budget and the linchpin of regional development—were diverted from southern to northern Nagano, where road building began in earnest. At the same time, prefectural support for the movement to revise land values in the Ina valley evaporated.

The first attempt at redress had taken the form of petitions from each Shimoina district head (*kuchō*) to the governor of the newly formed Nagano Prefecture. But Nagano Prefecture was a very different political entity from its predecessor. Its governor was an outsider, appointed by the central government and charged, not with representing the local will, but with enforcing Tokyo's policies. Moreover, his seat of administration lay in an area whose landlord class was less powerful than its counterpart in the south, and where no tax revolt loomed. Not surprisingly, no prefectural help was forthcoming to assist the incensed landlords of the Ina valley in their fight against Tokyo.

In 1877, the village leaders of lower Ina abandoned prefectural channels and began sending their representatives to Tokyo to lobby the land-tax bureau directly. Some satisfaction was obtained later that year, when

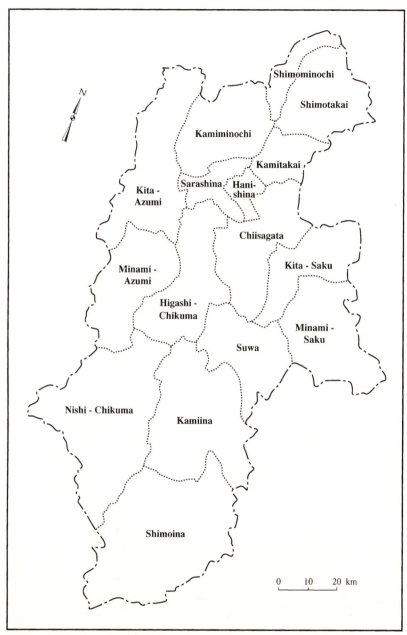

N

Shimominochi

Shimotakai

Kamiminochi

Kamitakai

Kita -
Azumi

Sarashina Hani-
shina

Chiisagata

Minami -
Azumi

Kita - Saku

Higashi -
Chikuma

Minami -
Saku

Suwa

Nishi - Chikuma

Kamiina

Shimoina

0   10   20 km

Map 24. The Counties (*Gun*) in Nagano Prefecture, Early Meiji.

nearly 8,000 yen worth of tax obligations within the Iida office's jurisdiction were canceled simply by eliminating field ridges, irrigation canals, and rock piles from the roster of taxable land. The same year, in the face of nationwide unrest, the government lowered the tax rate throughout the country from 3.0 to 2.5 percent of assessed land value. But lobbying continued for a reduction in Ina's land assessments, which remained high relative to those of its neighbors, and in 1880, when Ina County was divided in half, the newly created Shimoina County was named one of only four districts in the country where the government conceded the need to reopen the question of land prices. Meanwhile, however—angry at being ignored by the Nagano administration—a coalition of leaders from the southern counties began agitating to move the administrative offices to the geographical center of the prefecture, or, failing that, to secede altogether. This effort took shape against the backdrop of a broad movement for local self-government, including calls for wider prefectural assembly powers and attempts to change county and prefectural governorships from appointive to elective positions. These demands formed an essential plank of the People's Rights movement (*jiyū minken undō*) and the associated Liberal party (Jiyūtō) during the early 1880s. But on the issue of where to locate the Nagano prefectural capital, as on other questions of local political geography, the Liberal party found itself hopelessly divided into sectional camps, and in the end, the status quo was upheld.[11]

UEDA AND SUWA: THE NEW NAGANO POWERHOUSES

Supporting the new prefectural capital's effective usurpation of control over all of Nagano, despite sustained and sometimes militant protest by the residents of the southern half of the prefecture, were the political and economic strategists of the Meiji regime. From Tokyo's standpoint, the new regional capital of Nagano was a gateway to the rich rice and oil fields along the coast of the Sea of Japan. A commercial corridor was soon forged from Tokyo to Nagano, and the Chiisagata County town of Ueda—situated along the new route—experienced a marked growth spurt, surging past Shimoina both in sericultural output and as a transport and wholesaling center for interregional trade.

The greatest boon to Ueda was its inclusion on Nagano's first railroad, the Shin'etsu line: a direct link between Naoetsu, on the Sea of

11. For more on internal divisions within the Liberal party during these years, see Vlastos 1989:413–18. Shimoina's contradictory role in the Nagano secession dispute is discussed at the end of the present chapter.

Japan, and the Tokyo-Yokohama industrial complex (map 25). When this railroad was completed in 1890, the travel time between Tokyo and Nagano City was cut from six days to less than twelve hours, and a lucrative trade was established carrying Niigata rice, Hokkaido coal, and oil from the newly developed oil fields of the Sea of Japan coast to the nation's capital.[12] The early completion of this railroad enhanced the eastern Nagano trade corridor through which it ran, bolstering the fortunes of Ueda's merchant community. Before the line was pushed through, Iida's prices for such basic commodities as grains and fuel had been lower than Ueda's; afterward, the relationship was reversed.[13] To make matters worse for the Ina valley, before the Shin'etsu line was finished, the original project it was designed to serve—a proposed Nakasendō railroad through central and southern Nagano—was shelved for lack of funds. Instead, a less costly coastal line was constructed along the old Tōkaidō.

While its strategic location between Tokyo and Nagano boosted Ueda's commercial development, it was the Suwa basin that took the lead in the Nagano silk industry.[14] The convergence of three major inland trade routes just east of Lake Suwa (the Nakasendō, the Kōshūdō, and the Ina road; see map 7) had encouraged protoindustrial enterprises in the basin from an early date, with a primary concentration on cotton processing. Merchants from villages and towns around the lake shores had imported ginned cotton from Ise and Mikawa, putting it out to families throughout the area to willow and spin during the winter months. Already by 1875, the majority of landholdings in the Suwa basin were clearly insufficient to support a family through agriculture alone, demonstrating a widespread reliance on domestic industry for supplemental income.[15]

12. Hokkaido coal was first taken to Tokyo over this line in 1892; Sea of Japan oil, after 1902. Matsuzaki 1977:207; Nagano-ken 1971:449ff.

13. Examples include rice, wheat, soybeans, salt, soy sauce, *sake*, oil, and even firewood. Nagano-ken 1985:605ff.

14. Close ties with the Tokyo area may have boosted Ueda's commerce, but it had a more ambiguous effect on the region's industrial development. As in neighboring Gunma Prefecture, silk production in Ueda was dominated by powerful North Kantō merchant families committed to the putting-out system. This class of merchants was more concerned with its commercial dealings than with mechanizing production, so the established treadle (*zaguri*) technology proved remarkably tenacious in both areas. Nagano-ken 1971:155–59; Pratt 1991b:7–14.

15. In 1875, the average household in Hirano village (west of the lake) owned 4.9, in Shimosuwa 4.1, and in Kamisuwa only 2.5 *tan* of land (1 *tan* equals one-tenth of a hectare). A minimum of 5 *tan* was required to support a family from agriculture alone. Oguchi 1960:189. While much has been made of Suwa's protoindustrial legacy for its

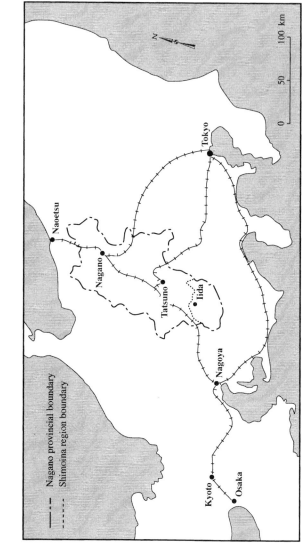

Map 25. The National Railroad Network in Central Japan at the End of the Meiji Era (1912).

With the demise of the domestic cotton industry and the rising demand for silk after 1859, the networks established for cotton processing appear to have been converted to silk reeling. Accumulation in the region was boosted when settlements near the lake began making improved treadle reeling apparatuses for sale to the surrounding basins. With a dozen fast-flowing streams descending toward the lake from the surrounding mountains, the Suwa basin was well placed to make use of waterwheels, raising output greatly during this early industrial phase. But the region first became nationally prominent as a filature center when Western mechanical reelers were imported in the 1870s, propelling central Shinano forward as the leader of the nation's silk industry.[16]

For Shimoina, this division of the spoils of early Meiji development between its northern neighbors offered little consolation. With Ueda gaining steadily in commerce and communications, and Suwa upstaging the rest of the prefecture in industrial technology, the Ina valley was doubly eclipsed. Despite the fact that its traditional industrial base remained strong until the recession of the mid 1880s, and despite its successes in expanding sericultural output, as a regional power, Shimoina was already falling behind. Masaki Keiji, one of few local scholars to acknowledge the extent of Iida's relative decline in the Meiji period, dates its origins to the aftermath of the Matsukata deflation.[17] Yet many available indicators suggest that a critical shift in the balance of regional power had begun during the previous transition years.

EARLY INDICATORS OF IIDA'S LAG

Transport and trade were among the first sectors where Iida was overtaken. Aside from the slow expansion of Tenryū River shipping (which carried only 5 percent of local cargo throughout the last quarter of the century),[18] Shimoina's transport infrastructure changed little during the first years of the Meiji era—so little, in fact, that the Ina valley lost its claim to being an important commercial channel between the mountains and the coast. By 1882, Iida carriers were

subsequent rise to the forefront of the filature industry, Edward Pratt (1991b) contends that the connection was less significant than has usually been supposed.

16. The Nakayamasha, founded in 1875, developed its own reeling machinery, combining principles of both the French and the Italian systems in a prototype for what would become known as "Suwa-style" reeling equipment. Oguchi 1960:189–92; Hirasawa 1953c:27.

17. Masaki 1978.

18. Kobayashi 1959:16.

Table 16 *Carriages, Wagons, and Carts in Shimoina and Chiisagata Counties and Nagano Prefecture, 1883*

	Shimoina	Chiisagata	Nagano Prefecture
Rickshaws	72	521	3,081
Horse-drawn carriages	0	13	114
Horse-drawn wagons	0	16	44
Ox-drawn wagons	0	21	64
Large two-wheeled carts	1	30	50
Small handcarts	103	2,014	12,668

SOURCE: Nagano-ken 1985:670–72, 714–15.

handling less cargo than their counterparts in Ueda, Nagano, Matsumoto, or Suwa.[19] Despite several attempts, the county had managed to build no permanent bridges over the Tenryū as of 1882.[20] On land, virtually the sole innovation of the period was the introduction of a handful of rickshaws and small handcarts, far fewer than Shimoina's neighbors had acquired. While horse- and ox-drawn wagons were beginning to make their appearance elsewhere in the prefecture—notably in east-central Nagano—Shimoina had not a single animal-drawn wagon in 1883 (table 16). A map of road improvements (map 26) is especially telling: as of 1885, all of southwestern Shinano was a cul-de-sac for wagon traffic. The only route by which wheeled vehicles could go from one boundary of the prefecture to another neatly delimited the new corridor of power running from Tokyo through Ueda and Nagano to the Sea of Japan.

Other indicators of the valley's stagnation relative to its neighbors can be found in the fields of communications and finance. The central post offices of Nagano, Ueda, and Matsumoto began offering savings accounts in 1875; comparable services were not available in Iida until 1878. The first telegraph line in Nagano Prefecture was opened in that year, passing along the same corridor from Tokyo to Niigata through Ueda and Nagano cities, and other lines followed in the early 1880s; Iida's hookup was the last in the prefecture, built in 1886. Seventeen local telegraph extension offices were established with local funds in the prefecture thereafter; while Suwa funded seven, and even Kamiina county three, Shimoina built only one. Most tellingly, Iida's national

19. Nagano-ken 1971:182.
20. Ibid.:176–77.

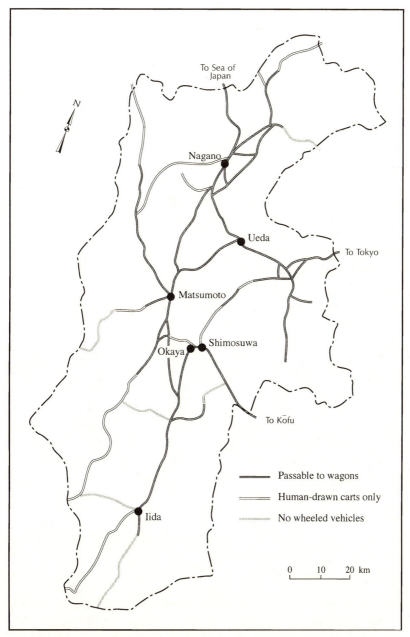

Map 26. The Road Network in Nagano Prefecture, 1885. (Adapted from
Nagano-ken 1971:175.)

bank was the last of the five to be founded in Shinano; it also was the most dependent on samurai funds, had only half the start-up capital of the others, and offered the lowest returns on deposits.[21]

Even in sericulture, Shimoina's involvement in the most profitable sectors of the industry lagged behind that of its neighbors. The county was a relative latecomer to silkworm-egg production, a source of significant capital accumulation in Ueda during the last years of the Tokugawa regime. By the time Shimoina's sericulturists were beginning to enter the egg market, Europe's silkworm blight had been brought under control and prices had plunged, leading many of the county's belated egg producers to financial ruin.[22] More seriously in the long term, the county's thread production consistently lagged behind cocoon output.[23] This allowed important industry profits to be captured by others, for rural producers were typically ignorant about price movements, pressured to sell their fresh cocoons within ten days, and (in many cases) indebted to manufacturers who had advanced silkworm-egg cards at the start of the season.[24] It was in the filatures, not the sericulture rooms, that silk fortunes were to be made.

In short, during the same years when the silk industry became firmly established in the county, Iida was assigned a lowly place both in the emerging industrial division of labor and in the new administrative hierarchy. As prefectural and national funds were diverted to their neighbors, local representatives were only gradually able to upgrade the county's financial and communications infrastructure, and Iida all but lost its one-time role as a trade entrepôt as well. This debilitating realignment of Shimoina's political and commercial context would be sealed by two developments of the later 1880s and 1890s: the Matsukata deflation and the building of the Chūō Railroad.

The Matsukata Deflation and the Iida Incident

The fifteenth year of the Meiji era, 1882, was an important turning point in the development of the modern Japanese econ-

21. Ibid.:172–73, 184–85, 261. To add insult to injury, the itinerary for the Meiji emperor's two visits to Shinano (in 1878 and 1880) skirted Shimoina.

22. Ibid.:151.

23. If prefectural statistics are to be believed, Shimoina in 1879 produced 4.5 percent of Nagano's cocoons, but only 3.8 percent of its reeled silk. Nagano-ken 1985:45ff.

24. Eng 1978:230–31.

omy. During this year, Finance Minister Matsukata Masayoshi's deflationary policies began to reverberate through the countryside, reining in a period of feverish inflation and speculative investment. The recession that followed triggered an avalanche of bankruptcies and foreclosures, hastening a major reorganization of the national economy in the 1880s and 1890s.

Scholarship on the Matsukata deflation has long been polarized around the issue of whether or not the deflation accelerated the differentiation of the Japanese peasantry. Here again, I would suggest that reconceiving the question in spatial as well as social terms allows us to transcend the existing stalemate. At least for Shimoina, the most important long-term effect of the Matsukata years was to shift power up the spatial hierarchy, strengthening existing tendencies toward central control. Particularly decisive were a series of firm governmental responses to an uprising plotted in the depths of the recession, known today as the Iida Incident.

DARK DAYS IN THE VALLEY

On October 21, 1881, at the crest of the expansionary wave of the early Meiji years, Matsukata Masayoshi replaced Ōkuma Shigenobu as finance minister. Matsukata's four-year tenure in the ministry would prove pivotal in modern Japanese history. To reverse the runaway inflation of the previous five years, he began a series of firm retrenchment policies, the centerpiece of which was the retiring of roughly 20 percent of all bank notes then in circulation in order to reduce the gap between paper currency and specie reserves. Matsukata also established a central bank to regulate currency and set fiscal policy, raised taxes on tobacco and *sake*, auctioned off a number of money-losing factories and mines, and instituted other policies of fiscal austerity. Since the land tax was fixed in yen (based on the recently completed assessments of land value), the resulting deflation raised the real value of government revenues at the expense of both landowners and tenants.[25]

These fiscal policies coincided with disastrous weather in much of Japan, from which Shimoina was not spared. June 1882 saw the worst flood in sixty years on the Tenryū; during the summer of 1883, the valley was desiccated by a forty-day drought. In 1884, cold wet weather

---

25. Ohkawa and Rosovsky 1965:65–66; Hirschmeier 1965:222. In Shimoina, it was estimated that local farmers in 1882 had to sell 50 percent more rice than before in order to raise the cash to meet their tax obligations. Nagano-ken 1971:178.

set in throughout summer and fall, bringing widespread floods and ruining the county's rice, barley, and tobacco crops. Finally, in 1885, heavy rains in April were followed by a freak snowfall in June and a flash flood in July, resulting in another nearly total harvest failure.[26] Meanwhile, the early 1880s also saw the revival of Italian silk production and a jump in the value of silver, both of which caused the price of silk to plummet.[27]

These events wracked Shimoina from one end to the other. As Matsukata's policies took hold in 1884, the market for the luxury goods around which Iida's economy had come to revolve fell off sharply. Output of wooden bowls for lacquering dropped from over 88,000 in 1883 to fewer than 12,000 in 1884; the following year, half of the lacquerware wholesalers, and three-quarters of the retailers, went out of business. Paper lost 30 percent of its value from 1882 to 1884, prompting nearly one hundred paper artisans to leave the region and forcing many more to turn to other forms of work. In other protoindustrial sectors too, artisans and outworkers alike could find virtually no demand for their skills.[28] The value of Tenryū River trade fell from over 30,000 yen in 1882 to less than 5,000 yen during the next twelve months, with local products declining in value faster than imports.[29]

Agricultural prices and wages plummeted as well, with the latter taking the more dramatic tumble. The price of rice—which had nearly trebled during the previous six years—dropped 38 percent between 1883 and 1884; returns to female on-farm labor declined by fully 46 percent in the same period.[30] The price of cocoons followed a similar course. Since silkworm maturation in 1884 was hampered by the cold weather, total silk floss output in Nagano Prefecture declined by nearly a quarter; yet the value of the relatively scarce cocoons harvested that year, far from rising, actually fell by half.[31] Daily necessities that could be made by ordinary rural households lost an even higher share of their market price. Homemade clothing and textiles fell to one-fifth their prerecession value, while charcoal—to which many desperate families

26. Masaki 1978:296, 374; Murasawa 1964, part 5:18; Seinaiji-mura 1982, 2:185.

27. Eng 1978:112.

28. In the lacquerware industry, for instance, wholesalers were reduced from 12 to 6, retailers from 29 to 8. The number of papermaking households in the county dropped from 458 in 1883 to 360 in 1885, while production dropped by half in the same period. Masaki 1978:296–302, 412–21.

29. Ibid.:395; Nagano-ken 1971:133.

30. Nagano-ken 1971:235–36; Nagano-ken 1985:634.

31. Cocoon prices fell from 4.50 yen per *kan* in 1882 to 2.60 yen per *kan* by 1884. Murasawa 1964, part 5:17–18; Toyooka-mura 1975:1050.

Table 17 *Owners, Owner/Tenants, and Tenants as Percentages of the Total Agricultural Population, 1884*

	Owners	Owner/Tenants	Tenants
Shimoina *gun*	30.9	44.5	24.6
Nagano Prefecture	36.8	41.9	21.3
All Japan	38.0	40.9	21.1

SOURCE: Nagano-ken 1971:243–44.

turned in the absence of other livelihoods—lost nearly nine-tenths of its market price in the space of a single year.[32]

Given the rising real tax burden, these figures translated into acute suffering for most of the rural populace. Local food stores were virtually exhausted by 1885. When a species of bamboo (*suzutake*) flowered in the mountains that year, families residing near the Tenryū banks undertook a two-day walk to the Tōyama valley to collect the edible seeds.[33] More ominously for the long term, many of the county's marginal farmers were forced to decapitalize in order to survive. Already by 1884, tenants comprised a significantly higher proportion of Shimoina's farm population than was true elsewhere in Japan (table 17); during the following year, the amount of land farmed by owner-operators in Shimoina fell by another 15 percent.[34]

In addition to losing their land, residents of the county were forced to decapitalize in other ways. The draft-animal population tumbled during the years immediately after the recession, a trend affecting cattle in particular, whose owners tended to be less affluent than their horse-owning neighbors.[35] Household possessions were sold off as well. Some families reportedly dismantled their furniture and even their homes to

32. In Ōdaira, a 3-*kan* bag of charcoal that fetched 30 sen in 1883 sold for only 3.5 sen the following year. Masaki 1978:296.

33. Ibid.

34. Owner-operators' acreage dropped from 60.1 percent of the total in 1884 to 51.7 percent in 1885 (Nagano-ken 1985:14). Similarly, the number of eligible voters dropped by 10 percent (from 4,833 to 4,340) between 1882 and 1883 alone, and another 4 percent the following year. Masaki 1978:298; cf. Nagano-ken 1971:233.

35. The total number of cattle in Shimoina peaked at 231 head in 1885, declining during the following year to 95, and by 1888 to a mere 30 head. It is significant that farmers killed or sold mostly male animals, while retaining their breeding females. Nagano-ken 1985:418. Comparable figures given for horses are difficult to credit. The total for the county is recorded as having fallen from 17,201 in 1885 to 10,142 in 1886, but as the 1884 total is only 9,341, it would seem likely that the 1885 figure is a mistake. See Nagano-ken 1985:425.

feed their kitchen fires; public auctions mounted; and the number of items pawned in the county increased tenfold.[36]

Nor was decapitalization limited to the marginal farming class. Equally hard-hit were the filaturists who had entered the business in the expansionary years immediately preceding the crisis. Few reeling plants in Shimoina posted profits in the mid 1880s, when three bad years in a row broke many small producers.[37] Even those operators who remained in business were largely drained of cash, so that when the cocoon-buying season came round in 1884, few had funds to purchase supplies. With the reduction in circulating currency and a general tightening of credit, loans had become much more difficult to obtain. The county's leading silk investor, Hasegawa Hanshichi, managed to secure a 20,000-yen government loan—but only because he was able to find a worthy former samurai (the bannerman who had ruled his hometown) to join his firm as a partner.[38] Most operators were not so lucky. Shimoina County as a whole saw its filature employment fall by 46 percent, as half of its firms went under in five short years.[39]

THE IIDA INCIDENT

The well-documented class differentiation of the rural populace was mirrored by a widening division within the recently formed Liberal party. Nagano Prefecture had long been a regional stronghold of this reform-minded organization,[40] and the local branch of the party—dominated by landlords from the Iida area—had focused its initial energies on lowering the land tax. In the course of that struggle, a local village headman named Mori Tahei had become converted to the party's broader agenda. In 1880, Mori joined the landlord-dominated Shōkyōsha, an organization of People's Rights activists from throughout Nagano Prefecture who were joining forces to push for the

36. Nagano-ken 1971:233; Nagano-ken 1985:578.

37. Of twenty-seven filatures in Shimoina employing ten or more workers in 1879, fewer than half (twelve) survived to 1893. In Kanae village, the heart of the county's reeling industry, only one of six firms that had been founded between 1877 and 1882 survived to 1884. Tenryūsha 1984:142; Kanae-machi 1969:637.

38. Hirasawa 1952:178–79, 182.

39. Masaki 1978:299. The county had supported 24 reeling plants employing 665 workers in 1878. The total number of filatures in the prefecture dropped from 488 to a mere 100 in the same period; Hirasawa 1952:138.

40. When the Liberal party was founded in 1881, Nagano (with 167 members) was one of only eight prefectures to boast more than 100 members. Haga 1977:278.

opening of a national assembly. The following year, he decided to bring the struggle home by founding a newspaper to gain support for the party's platform among the Ina valley's reading public. As Mori wrote at the time,

It is said that we are living in a new era, but the only thing new is that shogunal government has been replaced by imperial rule. . . . In regions such as ours, the people have not been enlightened, and popular rights have not arisen. . . . They do not even know what "rights" and "freedom" mean; indeed, it would not be too harsh to compare them to frogs in a well. It is for this reason that my colleagues and I have decided to publish the Deep Freedom Mountain News [*Miyama jiyū shinbun*].[41]

Although the paper folded within a year, its roster of local stockholders was a who's who of the young Shimoina elite. Landlords, filaturists, *sake* brewers, moneylenders, and paper and cotton merchants—including some past targets of peasant protest and not a few future politicians—were among the local notables to support the enterprise. These were the same men who had led the movement to lower the land tax, and many of them (or their fathers) were nativist scholars as well. Notably, only four of the entire group were over forty at the time.[42]

If Mori's backers betrayed the local Liberal party's class bias, their interests were by and large reflected in the newspaper's content as well. The paper's official platform was moderate and reformist, calling for more local self-government and advocating the teaching of constitutional democracy in the schools. But two columnists who published periodically in the *Miyama jiyū shinbun* articulated a more radical version of popular rights, going beyond the call for a national legislature to excoriate the government for administrative abuses and for repressing free speech. One of the two, Sakada Tetsutarō, published a provocative series comparing contemporary Japanese discontent with historical incidents of popular protest that had resulted in bloody revolutions. The installment of August 19, 1882, entitled "Jiseiron" ("On the Spirit of the Age"), contended that the will of the people

is not a mere human creation but a veritable force of nature. One may desire to struggle against such a force, but with mere human power it cannot be defeated; moreover, he who attempts such a struggle is bound to injure himself in the process. . . . Consider the great uprising of the French Revolution in the eigh-

41. Quoted in Shinshū Minken Hyakunen Jitsugyō Iinkai 1981:44.
42. Murasawa 1964, part 1. Out of 809 *Miyama jiyū shinbun* supporters, 419 were from the Ina valley.

teenth century, or the reign of Charles I in England, when the precious body of the sovereign became as [evanescent as] frost under the merciless blade and was no more. . . . Or turn your eyes to the last days of the Tokugawa Bakufu, a government that had survived for over two hundred years. Although it had many loyal retainers scattered in every province throughout the land, as soon as its forces were defeated at the battle of Toba, the shogunate fell from power and Japan entered into the era of imperial restoration. All of these rulers trifled with their power, pursuing only their own selfish interests while allowing their officials to tyrannize the people, who quickly tired of this loathsome treatment and responded to calls for renewal. . . . Be warned: whether across the ocean or here at home, whether now or in the past, every uprising against a state has without exception been provoked by a government following its own selfish ways, disdaining the popular will.[43]

Whether or not Sakada's editorial intended to justify popular violence against the government, it certainly hinted at a more radical course of action than that envisioned by Mori Tahei and his landlord and industrialist backers.

The recession induced by Matsukata's deflationary policies brought similarly extreme views to the fore within the Liberal party and the People's Rights movement. While an older, more conservative faction feared that Mori and his supporters were rousing the populace against the government at a time when civil order was dangerously threatened, less well-to-do farmers and leftist intellectuals grew disillusioned with the cautious politics of even the more progressive wing of the party.[44] From the latter group, in the depths of the recession, a socialist workers' organization emerged: one whose leadership ultimately turned toward a militant, underground struggle.

In Shimoina, this struggle was led by a photographer from eastern Nagano named Sakurai Heikichi. Sakurai came to Iida in 1884 and soon founded a proletarian party called the Aikoku Seirisha, or Patriotic Association for Just Government. Some 1,500 to 2,500 men and women from the industrial and transport towns in the Iida vicinity joined Sakurai's group, including tobacco cutters, tenant farmers, and laborers in the paper industries. Their activities focused primarily on self-help and democratic education. The Aikoku Seirisha planned mutual disaster aid, forbade its members to visit brothels, collected and disseminated peo-

43. From the installment published in *Miyama jiyū shinbun*, no. 69 (August 19, 1882). Reprinted in Masaki 1978:292–93.
44. On developments within the movement, see Nagano-ken 1971:95–99 and Shimoina Chiikishi Kenkyūkai 1982:28–33.

ple's rights pamphlets, and held regular seminars on coping with debt, illness, and other practical problems. Significantly, although the radical columnist Sakada joined forces with Sakurai, none of the stockholders in the *Miyama jiyū shinbun* participated in the founding or activities of this workers' organization.[45]

As the recession deepened in late 1884, the leadership of the Aikoku Seirisha decided the time had come for open rebellion against the Meiji regime. Through his co-conspirators, one of whom had served as a school teacher in rural Shimoina, Sakurai is believed to have been aware that Shimoina County had an unusually high concentration of rifles in private hands—more in fact than any other county in Nagano Prefecture.[46] Inspired by news of a massive uprising in Chichibu on October 31,[47] Sakurai and a handful of fellow activists from nearby regions conceived a plan for national revolution. First, they would print up and distribute leaflets in Tokyo calling for open rebellion; then they would stage the simultaneous seizure of local government offices in the cities of Iida, Okazaki, and Nagoya. The local strategy was to rally the Aikoku Seirisha members to attack the Iida police station and free all prisoners in the local jail, who were expected to join forces with their liberators. Thus fortified, the group planned to advance on the county offices and break into its coffers, stealing the proceeds from the much-detested *sake* tax. This money would be used to purchase food from urban and rural supporters for supplying a mountain redoubt above Iida, where the local band would wait for its Nagoya counterpart. Finally, the rebels would march to the Kantō plain, joining like-minded workers from throughout the nation in taking control of the capital. But none of this took place. On the eve of the coup attempt, the plot was discovered, and Sakurai was arrested. At a quickly convened trial in Matsumoto, the three instigators were found guilty of treason, and the entire Aikoku Seirisha organization was ordered dismantled.[48]

45. In Nagano Prefecture, this radical faction was led by individuals from Saku and Matsumoto counties (in eastern and central Nagano, respectively), where Matsukata's policies had been felt even more keenly than in Shimoina. For more on their activities, see Murasawa 1977–78.

46. Shimoina's residents possessed 3,828 rifles at the time, or nearly 50 weapons per village. Some had been licensed to hunters since the Tokugawa period, but most derived from the Iida domain armory, which was auctioned off in 1870 and 1871 (first to raise funds for the redemption of debased currency, and later simply in the process of liquidating the domain). Masaki 1973.

47. On the Chichibu Incident, see Bowen 1980:49–67 and passim; Bix 1986:211–86.

48. Masaki 1978:315ff.

THE SIGNIFICANCE OF THE RECESSION AND THE IIDA INCIDENT FOR REGIONAL DEVELOPMENT

The Matsukata years have become the focus of some of the most intense polemic in Meiji historiography. E. Sydney Crawcour sums up the prevailing macroeconomic view when he describes the effects of the Matsukata deflation as "depressing, though salutary"; while the recession "brought severe hardship to Japan's traditional industry," he contends, the recovery after the mid 1880s was "fairly rapid."[49] But for scholars who study rural society and agrarian protest movements, the benefits of the recession are more hotly debated. Most emphasize not only the short-term hardship to poor and middling farmers but also its social legacy: "a concentration of landownership in fewer hands, absentee landlordism by the 1890s, diminished face-to-face relationships between owners and tenants, a debt spiral that made it difficult to escape tenant status, and a gradual weakening of community ties and customary contacts."[50] By way of addressing the legacy question, a brief assessment of long-term trends in Shimoina is in order.

From a statistical point of view, the significance of the deflation for class differentiation within Shimoina appears to have been surprisingly minor. With the extensive development of protoindustrial domestic production in the Iida vicinity since the eighteenth century, tenancy rates were already high well before Matsukata's policies took hold.[51] While it is clear that some remaining independent farmers were undermined during the deflation years (as reflected in the decline of owner-operated land from 60.1 percent of the total in 1884 to 51.7 percent in 1885), this trend was relatively short-lived. The share of owner-cultivated land climbed back steadily to exceed the previous level just three years later, reaching 60.4 percent in 1889 and remaining above 60 percent throughout the 1890s. Even three decades later in 1921, after tenanted acreage had begun to expand once again, more land remained in the hands of

49. Crawcour 1988:388, 414.

50. Havens 1974:35. For a similar interpretation, see Vlastos 1989:419–20. Richard Smethurst has recently articulated a stridently opposing view, seeing increased tenancy after the 1880s as a by-product of reclamation and therefore a sign of "new opportunities for poor farm households to till larger farms," rather than "the purportedly invidious effects of the new tax system and the market economy." Smethurst 1986:57–73. For equally strident responses to Smethurst, see Bix 1987, Nakamura 1988, and Nishida 1989.

51. Furushima Toshio, in a study of mid Tokugawa and early Meiji landholding patterns in the centrally located village of Igara, found that as early as the later 1700s, nearly half (48 percent) of the village's households owned 3.3 *tan* of land or less; by 1874, 37 percent owned less than a single *tan* of land. Since a minimum of 5 *tan* were said to be

Table 18 *Owners, Owner/Tenants, and Tenants as Percentages of All Farm Households, Shimoina County, 1884–1895*

	Owners	Owner/Tenants	Tenants
1884	30.9%	44.5%	24.6%
1895	32.2	42.9	24.9

SOURCE: Nagano-ken, ed., 1985:14.

those who tilled it (56.0 percent) than had been true in the depths of the Matsukata deflation.[52]

Taking a different measure, tenants as a percentage of all households increased only incrementally during the following decade; the ranks of independent farmers actually saw greater proportional increases (table 18). Also, relatively few households in Shimoina lost land because of inability to meet tax obligations during this period. While tax defaulting in Nagano Prefecture as a whole jumped abruptly in 1884, the overwhelming majority of the affected households were from central Nagano, not from the Shimoina area.[53]

More important than this temporary rise in tenancy for Shimoina's long-term development, I would submit, was the crippling of the region's midlevel merchant-landlord elite. This class, as we have seen, had taken advantage of unprecedented, but risky, investment opportunities during the preceding decade. In the fallout from the recession, a number of households of headman rank in the villages near Iida were forced to sell all or most of their land. In Kamisato village, for instance, two of the six leading landlords of 1876 had lost all their land by 1885; another had his holdings reduced to one-fifteenth of their 1876 value by the same year. Fully half of the large landlords from this village were thus effectively eliminated in the space of a decade.[54]

The winnowing of the local elite class was also reflected in the demise of village-level financial and industrial enterprises, many of which had been established by the same families. A score of small lending companies had been founded during the inflationary wave of 1877 to 1882;

---

required to support a family from agriculture alone, this suggests that fully ten years before the Matsukata deflation, at least a third of the residents of this Iida-area village were earning their primary subsistence by means other than as owner-cultivators. Furushima 1956:52–60.

52. NSTI:25–28.

53. Ninety-one percent of all defaulting households (1,196 of 1,311) were residents of the Matsumoto plain. Nagano-ken 1971:233.

54. Kamisato-machi 1978:1103.

most of these went bankrupt during the subsequent deflation and never recovered.[55] An untold number of mutual-aid revolving-fund societies collapsed as well. The small entrepreneurs who were most affected were usually landlords who had mortgaged all or part of their land during the preceding years to obtain credit for relatively risky investments.

A similar process was apparent in the silk industry. Here too, statistics dramatize capital concentration after the recession. From 1883 to 1893, the proportion of large filatures in Shimoina (those with twenty or more basins) rose sharply, from 50 to 73 percent of the total. In fact, in 1885, a single mill owner accounted for over half of all production and employment in the Shimoina filatures. But these numbers, too, must be seen in light of the unprecedented boom in small investment that preceded the recession. Most of the small factories eliminated during this decade were very new operations, and many new ones would soon arise to take their place. By 1893, the proportion of large mills stood at exactly the same level as it had in 1880.[56]

More than accelerating class differentiation, what the Matsukata deflation and its aftermath accentuated was a particular form of sociospatial differentiation. Iida grew stronger relative to the rest of the region, while Shimoina as a whole was weakened relative to both Nagano and Tokyo. The enduring effect of these developments, I would argue, was to boost the multitiered centralizing trends already at work within Shimoina's economy. The primary sinews of an increasing local-center integration were few but powerful: the banking system, the police, and the telegraph.

On the one hand, as village elites languished, the more powerful Iida investors were able to extend their reach over the county's hinterland. With village-level credit systems falling on hard times, Iida's 117th National Bank—which was now closely supervised by the new Central Bank in Tokyo—emerged as the uncontested financial leader in the county. From 1879 to 1897, its reserves grew by over 400 percent, from 50,000 to 220,000 yen; by 1920, after a number of mergers with smaller banks, the 117th commanded four million yen and operated twelve branches in the Ina valley.[57]

At the same time, central power was directly augmented through a buildup in police forces and a crackdown on political organizations.

---

55. From 1877 to 1883, twelve financial institutions had been formed in Iida, and eight more in the nearby villages, with an average capitalization of only 4,855 yen. By 1886, only seven of the twenty were still in operation. Masaki 1978:296, 300–301.

56. Hirasawa 1952:138.

57. Kobayashi 1953.

Expenditures on police had already increased from 9.4 percent of the prefectural budget in 1879 to 15.5 percent in 1882. In response to a sharp rise in crime during the recession years of the mid 1880s,[58] as well as to the threat of political destabilization, the governor of Nagano Prefecture in 1884 announced a plan to increase the force by another 10 percent (adding 50 men to the existing 550). As this was a time of sharply declining tax revenues and program cuts, the prefectural assemblymen fought the governor on this issue for two years, but with only limited success. Police expenditures rose to over 19 percent of the prefectural budget in 1883 and 1884, and came down only one percentage point in 1885 and 1886.[59]

Other measures taken to forestall plots like the one uncovered in the Iida Incident augmented Tokyo's power as well. In 1885, immediately after the aborted uprising, the government pushed a telegraph line through from Tokyo to Iida, a potent symbol and indeed a potent instrument of the tighter supervision that would characterize center-local relations during the coming decades. Meanwhile, on the grounds that Aikoku Seirisha members had supported the treason attempted by their leaders, the state set about dismantling the People's Rights movement in the valley.[60] Torn asunder nationally by the same kinds of ideological divisions that had surfaced in Shimoina, the Liberal party in any case dissolved itself in 1884. Nonetheless, advocates of constitutional democracy were put under police surveillance, and teachers' associations with socialist leanings were forcibly reorganized under bureaucratic auspices. For three years, no political rallies were held in Shimoina County.[61]

In short, the Matsukata deflation and its aftermath boosted a sociospatial tendency that had already begun through the financing of silk production: the centralization of control over Shimoina's political and

58. The number of prisoners in the Iida jail increased by 350 percent during the deflation because robberies and break-ins rose dramatically. The area's suicide rate climbed as well. Masaki 1978:303; Nagano-ken 1971:233.

59. Nagano-ken 1971:233ff. For a detailed history of the Meiji police, see Westney 1987, ch. 2.

60. The extent of public support for the plotters in the Iida Incident remains controversial. For many years, the planned uprising was interpreted as a broad-based movement comparable to that in Chichibu. But Masaki Keiji argues that, unlike the genuine popular protests of earlier and later years, this attempt to overthrow the Meiji regime never had the support of more than a tiny coterie of intellectuals. In his view, the much-cited claim that 200 or more members of the Aikoku Seirisha supported the Iida plot is baseless, deriving from a later attempt by the government to discredit the entire organization. Masaki 1978:315–68; Masaki 1974.

61. In the resulting political vacuum, disaffected residents increasingly joined the Maruyama sect, a millenarian Buddhist organization that channeled anti-establishment

economic development. By weakening midlevel elites throughout the region, while strengthening a variety of Tokyo-centered financial and administrative institutions with regional headquarters in the county seat, the recession of the early 1880s bound Iida and its hinterland into a tighter geopolitical hierarchy, locally administered from the former castle town but increasingly directed from Tokyo. The lasting effect of the deflation and the events it set in train was to transfer power upward through this hierarchy, enhancing central control at both the intra- and interregional levels.

Further Realignments: Electoral Politics and Regional Rivalries

The last major shift in Shimoina's political field during the first quarter-century of the Meiji era involved the beginning of popular elections. Continued agitation by the People's Rights movement finally persuaded the oligarchs to create space for limited popular participation in the political process in the later 1880s. A constitution was handed down to the people in 1889, providing for the creation of a national assembly that would open in the following year.[62]

The resulting electoral maneuvering only reinforced southern Shinano's disadvantageous position relative to larger power centers on the prefectural and national map. It also created a climate of intense regional rivalry, channeled in the Ina valley into an aggressive railroad lobby, which dominated local politics for half a decade. Like the land-tax resisters of the 1870s and the coup plotters of the 1880s, Shimoina's railroad boosters were ultimately unsuccessful in their bid to redirect national resources according to their own priorities. Yet these political defeats must be seen as more than the inevitable ending to ill-starred efforts at local self-determination. Divergent as the aims and methods of these groups clearly were, all three movements not merely reflected but

---

sentiment into less overtly political forms of expression. By the end of the deflation years, Shimoina had over a thousand Maruyama adherents, more than any other county in Nagano or neighboring Aichi Prefecture. Masaki 1978:378; Haga 1977:287. On the political context of the rise of new religions like the Maruyama sect, see Harootunian 1989:215–31.

62. On the Meiji constitution's limited provisions for democratic participation, see Beasley 1989:651–64.

ultimately helped reshape the geography of control over the valley's subsequent development.

SHIMOINA AND THE RISE OF ELECTORAL POLITICS

Even as stronger hierarchical bonds were being forged between Tokyo and Iida, complementary shifts were taking place among the counties in Nagano Prefecture. Matsukata's insistence on fiscal austerity had translated into a sharp decrease in national resources for local development projects, just as electoral politics were being initiated at the prefectural and county levels. As a result, the local-center orientation of the first Meiji decade's political mobilization (the effort to win concessions on the land tax from Tokyo) soon gave way to intercounty struggles for prefectural revenues. Regional rivalries surfaced at other levels of the administrative hierarchy as well, pitting hamlet against hamlet, village against village, and southern against northern Nagano. As soon as they were formed, the newly constituted village, town, county, prefectural, and national assemblies became the sites of recurrent territorial struggles.

Elections to these councils began at various points between 1879 and 1891, quickly emerging as the most visible arena in which the interlocking hierarchies of region, class, and gender were displayed and contested. At the higher levels, centrally located elite men dominated the machinery of electoral politics. Only male heads of household who were at least twenty-five years of age and capable of meeting varying property qualifications were entitled to vote.[63] The town and village assemblies dominated by these property owners were able to institute highly regressive taxes to finance the public works and education needs that were increasingly delegated to local government.[64] Moreover, at the national level, not only was the enfranchised population a minority of the whole, but poor areas were decisively underrepresented. While the electorate for the national assembly encompassed just over 1 percent of the popu-

63. The property requirement for enfranchisement, measured in direct national taxes paid annually, ranged from two yen for voters in village and town elections to ten yen for prefectural assembly elections and fifteen for the national assembly. Steele 1989.

64. Most oppressive was the household tax, which would become the single most important source of local government revenue (more than doubling between 1900 and 1909). Pyle 1973:57. Nonetheless, this was not the sole tax that fell more heavily on the small and middling farmer than on the landlord class. For a compelling study documenting the increasing regressiveness of the tax system as a whole after the Restoration, see Chambliss 1965.

lation nationwide, for instance, it included less than 0.9 percent of the populace in Nagano Prefecture. The situation was most extreme in the Kiso valley, where only 0.01 percent held the franchise. But even in Shimoina, five mountain villages claimed not a single person eligible to vote for the national assembly.[65]

The county assemblies (*gunkai*) were a particular stronghold of the landlord class. In 1890, when the *gun* began to elect their own representatives (although their administrative heads [*guncho*] continued to be appointed), a bicameral legislature was created for the county assemblies. Any twenty-five-year-old male head of household who had resided in the county for two years, and who paid two yen in national taxes, was eligible to vote for village and town assemblymen, who in turn elected lower-house representatives to the county assembly from their midst. But only men owning at least 10,000 yen worth of property were entitled to vote or run for the six upper-house positions. In 1891, the year of the first elections, the qualified voters and candidates for these six positions in Shimoina totaled nine persons; by 1897, that number had fallen to seven. The geographical concentration of these men, Shimoina's wealthiest citizens, was telling. Of the first six elected to the assembly, two were from Iida, three from its immediate satellites (Matsuo and Igara), and one from an agriculturally rich village in the northern valley floor.[66]

Representatives to the prefectural and national assemblies were drawn from the same body of men who ran the county assemblies, and they reflected the same biases. In Nagano, most were landlords connected with the Liberal party, and after the deflation, their average landholdings increased notably. Moreover, in these higher assemblies, regional representation could be as skewed as class representation. In the upper house of the Nagano prefectural assembly, fourteen of fifteen delegates elected to the first session were from the northern and eastern counties.[67]

65. Shimoina at the time had 1,260 voters eligible for national elections, or 5.5 percent of the 22,998 household heads; in Nagano Prefecture as a whole, 10,641 men were eligible to vote. The five mountain villages lacking representation were Seinaiji, Namiai, Kado, Kizawa, and Yaegōchi. Masaki 1978:379.

66. Masaki 1978:385. Beginning in 1878, the *gun* served as an intermediate unit of administration, taking over the functions of the large *ku* or census districts. (See Umegaki 1988:136–39 on the role of census districts in early Meiji political integration.) The county assembly system was changed in 1899 to direct elections, with no landlord house; assemblies at the county level were eliminated altogether in 1921. Nagano-ken 1972:48–55; Ōkuma 1910:330–33. For a discussion of the county assemblies as bastions of right-wing politics, see Najita 1967:45–50; for an illuminating essay on the wider context of Meiji conservatism, see Pyle 1989.

67. Tsukada 1974:238–39. Over the next thirty years, with the phenomenal growth of the silk industry, the balance shifted decisively to central Nagano, and from landlords

It was in this context that Shimoina's identity as a political region was forged. As delegates to a deliberative body that was clearly stacked in favor of their northern neighbors, Shimoina's elected representatives saw themselves as beleaguered crusaders, fighting on behalf of a clearly defined geographical entity. At the same time, their new positions gave the uppermost tier of the local elite an unprecedented power to articulate where the region's interest lay, and to do so in a way that would simultaneously serve their personal interests as well. The project on which they fastened their highest hopes—and the failure that ultimately came to symbolize their diminished power more than any other—was the effort to secure a rail line through the county.

EARLY PLANS FOR A NAKASENDŌ RAILROAD

The first quarter-century after the opening of the ports had seen the rise of a new commercial corridor through eastern Nagano Prefecture, rerouting trade between the coast and the interior on paths that bypassed the Ina valley. With the spectacular rise of Suwa, however, the government announced plans to build a rail line from Tokyo to Nagoya through Nagano Prefecture. That the proposed National Railways Central Trunk Line (Kokutetsu chūō honsen) would pass through the Suwa basin was a given; southwest of Suwa, however, the track might be laid either through Kiso or through the Ina valley. As the latter option began to be seriously investigated, the Ina valley's hopes for maintaining a strategic role in regional trade were revived. For a brief, exhilarating moment in the early 1890s, Iida's merchant community saw a chance to recover the city's slipping share of wholesaling, and to stimulate local investment in the budding sericultural economy, if only the valley could be chosen as the route for the Chūō Railroad.

Japan was just emerging from a second recession in 1892 when the Chūō plan was broached. The idea was not a new one; similar intentions had been voiced more than fifteen years earlier, and 20 million yen in construction funds had been appropriated for the proposed Nakasendō railroad in 1883. But recession had hit the following year, and the project had been shelved for lack of funds.[68] It was silk interests in the Suwa basin that finally succeeded in pushing the government to revive its com-

---

to silk industrialists: by 1918, fourteen of the fifteen were filaturists, all but one of them from Suwa. Shimoina Chiikishi Kenkyūkai 1982:42–43. For trends in landholdings of Nagano representatives, see Nagano-ken 1971:102.

68. A private line linking Suwa with Kōfu and Matsumoto was also planned in the 1880s, but aborted in the recession of 1890–92. Harada 1983:195ff.

mitment to a rail line linking the central Japanese Alps with the port of Yokohama.

Suwa industrialists had emerged from the recession in a comparatively strong position. The small-scale mill owners from around the lake had banded together in 1879 to form a strong producers' association, coordinating purchases, sales, and shipping—a partnership widely credited with having allowed this area to survive the Matsukata deflation with minimal losses.[69] By the late 1880s, as production came back on line, the basin's factories had simply outgrown the local supplies of firewood, charcoal, cocoons, factory workers, and grain, all of which were being imported from an ever-widening hinterland. Cocoons alone were packed in from nine different prefectures—all on horse- and ox-back, and most during the plum rains of early summer, when the roads often washed out. In addition to the spoilage, delays, and high transport costs of this inefficient system, mill owners were forced to make regular donations to the prefectural coffers for upkeep of the heavily used passes into the basin.[70] Impatient with these expenses, but unable to raise the capital to build private railroads on their own initiative, Suwa's industrialists mounted a sustained campaign to bring a national line into the basin.

SHIMOINA FIGHTS FOR THE RAILROAD

In 1891, they succeeded. Just five years after abandoning the original Nakasendō project, the central government announced that it had indeed revived plans for a rail link from Tokyo to Nagoya through the mountains of central Shinano. On hearing this news, elected representatives from both Ina and Kiso immediately went to Tokyo and began lobbying to bring the railroad to their respective valleys.

Spearheading the railroad fight for Shimoina were Itō Daihachi, Shimoina's sole representative to the national assembly, and Ihara Gorobei, one of the six wealthy landlords elected to the county assembly's upper house the previous year. These two men represented the pinnacle of the local elite. Itō was the largest landowner in Igara village, Iida's neighbor to the south. Ihara, one of the nine richest landowners in the county, was also a lacquer dealer and the top taxpayer in Iida.

---

69. Ebato 1969:14–15, 74–75; Horiguchi 1972:157–58.

70. Harada 1983:189, 194; Nagano-ken 1971:456. Hirano village (in present-day Okaya City), the heart of the filature district on the western shore of Lake Suwa, was importing cocoons in 1885 from Saitama, Gunma, Ibaraki, Niigata, Fukushima, Yamagata, Shizuoka, Aichi, and Gifu prefectures.

Both Liberal party politicians, they had also been shareholders in the *Miyama jiyū shinbun*. In 1890, a year of poor harvests and rice riots throughout the country, both had donated relief funds for the Iida area from their private fortunes.[71]

When surveyors found that the Kiso valley route would be more direct than the Ina alternative and would cost considerably less to construct (13 versus 23 million yen), these men and their followers vowed to win the railroad for Ina regardless, and the age of serious regional boosterism dawned for Shimoina. Merchants organized themselves into promotional associations, the county assembly set aside a generous portion of its meager budget to support lobbyists in Tokyo, village assemblies voted to offer grants of land and labor, rallies were held, editorials published, and poems composed, all to persuade the parties concerned of the fortunes that were sure to accrue—to the nation as well as the region—should the railroad come to Ina. In 1892, in one of the more memorable contributions to the collective effort, a student group performed a drama in Iida's newly constructed auditorium (the grandly named Akebonoza, or "Hall of Dawn") entitled "The Opening of the Ina Railroad." In the play, a group of village youths collecting donations for the railroad movement call on a rice merchant named Hirata, who gives only a minuscule donation, contending that if the railroad goes through, the price of rice will fall and he will lose money. The youths berate him and leave. When the railroad is completed, however, Hirata decides he would like to ride to Tokyo, but the youths refuse to let him on board the train, leaving him to repent his earlier stinginess.[72]

Beginning in 1891, the Ina valley boosters concentrated their energies on compiling a comparative survey of Ina and Kiso as routes for the Chūō Railroad (*Chūō tetsudō Ina, Kiso ryōsen no hikaku chōsa*). This document was a glorious piece of regional propaganda, masquerading as an unprejudiced assessment of the economic potential of the two valleys. Its authors claimed for the Ina valley "five times the arable, ten times the population, and fifteen times the wealth" of its neighbor to the west. Through a lengthy series of calculations, they went on to demonstrate that any extra costs incurred during construction would quickly be offset by superior profits should the railroad be routed through the Ina valley. The culmination of their argument, concerning the potential for further development of silk, is worth reprinting at length.

---

71. Hirasawa 1967:13; Shimoina Chiikishi Kenkyūkai 1982:95–96.
72. Shimoina Chiikishi Kenkyūkai 1982:23.

Present cocoon output in Kamiina and Shimoina counties together averages 51,000 *koku* per year. At 23 yen per *koku*, this amounts to 1,173,000 yen. As this figure indicates, the sericulture industry in Ina is already of great proportions. Yet as mentioned above, this figure is kept from being even higher by the inconvenience of transport, which makes it necessary to keep many wild lands, which could otherwise be converted to mulberry fields, planted to grass in order to supply green manure. If a railroad were opened, [decreasing the cost of imported fertilizer and] allowing part of this grassland to be converted to mulberry fields, production would greatly increase. By contrast, the cold climate of the Kiso region has kept its cocoon production low, and we believe that the same conditions will prevent any hopes for future development in Kiso from being realized.

These are the conditions of cocoon production in the Ina region. Turning to a comparison of silk-thread production, the situation almost goes without saying. Ina's annual thread output averages 5,210 bales, worth 1,667,200 yen (at 320 yen per bale), and another 2,386 bales of waste silk worth 429,480 yen (at 180 yen per bale). The quality of the product may be surmised from the consistently high prices it fetches on the Yokohama market. Whatever the reasons for the excellence of the locally reeled thread, the fact that filaturists are now importing cocoons from as far away as Tohoku and Kyushu to process in the valley suggests that there is something special about Ina silk.

Yet the Ina silk industry has not yet reached its zenith, and we may anticipate even greater prosperity in the years ahead. Despite its other advantages, the valley simply cannot expand its output at this time because of a paucity of raw materials. At present, the cocoons being imported into these counties must be first brought by train to the nearest station; then they must be taken by porter or packhorse another fifteen to fifty or even sixty miles, delaying delivery and creating shortages. As a result, the filatures presently have no choice but to shut down for over half of the year. If these raw material shortages were overcome, and Ina silk producers were able to exploit their potential to the full, silk production would double over its present level. And how can one increase the supply? By constructing a railway and eliminating inconveniences of transport. If this were to happen, supply would meet demand, and the silk industry would be able to operate year-round without work stoppages. The result would be an increase of net production to over 4,336,000 yen per year.

. . . [Previous calculations indicate that] putting the rail line through Kiso would generate 42,600 yen in revenues, from increased production of lumber as well as from lowered transport costs. An additional 1,591,600 yen stands to be gained from agricultural land clearing. Putting the line through Ina, on the other hand, would generate over 4,237,600 yen from agricultural clearing . . . in addition to over 4,336,300 yen in silk-thread production. Thus, the final comparison between the two lines indicates that routing the railway through Ina would generate over 6,939,700 yen more in profits annually than would accrue if the line were put through Kiso.[73]

73. Chūō Tetsudō Inasen Iin 1894. Some figures have been altered to correct inconsistencies in the original, which is available on file at the Iida City Library. Large sections of the text are reproduced in Murasawa 1959. See also Sekishima 1972.

A dozen prominent local merchants and politicians, led by Ihara, made the four-day trek to Tokyo to assist Itō in presenting this document to the national assembly, taking with them a suitably distorted topographic map designed to make the Kiso valley appear narrower and the Ina valley wider than was in fact the case. But at least two forces were working against them. One was the stubborn issue of cost; the other, the interests of the Imperial Household. While the Kiso valley assuredly did lag behind Ina in agricultural and industrial potential, it included a vast expanse of woodland whose development was of keen interest to the court faction in the capital. The Kiso forest reserves had been declared the property of the emperor, and the Imperial Household Ministry was intent on maximizing its profits from proposed logging. Whether bowing to these or other interests, the national assembly voted on May 23, 1894, to route the Chūō line through the Kiso valley, and the last hopes for reviving the Ina valley's role as a trade corridor were extinguished.[74]

LESSER SPOILS: ROAD IMPROVEMENTS TO 1895

Local historians typically portray the railroad boosters as valiant crusaders for the region, and their failure as a major setback for Shimoina. One scholarly article on the movement, written in 1964, opens by declaring: "The efforts of our forebears to connect Iida to the national transport grid began in Meiji 26 [1893] and are only now—seventy years later—bearing fruit. Truly this is wonderful—the dawning of a new day for Iida. Let us recount the history of our self-sacrificing predecessors [*senpai*] who began this historic movement." And a popular history of the region, after recounting the story of the railroad movement, similarly concludes that "although of course they ultimately failed, we must ever remember their example of selfless love for their homeland."[75]

A favorite hero of such accounts is Itō Daihachi, the county's national assemblyman at the time, who is eulogized for salvaging what he could for the region through a clever, underhanded ploy. Itō allegedly pre-

---

74. Sakashita 1964–65; Murasawa 1959. Furushima Toshio disputes the importance of the Kiso timber connection, contending rather that military interests—specifically, the desire to be able to move large numbers of troops through the countryside in relative secrecy—were more important in the final decision. Personal communication, May 27, 1990. In any case, not everyone saw the resulting decision as a victory for Kiso. Shimazaki Tōson, a native of Kiso, lamented the destructiveness of the new railroad in his novel *The Family* (1976:284–85). For a sensitive discussion of Tōson's ambivalence toward modernization in the countryside, see Dodd 1993:44–59.

75. Sakashita 1964–65, part 3:19; Iida Bunkazai no Kai 1969:279.

pared a feast for the surveying crew that was sent to determine where the rails would be laid between Suwa and Shiojiri. After assuring that the engineers were drunk, he had them carried in palanquins on the circuitous route through Tatsuno. The surveyors approved the plan unawares and returned to Tokyo, where the national assembly appropriated funds for the project before the trickery was found out. The decision was never reversed, and the Ina valley ended up with an accessible station on what was celebrated locally as the Daihachi Loop (see map 25).[76]

But this small victory, however colorful, offered poor consolation for having lost the national line altogether. Through the next twenty years, the Ina valley's transport network would see only modest improvements, as road work was carried out on the various routes connecting Shimoina to the advancing railhead in the neighboring Kiso valley. Ihara Gorobei's son spearheaded the effort to fund a private railroad from Tatsuno to Iida, but without national support, nearly three arduous decades were required just to complete this short stretch of track. In the meantime, the focus shifted to the less glamorous chore of securing road improvement funds.

Maintenance allowances for major roads had been provided in the national budget until 1880, when responsibility for all prefectural roads and bridges fell to the prefectures themselves. At that time, the governor of Nagano determined what he believed to be the seven roads in the prefecture most in need of upgrading, among which was the Ina road. Half of the expenses would be paid out of prefectural taxes; local notables in the affected regions were asked to donate one-sixth, and the remainder was to be supplied from the national treasury. Construction was delayed while the government deliberated over the Nakasendō railroad, but with that project temporarily shelved in 1886, the deferred projects were reinstated.

Completing improvements on the Ina road was sufficiently important to Shimoina that the area's elite broke ranks with their counterparts elsewhere in southern and central Nagano over the issue of dividing the prefecture in two. When the attempt to move the prefectural offices back to Matsumoto proved fruitless, leaders from other counties of southern Nagano began agitating to secede from the prefecture, but Shimoina representatives and the powerful Iida Chamber of Commerce staunchly opposed such a move. In their view, to divide the prefecture before the promised road improvements had been completed would only increase

76. Sakashita 1964–65, part 3:16.

local taxes. Led by the intrepid Ihara Gorobei, the Shimoina group submitted a petition against splitting the prefecture, aiding northern interests who hoped to preserve the status quo. The secession movement ultimately failed, and the road project proceeded. By 1893, improvements had been completed in both Nagano and Aichi prefectures, and horse-drawn wagons were able to enter Shimoina from the south, opening a new chapter in the region's economic development.[77]

Conclusion

The first contention of the foregoing narrative, implicit in its point of view, is that to understand Japan's metamorphosis into a unified state we must ask how change was experienced at the local level.[78] Centralization was by nature a differentiating process; creating a new geopolitical cartography meant designating cores and peripheries, elevating certain places at the expense of others. As a result, assimilation into the emerging state had very different implications for different places. Shimoina's relatively privileged position in the short-lived Chikuma Prefecture, and its loss of clout when subsumed into the larger Nagano Prefecture, provides an apposite reminder of how variable the fortunes of particular locales could be.

The same example underscores a second theme that has been sounded here: the extent to which the formation of the new order was both contingent and contested. The mandate to create new units of local administration may have been handed down from above, but the inclusiveness of any one prefecture's territory and the location of its capital were open to impassioned debate. Similarly, while the ruling clique might determine high-order priorities in the extension of the railroad network, the precise routing of any given stretch of track was a matter of intense local interest and lobbying effort. Railroads could break much more radically with physical geography than could earlier transport systems, making their routing unusually manipulable; at the same time, because local investors recognized in them "the chief device for introducing a new capitalist logic" to the region,[79] rivalry for securing

77. Nagano-ken 1971:302–3.

78. Others who have argued for a view from the provinces include Gary Allinson (1975:10) and Neil Waters (1983:25); see Wigen 1992:4–5.

79. The quotation is from Cronon 1991:81. On the importance of railroad boosterism in nineteenth-century America, see ibid., ch. 2.

train access was fierce. It was in this competitive context that regional identities were forged.

A third observation has to do with the paradox of scale. At the regional level, there is no doubt that the dislocations traced here were debilitating in tangible ways. Slowly but surely, the scope of local self-determination was reduced until, by 1895, the Ina valley had been diminished to the status of a pawn in a large and hierarchically structured geopolitical universe. Yet if Shimoina as a whole fared poorly relative to its neighbors in each of these three rounds of national consolidation, the same cannot necessarily be said of individual communities within the county, any more than it can be said of individual citizens. In particular, the same geopolitical restructuring that disadvantaged Shimoina as a whole, relative to other counties in the prefecture, privileged Iida relative to its local hinterland. Under the new system, the terrain that fell within Iida's jurisdiction was more than four times larger than the former Iida domain. And designation as a county seat gave the former castle town not only administrative hegemony over the newly encompassed territories, but financial and communications tools by which to assert increasing economic hegemony over them as well. In short, the implications of redrawing the political map were more ambiguous than a simple interregional comparison can suggest. For the historical geographer, it is essential to sort out those ambiguities on different spatial scales.

These findings in turn relate to a final theme running through this chapter: the observation that regional rivalry was pressed into the service of social order. While ruthlessly suppressing working-class and socialistic movements such as the Aikoku Seirisha on the grounds that social harmony was imperative for national survival, the new regime did little to suppress conflict between regions. In a word, geography emerged as the permissible language of difference in Meiji Japan.[80] To be sure, as Carol Gluck has observed, "the rhetoric of national unity was constantly wielded as a talisman against the conflict of local interests."[81] Yet by the structural logic of Meiji politics, territorial contests were tolerated and even promoted, emerging as arguably the most legitimate focus of competitive social mobilization. And whether by design or by de-

---

80. Gender was, of course, another important lexicon of difference, but one that could alternately assume condoned or subversive inflections. For a probing analysis of official Meiji rhetoric on women, concluding that its essential message was to extoll a "cult of productivity" (in contrast to the American "cult of domesticity"), see Nolte and Hastings 1991; on attempts to generate an alternative discourse on women, see Sievers 1983.

81. Gluck 1985:38.

fault, regional conflict served the regime well. Although the kind of evidence presented here does not speak to the question of intent, it does suggest ways in which regional rivalries proved beneficial to the ruling elite, particularly to those involved in the newly founded parties.[82] The need to compete in regional blocks for scarce resources created a politics of patronage, where local leaders could reasonably argue that it was in the best interests of everyone in the community—women as well as men, young as well as old, tenants as well as landlords—to rally behind its richest and most powerful men. Yet in practice regional rhetoric often served as a mask for personal and class interests; as numerous critics recognized, projects put forward in the name of the common good rarely promised equal dividends for all of Shimoina's citizens.[83] As prominent as the new regional rivalries became, the landscape they overlay—and helped to reshape—was already complexly divided along other lines.

82. "The center and the provinces were now linked by channels of mutual demands brokered by a local elite who were increasingly tied to the parties," Carol Gluck observes of the period after the Russo-Japanese War (Gluck 1985:236). In the case of Shimoina, I would argue, this was true more than a decade earlier.

83. To cite but one example, in 1882 (the first year of the Matsukata recession), when leading filaturists throughout Nagano clamored for road improvements, their critics countered that the six expensive projects under consideration would actually benefit only 7 to 8 percent of the prefecture's people. Nagano-ken 1971:178.

CHAPTER SEVEN

Precarious Prosperity

Industrial Restructuring and Regional Transformation, 1895–1920

By the early 1890s, the major institutions of the modern Japanese state had taken shape. The treasury was on a solid footing, the administrative hierarchy had stabilized, and the rudiments of a constitutional democracy had been put into place. As the global economy swung out of recession later in the decade, Tokyo had at its disposal a new set of political levers with which to mobilize the country's resources. The next round of economic expansion, lasting for roughly twenty-five years, would see Japan transformed into an industrial and imperial power.

The first landmark in that transformation was a military victory in the brief but bloody Sino-Japanese War. Among several favorable provisions in the Treaty of Shimonoseki, which ended the war in 1895, none was more important than a punitive reparations payment demanded from the Chinese. The indemnity amounted to a remarkable 360 million yen (200 million taels), the equivalent of 4.5 times the Japanese government's annual budget for 1893. Beyond reimbursing the treasury for all expenses incurred in the invasion, this massive infusion of cash permitted Japan simultaneously to go on the gold standard and to begin building up its military and industrial base without resorting to major overseas loans. The following decade witnessed a spate of shipbuilding and railroad construction, as well as a proliferation of textile factories: hallmarks of Japan's first industrial revolution.[1]

---

1. On the scale of the indemnity, and its benefits for Japanese heavy industry, see Gordon 1991:70; on 1895 as a watershed in Japanese economic development, see Duus 1988:4–5.

Particularly important for central Honshu was the continued rise of the filatures. Although the European market for Japanese floss had been largely foreclosed once France and Italy recovered from the silkworm blight of the 1860s and 1870s, a new market was quickly arising in the United States. Legislation to protect the American textile industry was passed after the Civil War, shifting demand away from finished silk cloth and toward imported floss. Once again, the Japanese were in the right place at the right time. Especially after the U.S. transcontinental railway was completed in 1884, slashing shipping costs from Pacific ports to mills on the Atlantic seaboard, European producers proved unable to compete with low-priced Asian silk thread in New York. Chinese manufacturers, meanwhile, suffered from a different set of difficulties, growing out of their country's domestic problems. Political fragmentation and repeated military incursions limited the mobility of inputs, hampered diffusion of technology, rendered the state unable to regulate producers, and subjected shippers to grossly inequitable European trade and tariff practices. By contrast, Japanese filaturists not only profited from aggressive state assistance but managed to bypass brokers and get cocoons directly from peasants, resulting in savings their Chinese competitors could not match. This combination of advantages yielded tangible results; from 1867 to 1894, annual Japanese raw silk exports to the United States had already increased a remarkable seventyfold. By the turn of the century, Japanese exporters were supplying the bulk of this important market.[2]

Helped in part by the foreign exchange earned from silk sales in the United States, the Japanese were able to launch a second successful military campaign after the turn of the century, this time against Russia. Japan's 1905 victory in the Russo-Japanese War was won only at enormous social and financial expense; the fighting claimed a staggering eighty thousand lives and cost two billion yen, or eight times as much as the Sino-Japanese conflict, while bringing the Japanese neither territorial gains nor indemnity. Yet it carried considerable symbolic weight, signaling to other Asians—and indeed to subject peoples around the world—that the Europeans were not invincible.[3] The implicit promise of a new order, however, was soon betrayed by Japan's own continuing colonial exploits. Over the next fifteen years, the emboldened Japanese

2. For a concise comparison of the conditions under which the Chinese and Japanese silk industries developed, see Eng 1978:218–39. An account of one early Japanese trader's adventures in the New York silk market may be found in Reischauer 1986:207ff.

3. On Japan's role as a model for anticolonial movements elsewhere in Asia, see Jansen 1954.

army and navy would move on to seize Korea and a number of former German possessions in the Pacific, assembling a formidable empire by 1920. Although such aggression was costly, most of these colonies were soon forced in one way or another to contribute to the Japanese economy, whether by supplying raw materials for Japan's industries, markets for its manufactured goods, food and fertilizer for its farms, or an emigration frontier for its burgeoning population.[4]

Behind all of these exploits lay a remarkable metamorphosis of the domestic countryside, nowhere more evident than in the southern Japanese Alps. On the one hand, military mobilization had been paid for by the sweat and blood of the rural populace.[5] Shimoina residents had to shoulder nearly 7,000 yen in additional taxes to finance the Sino-Japanese War alone; for the fight against Russia, the Japanese government drafted more than 2,500 local men, and imposed special surcharges that put local tobacco growers and domestic weavers out of business.[6] Tokyo also raised the land tax twice—inadvertently doubling the size of the national electorate in the process.[7] But at least as important as military mobilization was the more general redeployment of regional resources. In a very tangible sense, the work of young women in sericultural rooms and filatures up and down the valley helped finance Japan's colonial exploits. Silk accounted for more than a third of the country's exports at the turn of the century, earning hard currency that paid for machinery and armaments from the West. Clearly, there was more than empty rhetoric to the company song that declared, "Boys to the army/ Girls to the factory / Reeling thread is for the country too."[8] And far from subsiding after the costly victory over Russia, the pace of industrialization only accelerated further. As the Japanese economy moved through a postwar recession and finally into another expansionary cycle during World War I, the financial and geopolitical infrastructures put in place during the previous quarter-cen-

4. On the economic relations between Japan and its prewar empire, see Myers and Peattie 1984, part 3, and Duus, Myers, and Peattie 1989, part 1.

5. On the costs of the Russo-Japanese War for rural communities across Japan, and the significance of military conscription for rural social relations, see Waswo 1988:562–66.

6. Nagano-ken 1971:522; Shimoina-gun 1907:76. Although only 208 of the 2,526 local men drafted for the latter conflict actually died in battle, thousands of rural families were deprived of their loved ones—and of their strongest male laborers—for the duration of the war.

7. Fewer than 763,000 men were qualified to vote before the war surcharge was imposed, compared with over 1.5 million afterward. For contrasting views on the significance of this doubling of the electorate, cf. Mitani 1974:7–21; Banno 1983.

8. Quoted in Tsurumi 1990:92. Western scholars have been slow to recognize the extent of women's contributions to Japan's growing economic prowess during these years; for a provocative review of the literature, see Marburg 1984, ch. 4.

tury were wielded to turn Shimoina inside out for timber and silk. The result was a new landscape of production and exchange—one where heady profits in the short term masked a growing vulnerability in the long term.

The font of Shimoina's prosperity—and the source of its vulnerability—was the region's increasing commitment to sericulture. By 1920, Nagano Prefecture had become a "silk kingdom" (*seishi ōkoku*), and silk had come to dominate the Ina valley's economy by any conceivable measure. Mulberry plantations occupied nearly half of Shimoina's arable land, and filatures occupied 60 percent of its industrial work force. The same proportion of all households in the county raised silkworms, producing on average more than 280 kilograms of cocoons each per year. Throughout Nagano Prefecture, the silk industry accounted for over 70 percent of all bank loans made in the 1910s. And in 1920, when silk prices reached their prewar peak, the net value of the region's cocoon harvest was an astonishing 2.3 times the sum of all other produce from the land.[9] Meanwhile, the dramatic growth of cocoon production and reeling capacity had spawned subsidiary activities throughout the region. While thousands of families in the hinterlands and on the valley floor alike found employment supplying silk producers with inputs or processing their by-products, Iida's merchants flourished by catering to new consumer markets in the county.

But the profits that drove this economy would prove ephemeral, leaving the southern Japanese Alps open to devastation when the market collapsed in succeeding decades. By 1920, wholesale conversion to timber and silk had weakened the Ina valley's once-robust economy in four distinct ways: its environment was degraded, its export base had contracted, its population had become increasingly dependent on imported food, and even its core communities were steadily losing high-level processing to other areas. On close inspection, the valley's celebrated industrial transformation was surprisingly shallow, concentrated in a single industry and based on intensive use of low-wage labor.

Many of these developments were captured in the Statistical Abstract for Shimoina County, *Nagano-ken Shimoina-gun tōkei ippan* (NSTI), a 1921 economic census published in 1923. This document, prepared by officials in Iida, affords a priceless snapshot of Shimoina as silk prices reached their prewar zenith, with sections enumerating the county's population, crops, commerce, finance, industry, forestry, animal husbandry, wages and prices, postal and telegraph services, roads, schools, local government, and more. On many of these topics, separate statis-

---

9. The figure on prefectural bank loans going to the silk industry is from Nagano-ken 1972:170. All other statistics derive from NSTI.

tics were given for each of the forty-one villages making up Shimoina at the time. As mistakes and discrepancies were not completely eliminated, the data must be cross-checked continually, both internally and with other sources. But used with discretion, the abstract provides a valuable picture of the new economy of production. In the following discussion, a wide variety of statistics from this census, bolstered by supplementary information from prefectural and village sources, are adduced to reconstruct Shimoina's economic geography at the height of the silk boom. The chapter begins with an attempt to recapture the perspective of the time, documenting the prosperity that dazzled visitors and local boosters alike. Only later are the shadows filled in, as the analysis turns to the more sobering question of the region's growing precariousness.

A Vital Landscape

Despite the gloom that pervaded the Ina valley in 1894 when it was announced that the Chūō Railroad would be routed through Kiso, twenty-five years later Shimoina's economy was steaming ahead. Fueled by strong markets for timber and silk, the county's population had grown nearly 30 percent since the turn of the century, and the net value of local goods and services had grown even faster; most residents now had access to schools, banks, electricity, and improved roads or cableways. These infrastructural investments had extended Iida's reach over its administrative hinterland; they also brought income from throughout Shimoina into the county seat, where restaurants, shops, inns, and brothels were doing a brisk business. Most propitiously, funds had finally been raised to extend a private rail line down the valley from Suwa.[10] In 1920, the opening of direct train service to Iida was only three years away. To local boosters, the good times were just beginning; the Ina valley's potential had only begun to be developed.

A BOOM FOR THE HINTERLANDS

One of the primary enterprises driving Shimoina's prosperity after the turn of the century was a wave of timbering in the region's

---

10. Shizuoka Tetsudō Kanri Kyoku 1955; Nagano-ken 1972:64–65, 146. On the chronic difficulties of local fund-raising efforts, see Matsuzaki 1977:246–48; on national investment patterns in the light rail boom of the early 1910s, see Wakuda 1981:54–55.

mountainous hinterland. Three developments had converged to make Shimoina an attractive target for commercial loggers during these years. First, timber demand was up nationwide, as the rapid economic growth throughout Japan in the closing decade of the 1800s triggered a sustained construction boom. At the same time, after a long period of hesitation, the government had moved in the early 1890s to establish clear title to all timber reserves. In Shimoina, most woodland on the terraces around the valley floor had long since reverted to individual or village ownership; in fact, thanks to the long and well-documented history of commercial forestry in the Ina valley, an unusual number of individuals and communities in Shimoina successfully defended private ownership claims. Nonetheless, extensive tracts of prime forest in the rugged, lightly populated tributary valleys south and west of the castle town were now claimed by the state.[11]

The final development that prompted foresters to penetrate these previously isolated areas was the upgrading of the region's roads. In the 1890s, the principal packhorse routes were widened to between five and six meters, and their steeper sections were replaced by winding segments designed to avoid gradient extremes. With this combination of breadth and slope adjustments, horse-drawn carriages (*basha*) and cargo wagons (*nibasha*) appeared in the county for the first time. It was a far cry from acquiring a national railroad, but these modest vehicles increased the competence of the local transport system sevenfold: enough to create radical changes in the economy of Shimoina.[12]

In 1894, as soon as wheeled vehicles could travel directly from Iida to Nagoya, professional lumbermen began streaming in from the south to fell the county's prime timber. Over the next twenty-five years, following the steady progress of road improvements, commercial loggers—joined by displaced packhorse drivers—pushed north and east through the county's mountainous rimlands.[13] While it lasted, the lumber fron-

---

11. In 1890, all wooded acreage to which individuals or local governments could prove ownership was designated "people's forest" (*min'yūrin*). The largest subcategory, made up of prefectural, town, village, and hamlet forests (collectively known as *kōyūrin*), consisted mostly of former commons or village-owned lands. Remaining woodlands, including most former Bakufu and domain holdings, were declared *kan'yūrin*, or state-owned forest (a category encompassing both imperial [*goryōrin*] and national [*kokuyūrin*] forest lands). Nagano Prefecture led the nation in quasi-private *kōyūrin* acreage. Nagano-ken 1971:420; Nagano-ken 1978:11–13, 24; T. Satō 1981; Tsutsui 1974b.

12. Carriages and wagons could carry up to 200 *kan*, pack animals only 30. Sekishima 1960. On the sequence of local road improvements, see Osawa 1966.

13. Many former *chūma* drivers became lumberers during these mid Meiji years. Whereas a horse cost only 40 yen in the early 1890s, a cart cost 200 to 300 yen, an investment few packhorse drivers could afford. Miura 1988:127.

tier created a frenzy of commercial opportunity for local merchants and lucrative employment for resident laborers, who helped fell, transport, and process the logs. Just four years after improvements on the Seinaiji Pass were completed, for instance, nine local lumber mills were operating at capacity, employing over 150 local men.[14] Population surged along the timber frontier (map 27); in remote southeastern Shimoina, an encampment of 1,500 men sprouted in the midst of a local population numbering only 4,500. Since the targeted areas typically supported little agriculture, more than one local businessman made his fortune procuring supplies for this newfound market.[15]

SERICULTURE AND FILATURE

More steady than forestry, however, were sericulture and filature and their numerous spin-offs. Silkworm-rearing techniques had been systematically improved—and their advantages persistently advocated—by the local chapter of the national agricultural association (Dainihon nōkai), founded in Nagano in 1886. As farmers struggled to recover from the recession of the early 1880s, the bureaucrats and large landowners who ran the agricultural co-ops had coordinated a massive move into sericulture; with the decline of the packhorse trade, even horse barns along the Ina road were converted to silkworm-rearing rooms.[16] By 1921, raising silkworms had become a nearly ubiquitous employment in rural Shimoina. Fifty percent of the county's households—and 60 percent of those living outside the urban/commercial core—produced spring cocoons; an even higher percentage raised the summer or fall silkworms that had begun penetrating the county in the 1890s. In every rural village, between 40 and 90 percent of all households were engaged in sericultural activity at some time during the year.[17]

Silk-producing areas experienced rising populations, too, albeit in less dramatic numbers than the lumber camps. The growth of sericultural employment within the home enhanced the potential economic contributions of children while enabling rural households to support more offspring if they so chose. It is surely no coincidence that average house-

14. Seinaiji-mura 1982, 2:21, 279.
15. To support the six hundred woodsmen hired by Ōji Paper in rugged southeastern Shimoina, for instance, an Iida merchant contracted to have twenty bags of rice a day polished and carried in to the Tōyama valley. Shimoina Chiikishi Kenkyūkai 1982:56.
16. Shimoina Kyōikukai 1962:54.
17. NSTI:48–53. On the government's role in establishing agricultural cooperatives and experiment stations during this period, see Nagano-ken 1971:406–8.

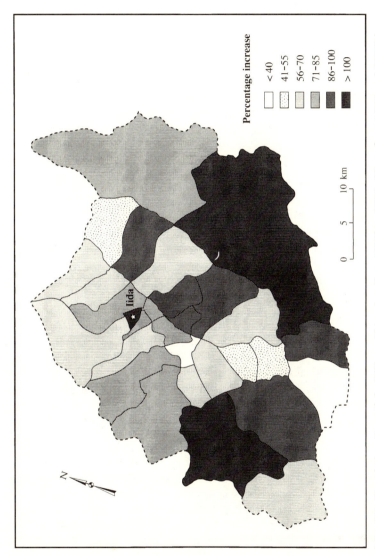

Map 27. Population Growth by Village, Shimoina *Gun*, 1875 to 1921. (Data from NCS and NSTI.)

Percentage increase

< 40
41–55
56–70
71–85
86–100
> 100

hold size in rural Shimoina, after declining during the Tokugawa period, climbed once again through the silk-boom years (rising from 4.8 in 1876 to 5.5 in 1921). It is also suggestive that sericultural centers generally supported larger than average families (map 28) and relatively low ratios of men to women (map 29). These villages also posted impressive net population gains, although their growth was temporarily surpassed by that of mountain villages where logging crews were in residence when figures were collected in 1921 (see map 27).

Most significant from the local perspective was that the county's filatures grew as well, despite strong competition from the ever-expanding factories headquartered in the Suwa basin. During the four decades that spanned the turn of the century, local thread output rose a remarkable seventeenfold, making Shimoina one of only three Nagano counties outside Suwa whose share of thread production increased between 1882 and 1921.[18] It was these silk mills that formed Shimoina's industrial base, uniting the northern two-thirds of the county into a single production complex (map 30).

One of the more positive features of the Ina valley's textile-based industrialization was the extent of local ownership. Admittedly, local merchants had to borrow heavily to fund each round of cocoon buying, and some local filature growth was directly financed by large outside firms. Shimoina's largest mill in 1914, for instance, was a branch operation of the Katakura concern, Suwa's leading filature company. But outsiders never owned more than a fraction of Shimoina's reeling capacity; most of the county's filatures throughout this period were locally owned.[19] One reason local ownership persisted into the early twentieth century was that economies of scale were not as yet decisive in the industry. The size of the local mills did grow over time; by 1921, the average filature in the county had eighty-eight basins, triple the comparable figure for 1893. But well-run small firms remained competitive even as the average plant grew in size. Small firms were helped by the relative affordability of waterwheels (the main power source in Japanese filatures

18. Shimoina's share of prefectural thread output rose from 5.8 percent in 1882 to 8.9 percent in 1919. More impressive, Shimoina's absolute growth in annual floss output (126,677 *kan*) was more than twice that of the other two counties combined. Computed from figures in Nagano-ken 1971:155, and id., 1972:105.

19. In 1920, Katakura owned twenty-three mills in Japan, with nearly 12,000 reeling basins, accounting for over 10 percent of the nation's reeling capacity. But this massive firm was forced to retreat in succeeding years as the smaller mills of the Ina valley banded together to form the Nanryūsha, one of Japan's most successful producer cooperatives. On the local silk co-op movement, see Tenryūsha 1984; Hirano 1990.

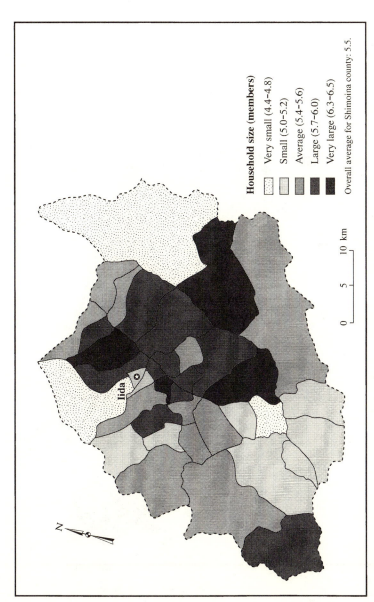

Household size (members)

Very small (4.4–4.8)

Small (5.0–5.2)

Average (5.4–5.6)

Large (5.7–6.0)

Very large (6.3–6.5)

Overall average for Shimoina county: 5.5.

Iida

N

0   5   10 km

Map 28.   Average Household Size by Village, Shimoina *Gun*, 1921. Resident population. (Data from NSTI.)

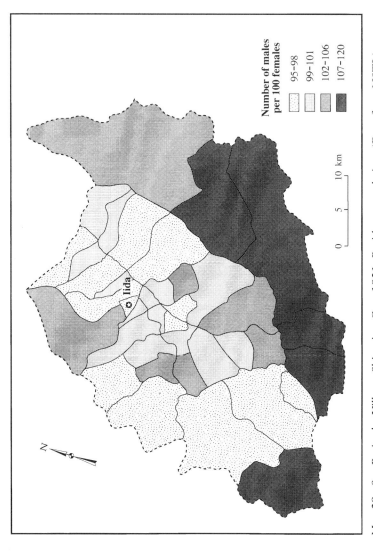

Map 29. Sex Ratios by Village, Shimoina *Gun*, 1921. Resident population. (Data from NSTI.)

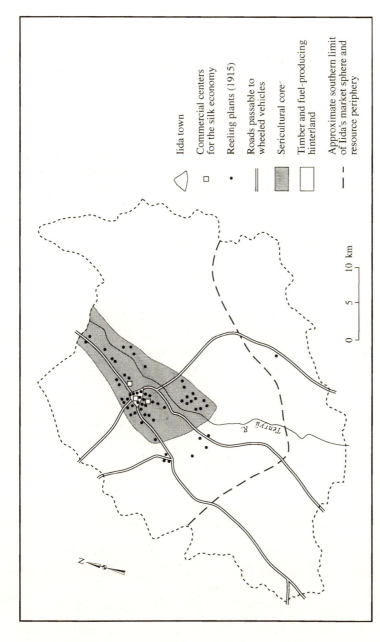

Map 30. The Territorial Division of Labor in Shimoina *Gun*, ca. 1920. (Data on the location of commercial centers for the economy, from Matsuo-mura 1982:604 and Kamisato-mura 1978:1099; on the distribution of reeling plants, from Tenryū sha 1984:193. Other data compiled from NSTI.)

Iida town

☐ Commercial centers for the silk economy

• Reeling plants (1915)

Roads passable to wheeled vehicles

Sericultural core

Timber and fuel-producing hinterland

Approximate southern limit of Iida's market sphere and resource periphery

Tenryū R.

N

0   5   10 km

throughout this period), which could be profitably employed even in a ten-basin plant. They also benefited from the early introduction of electricity, which was first generated in Iida in 1899.[20]

Boosted by steady improvements in their equipment, as well as by the longer working hours that electric lighting made possible, labor productivity in all Shimoina filatures rose considerably. In the six short years after Matsukata, annual output per worker nearly trebled; in the following three decades to 1921, productivity gains were nearly as impressive. While Shimoina's filature work force increased nearly sevenfold, output grew more than tenfold (table 19).[21]

In short, while the dominant firms in the industry clearly grew in size, small local filatures managed to function profitably alongside them throughout the period under consideration. It was primarily to these small-scale operators that Shimoina owed its overall industrial growth after the turn of the century.

SILK SPIN-OFFS: BACKWARD LINKAGES

In addition to the income directly generated in sericulture and filature, silk production spawned numerous backward and forward linkages as well. Two minor input industries noted in local sources were the making of hardwood spools for the filatures and the weaving of silkworm shelving and cocoon baskets from bamboo and straw. Lathe-turners (*kijishi*) in the Seinaiji area fashioned the hardwood spools called *giri* on which silk was reeled, marketing them not only to Iida but throughout Japan.[22] In addition, more than 600 Shimoina households were engaged at least part-time in weaving silkworm racks, giving this sector the largest work force of any craft industry in the county.[23] But still larger industries sprang up to supply lime-based fertilizer (*sekkai*) for the mulberry fields and wood fuels for silk processing, providing clear

20. Minami Ryōshin (1976, 1987) has argued that the swiftness with which electrification followed the introduction of steam generators in Japan allowed small factories to remain competitive with large firms. Overall, Shimoina seems to confirm Minami's hypothesis; very small filatures remained competitive throughout the boom years. The fitful progress of local electrification is recounted in Chūbu Denryoku 1981, Shimoina Chiikishi Kenkyūkai 1982:58–59, and Ishikawa 1980:111.

21. Masaki 1978:401; Hirasawa 1952:205–6.

22. Tate 1979:7–9; Seinaiji-mura 1982, 2:279.

23. Shimoina produced 3,720 *koku* of bamboo goods in 1912, along with 14,870 straw items (including sandals, horseshoes, mats, and bags for charcoal). Kobayashi 1959:63; Ishikawa and Takeuchi 1986:16. See also Takagi-mura 1979, 2:361.

Table 19 *Shimoina Filature Employment and Productivity Gains, 1893–1921*

	1893	1921	Percentage Increase
Mills	48	91	89
Basins	1,352	8,041	495
Basins per mill	28	88	216
Workers	1,384[a]	9,130	560
Thread output (*kan*)	11,239	121,643	982
Output per worker (*kan*/yr)	8.1	13.3	64

SOURCES: Hirasawa 1952:200; NSTI:67–72. Employment figure for 1894 from Masaki 1978:404.

[a]Employment figure for 1894.

evidence of the new economy's power to generate backward linkages even in the Shimoina hinterlands.

Mulberry culture created a prodigious appetite for lime to neutralize Shimoina's predominantly acidic soils. Beginning almost immediately after the opening of Yokohama, limekilns were put into service throughout the eastern portion of Shimoina County, with Chiyo village emerging as the undisputed center of production in the county. This generated considerable employment in areas of limestone deposits, where local residents began mining, baking, and transporting the finished product in unprecedented quantities and over unprecedented distances. Significantly, the local corporation founded to oversee lime production in Chiyo drew its primary stockholders from Matsuo, Igara, and Ichida villages: all major centers of silk-mulberry cultivation.[24]

The rising demand for lime in turn helped create a market for charcoal, the single most important backward linkage supported by the silk industry. Limestone had to be treated at high temperatures, usually in stone kilns, before it could be applied to the fields. As the 1906 correlation of high charcoal prices with liming districts testifies (map 31), some of the most intense fuel demand in Shimoina was associated with the county's limekilns. But lime demand was only one way in which the silk economy increased Shimoina's regional market for fuel. For instance, by the turn of the century, silkworm-rearing rooms were increasingly provided with wood-burning stoves; maintaining a warm, even temperature had been shown to hasten maturation of the larvae, improve their

24. Sources on the Chiyo lime industry include Shimoina-gun 1911; Chiyo-mura 1965:13; Tenryūsha 1984:212–13.

productivity, and reduce turnover time between cocoon harvests. In 1921, farmers may have used the equivalent of one-sixth of the region's commercial charcoal just to incubate their silkworms.[25] Filature operations also required fuel, both for drying the fresh cocoons and for heating the water in which they were immersed before reeling.[26] This industrial demand was augmented by general consumer needs, emanating from the households of Shimoina's rapidly growing population. Since rail transport would not reach Iida until 1923, fuel imports from outside the county were impractical during the years under consideration, creating a steady market for locally produced kindling and charcoal. In addition, the growing urban population along the industrializing Pacific coast pushed the fuel hinterland of the Tōkai northward, exerting a strong pull on charcoal produced in the southern Shimoina fringe.

As these pressures began to be felt in the mountains rimming the county, charcoal production became a major commercial industry. The great advantage of charcoal over firewood, of course, was its higher ratio of value to volume. Weighing only 15 to 30 percent as much as the wood it was made from, and selling for three times as much as firewood by weight, charcoal could more readily bear the shipping costs from the outlying hills. As a result, while firewood continued to play a very important role in overall fuel consumption for some time, *commercial* fuel forestry in the county was heavily slanted to charcoal.[27]

Charcoalers found suitable hardwoods at two distinct types of sites. Former meadows, which were rapidly being converted to woodlots, were one target area. Since the peak labor demand associated with sericulture coincided with the traditional spring grass-mowing season, many households were choosing to forgo their rights to green manure, purchasing

25. The quantity of fuel consumed in heating the rearing rooms (a practice that began in the early 1880s) is reflected in farm budgets. One northern Nagano household in 1920, in order to produce 350 kg of cocoons, required three bags (totaling 50 to 60 kg) of charcoal, plus three horse loads (340 to 380 kg) of firewood (Nagano-ken 1971:116). If this may be taken as a rough index of the ratio of fuel consumption to cocoon output in the countryside, Shimoina in 1920 may have required on the order of 80 to 100 metric tons of charcoal, and another 550 to 600 metric tons of firewood, for sericultural production alone.

26. For best results, cocoons had to be dried over slow fires, usually in large drying rooms or ovens. (As noted above, steam treatment and salt solutions also worked, but they damaged the silk.) The second fuel-intensive step, heating the water in the reeling basins, was necessary to dissolve the glutinous coating binding the cocoons together. The changing technology of silk processing is discussed in Oguchi 1960 and Hirasawa 1960a.

27. Kawaji-mura 1988:292; Seinaiji-mura 1982, 2:16.28. Miura 1988:128. On the ecology of Japanese coppices, see Itow 1984.

commercial fertilizers instead. With the draft animal population declining after the turn of the century as well, grasslands throughout Shimoina began reverting to scrub forest.[28] At the same time, an abundance of appropriate wood for charcoal making could be found in the wake of the lumbermen, since timbering and fuel forestry were more complementary than competitive in their use of forest resources. Charcoalers preferred the smaller hardwoods that loggers routinely left behind; they could make use of the timbermen's rejected tops and branches as well.[29] The first specialized charcoalers in Shimoina in fact followed the lumbermen into the county from the south, near the turn of the century. At the peak of production, several hundred such workers were encamped in the forests of southern Shimoina, many with families in tow.[30] Packhorse drivers or teamsters made the rounds of the collection points, often fattening their shipping wages by operating simultaneously as brokers, buying cheap in the uplands and selling dear in the industrial belt of the valley or in the Tōkai region. The result was a new geography of trade (map 32).

The retail value of charcoal in both absolute and relative terms escalated dramatically during the 1910s (fig. 2). In the spring of 1920, when silk prices reached their prewar zenith, charcoal prices peaked as well, and charcoal surpassed lumber as the most profitable sector of forestry in Nagano Prefecture.[31] The value of one bag of charcoal, roughly a day's wage for the charcoaler, had by then reached 45 *sen*—only half as much as a day laborer in Iida could earn at the time, but quite high remuneration by the standards of mountain dwellers.[32] Not surprisingly, overall production levels had risen dramatically as well: Shimoina's annual charcoal output had tripled since 1906, until its value exceeded that of the county's lumber output by fully 50 percent.[33] Through its sustained

28. Miura 1988:128. On the ecology of Japanese coppices, see Itow 1984.

29. Lumberers in this area targeted fir, hemlock, cypress, and mature beech (*momi, tsuga, hinoki,* and *buna*), while charcoalers preferred oaks (*nara* and *shide*) and miscellaneous broadleaf species. Miura 1988:128, S. Satō 1966:42; cf. Saga 1987:196.

30. Although they entered from the south, the first professional charcoalers to work in Shimoina's mountains hailed originally from the coast of the Sea of Japan to the north. Their home district of Hokuriku, known for its high tenancy rates, exported female labor to Shimoina as well, including many of the prostitutes in Iida's flourishing red-light district (Shimoina Chiikishi Kenkyūkai 1982:39.) On local charcoal-production techniques, see Seinaiji-mura 1982, 1:258, Anan-machi 1987, 2:482, Tatsuoka Kōminkan 1985:3–6, and Mukaiyama 1959.

31. Nagano-ken 1971:142.

32. Ōshika-mura 1984, 2:229.

33. NSTI:92–93.

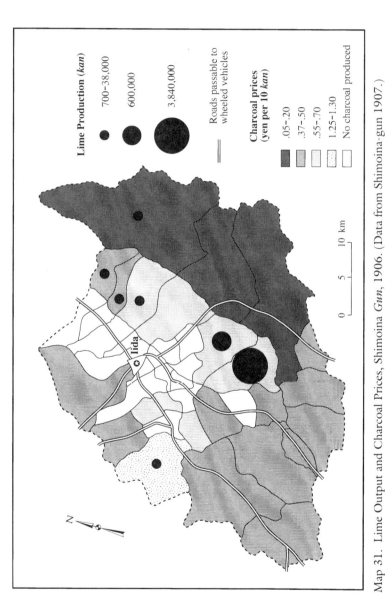

Lime Production (*kan*)

- 700–38,000
- 600,000
- 3,840,000

Roads passable to
wheeled vehicles

**Charcoal prices
(yen per 10 *kan*)**

- .05–.20
- .37–.50
- .55–.70
- 1.25–1.30
- No charcoal produced

Map 31. Lime Output and Charcoal Prices, Shimoina *Gun*, 1906. (Data from Shimoina-gun 1907.)

demand for wood fuels, the silk industry had brought relative prosperity to the valley's hinterlands.

FORWARD LINKAGES

Employment in all of these input industries—including production of silkworm shelving and hardwood spools as well as lime and charcoal—was concentrated in Shimoina's mountainous periphery. Areas near the primary silk centers, by contrast, took advantage of the copious by-products of sericulture to establish lucrative downstream or forward linkages.

The pupae that accumulated as the cocoons were reeled were one valuable silk by-product. Tons of putrifying silkworms made for a smelly, troublesome industrial waste. But they also constituted an organic treasure, fatty and rich in protein. After the silk fibers had been removed, the exposed chrysalides were collected from the boiling basins and crushed to extract their oil, which was highly prized for cooking. The pressings could then be sold as a premium fertilizer, along with the copious silkworm droppings. In fact, an agricultural experiment sponsored by the government in 1890 found silkworm pupae to be the most effective of several fertilizers tested for increasing rice production. Twenty years later, production of silkworm-based fertilizer had grown into a major commercial industry in Kanae village, Iida's neighbor to the south.[34]

Another alternative was to feed pupae wastes to carp, creating a new source of human food while fertilizing the paddies in which the fish were raised. In this inland valley, where most households were lucky to eat dried fish twice a month, fresh carp provided a novel and welcome treat. In 1873, when Hasegawa Hanshichi of Takagi village erected Shimoina's first modern filature factory, he also pioneered in feeding the pupae to fish. Fifty years later, aquaculture in Shimoina had emerged as a major business in its own right, generating over 100,000 yen in annual sales. Its geography shows its close ties to the silk industry; the villages in which carp ponds were to be found—Kamisato, Zakōji, Takagi, and Matsuo— were without exception Shimoina's major silk-reeling centers.[35]

Even more valuable than silkworm pupae was another by-product of sericulture: waste silk (*tamaito*), the coarse thread reeled from double,

34. T. Miyamoto 1967:217; Shimohisakata-mura 1973:687. As early as 1876, pupae-fertilizer output was recorded in four villages; by 1910, the industry generated 6,000 yen in sales in Kanae village alone. NCS; Shimoina-gun 1911.
35. Matsuzaki 1977:152ff.; Shinano Kyōikukai Shimoina Bukai 1934:301–25; T. Miyamoto 1967:215–17; Matsukawa-machi 1965:770.

punctured, or discolored cocoons. Although its export value was nil, the nubby, textured fabric made from these fibers was highly prized locally, and the rising volume of imperfect cocoons that accompanied the growth of sericulture as a whole expanded the basis for the thriving weaving industry in the valley. In addition, rural families typically retained some of their higher-quality cocoons to satisfy their own demand for silk fabrics. The combination of rising incomes and increased cocoon output caused commercial weaving to expand greatly in the valley during these years, as did the imposition of a loom tax during the Russo-Japanese War, which discouraged domestic producers. The net value of Shimoina textiles soared from 78,000 yen in 1917—already the highest of any county in Nagano, and nearly a quarter of prefectural production—to 391,000 yen just five years later.[36]

The rise of a professional weaving industry boosted the dyeing business in the valley as well. Fabric designers were invited to Shimoina from more advanced centers, and the industry began to generate highly specialized spin-offs. Originally, eighteen types of vegetable dyes (of which indigo was foremost) were employed, but in 1889, when a local dyer began to use modern chemical dyes and mordants, the industry was revolutionized. Iida-area dye-works serviced an area that gradually expanded to include central and eastern Nagano and even part of Gifu Prefecture.[37]

PURVEYORS TO NEWFOUND AFFLUENCE:
GROWTH OF THE RETAIL AND
SERVICE SECTORS

In addition to supporting lime, charcoal, and lesser input industries in the mountainous hinterland, and fostering silkworm-based fertilizer production, aquaculture, weaving, and dyeing in the valley's sericultural core, the silk economy also created the basis for a growing service industry. Iida in these years became known as a place to shop, eat, drink, and play. Although through-trade and wholesaling were on the decline, the merchant community of the former castle town throve by catering to the region's silk-enriched farmers and manufacturers. A similar phenomenon occurred in the former outposts as well; while long-distance drayage dropped precipitously with the completion of the railroad through the neighboring Kiso valley, merchant communities in all

36. S. Hayashi 1962:54, 133, 141; Shinano Kyōikukai Shimoina Bukai 1934:11–12, 307–15.
37. S. Hayashi 1962:168–73.

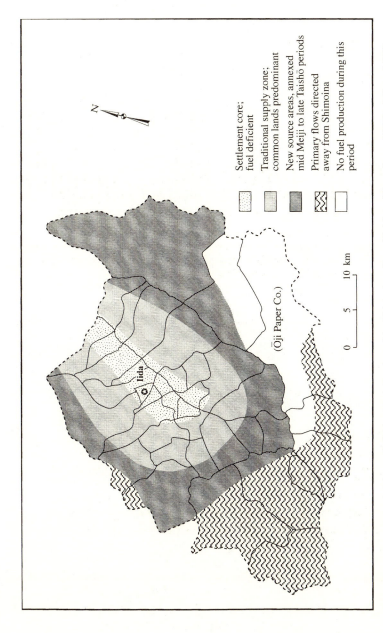

N

Settlement core;
fuel deficient

Traditional supply zone;
common lands predominant

New source areas, annexed
mid Meiji to late Taishō periods

Primary flows directed
away from Shimoina

No fuel production during this
period

(Ōji Paper Co.)

Iida

0   5   10 km

Map 32. Primary Trade Flows for Charcoal Produced in Shimoina *Gun*, Taishō. (Data from NSTI; Miura 1988; Shimoina Kyōikukai 1962; Seinaiji-mura 1982, 1:258; Sawada 1981:178–82.)

Yen

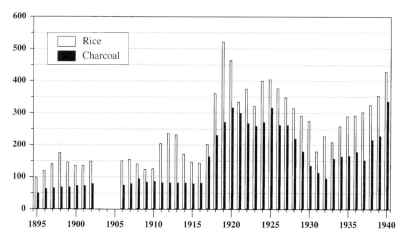

(a) Absolute Prices: Annual Variations. Rice prices are expressed as yen per 10 *koku*; charcoal prices, as yen per 100 *kan*. Annual averages have been calculated from four seasonal figures given for each year. (Data from Nagano-ken 1986.)

%

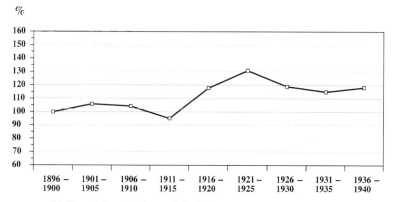

(b) The Value of Charcoal Relative to Rice, Five-Year Averages. Indexed to 1896–1900 values. (Data from Nagano-ken 1986.)

Fig. 2. Rice and Charcoal Prices, Iida, 1895 to 1940.

these commercial hubs maintained their vitality by switching from wholesaling activities to locally oriented retailing. The largest market town in the county grew up north of Iida in the former post station of Ōshima, now home to the only local bank outside of Iida; a close second was Komaba, an important packhorse center to the south.[38] Moreover, as sericultural income steadily increased the scope of the cash economy, permanent general-goods stores began to appear even in rural areas; by the turn of the twentieth century, virtually every village in Shimoina had witnessed the growth of a local commercial district. Most of the inventory of the new village stores would have been familiar to a packhorse driver in 1750: knives, sieves, crockery, lacquerware, used clothing, sewing notions, pharmaceuticals, tea, tobacco, dried fish, seaweed, confections, spices, and *sake*. More novel were such popular items as oil lamps, silkworm-egg cards, and Dharma dolls, considered good-luck charms for sericulturists.[39]

The importance of silk for Shimoina's growing business community far outweighed its representation in the roster of commercial operations. Only a minority of Iida's merchants served the silk industry directly; from 1881 to 1901, just fourteen silk-related corporations were formed in the county, to sell thread, cocoons, or silkworm eggs, to operate warehouses or cocoon markets, and to raise mulberry seedlings.[40] But the majority who did not become directly involved clearly benefited from sericultural expansion as well. The number of horse-drawn wagons (*niguruma*) employed in the county, to take but one instance, increased by over 50 percent in the first half of the 1910s; most of these operated in the flatter parts of the valley, where their primary cargo consisted of fertilizer and cocoons.[41] Indeed, a Tokyo newspaper later reported that in these booming years, before railroads or motorized transport reached the valley, Shimoina's rising entrepreneurial class as a whole was irreverently referred to as *bafun narikin*, or "horse-dung nouveaux riches."[42]

Nor was the growth of commerce confined to the service sector. Sericultural income contributed to a number of new manufacturing opportunities as well, particularly in the realm of food and drink. By drawing women's time away from food preparation, while simultaneously

38. Lesser marketing centers included the east bank's Shimohisakata, Ajima, and Hiraoka, the latter a gateway to the Tōyama valley. Iwashima 1967:76–77.

39. On peddling, see Takeuchi 1976:73–90. For detailed commercial records from the largest satellites, see Kanae-machi 1969:631 and Kamisato-machi 1978:1099.

40. Masaki 1978:409.

41. S. Satō 1966:43.

42. Tōkyō Nichinichi Shinbunsha Keizaibu 1930:306.

allowing them for the first time to earn a sizable cash wage, the growth of the silk industry in the Ina valley enhanced a number of local food industries. One of these was the confection business, established during the late Tokugawa period to cater to travelers and traders along the region's thoroughfares. The Ina valley bakeries found their local clientele significantly expanded in the 1890s, when rice crackers and cakes became a customary snack for the filature workers.[43] The confectionery industry in turn called for large quantities of flour, as did such newly popular fast foods as noodles and dumplings. Milling thus represented another regional food industry augmented indirectly by the silk boom.[44]

Changes in women's labor schedules, along with silk-based cash income, also enlarged the commercial market for fermented soybean paste, called *miso*. Most of the region's rural households had traditionally processed bean paste for themselves. In the later Meiji and Taishō periods, however, professional *miso* producers not only gained a regional market but discovered customers elsewhere in the prefecture and the nation (including the armed forces)—partly because of the discovery, credited to one Nakada of Iida, of a new technique that hastened fermentation (requiring only twenty days instead of the traditional one hundred) while permitting year-round production (the industry having previously been limited to the warm summer months). This breakthrough allowed manufacturers who had formerly been able to produce only one batch per year to increase their output fully eighteenfold without altering the size of their workshops. By 1918, Nagano Prefecture was producing nearly a fifth of all the soybean paste consumed in Japan, and a local *miso* manufacturer named Hara Gorobei had risen to become the second-ranked taxpayer in Iida. A lavish festival he hosted the following year (at a shrine dedicated to the patron god of *miso*, soy sauce, and *sake* brewers) made a public display of the industry's growing prosperity.[45]

A novel variant of another traditional soy-based food similarly became important to Shimoina's commercial economy during this period. This was freeze-dried bean curd (*kōri dōfu*), a light-weight, long-lasting form of an otherwise highly perishable staple. First developed in the 1850s, the market for this product expanded rapidly in the 1890s with the rise of an urban, industrial work force and the concurrent demand for easily prepared foods. In 1909, Shimoina manufacturers began shipping

43. On the surprisingly large share of their meager allowances that factory girls typically spent on confections, see Tsurumi 1984:7 and Kidd 1978:49.
44. Iida Shin'yō Kinkō 1976:89, 105; Kobayashi 1959.
45. Iwashima 1967:61–63.

their product to other cities in the prefecture; two years later, when the Chūō Railroad line reached the head of the Ina valley, soybeans began to be imported from Korea (a development of critical importance to the *miso* industry as well), and commercial outlets were established in the Osaka area. By the late Taishō period, Shimoina was Nagano Prefecture's top *kōri dōfu* producer, accounting for roughly a third of the prefectural total.[46]

In addition to sparking demand for prepared foods such as confections, *miso,* and freeze-dried tofu, the infusion of cash throughout the region during the early twentieth century created a vigorous market for alcoholic spirits. A specialty product of the region from pre-Meiji times, *sake* had been chosen as one of the first local items sent to Yokohama to test the export market. Although that experiment had failed, sericultural income eventually enabled the local populace to support a brewery sector in Shimoina that ranked second among the sixteen counties in Nagano Prefecture in total production. By 1921, Shimoina's annual *sake* production had risen to 30,000 *koku* (5.4 million liters), all of which was allegedly consumed within the county—yielding an impressive consumption rate of 174 liters per household per year.[47]

Finally, at the height of the silk boom, Iida supported a bustling community of taverns and brothels, centered in and around the town's newly established red-light district. By 1869, an unofficial brothel district had arisen in Tenmachō, the northern area of town where packhorse drivers congregated, but a licensed red-light district (*yūkaku*) was opened with great fanfare in 1882. The new "pleasure quarter," located in an exclusive neighborhood of former samurai residences called Nihonmatsu, initially housed ten brothels and twenty-one taverns. Merchants from the town, led by the local police chief and the prefectural assembly representative, had raised 30,000 yen to lobby the prefecture for its license, on the grounds that it would stimulate economic development.[48] A local newspaper, the *Ina kōhō,* published figures documenting the growth of Iida's red-lantern district through the years of the silk boom (table 20); no other documentation survives. Apprentice geisha, who are known to have been numerous in Nihonmatsu as well, were not counted in this compilation.

46. By 1921 the industry supported 192 laborers and generated over 54,000 yen in sales. Shinano Kyōikukai Shimoina Bukai 1934:211–16; NSTI:76.

47. Ishikawa and Takeuchi 1986:16; on local *sake* consumption, see Iwashima 1967:58–59.

48. Iwashima 1967:82–83; Murasawa 1983:7–8, 72.

Table 20 *The Growth of Iida's Official Red-Light District, 1882–1915*

	Geisha	Prostitutes	Inns	Taverns
1882	4	43	60	330
1883	18	64	—	—
1884	36	65	—	—
1886	42	62	46	449
1894	78	54	131	493
1910	113	—	—	—
1915 (est.)	380	500	—	500

SOURCES: Murasawa 1983:15; Iwashima 1967:83.

The key to the Nihonmatsu district's success was its proximity to the central warehouses where farmers came to sell their cocoons. Elderly Iida residents remember seeing twenty to thirty horses at a time tied up in the garden of a nearby temple while their owners dallied at the Nihonmatsu teahouses. Meanwhile, a second, informal entertainment district sprang up next to the cargo dealers' warehouses in a district known as Ōyokochō, where the main guests were not farmers but packhorse drivers and teamsters. These traveling men, notorious for being free with their cash, provided a steady clientele for the inns of the town. Many were jokingly called by the names of the geisha they were currently seeing. Each year during New Year's celebrations, the women who worked in Ōyokochō processed from their teahouses to the cargo dealers' shops, bearing gifts to thank them for their patronage.[49]

It is little wonder contemporary observers were struck by such a scene. Iida was awash in cash; from brothel owners to flour millers, the local merchant community was reveling in unprecedented prosperity. Difficult as its transition in the mid Meiji era was, Shimoina had not been left out of the nation's long wave of economic expansion. Nor had its economy been visibly simplified; in place of the intricate fabric of protoindustrial production, a new and equally intricate set of agro-industrial linkages had begun to weave the region together in new ways. Most impressive, despite the financial centralization of the early Meiji years, local capital continued to play an important role in the filatures that comprised the heart of the new complex. Shimoina's political and business leaders might well congratulate themselves on their role in the region's evident success and expect still greater things to come. But beneath the

49. Kobayashi 1959:20; Iwashima 1967:82–83.

bustling economic landscape of the early twentieth century lurked a darker reality. As the following decade would dramatically reveal, Shimoina's newfound wealth was a truly precarious prosperity.

Vulnerability: The Underside of Shimoina's Modernization

The most important weakness of the new regional economy was also the most intractable: despite the superficial variety of enterprises, the sources of Shimoina's prosperity could increasingly be traced back to a single commodity. Once the large timber was gone, it was to silk that Shimoina's consumers owed their unprecedented cash flow: not only farmers and filaturists, but workers in the subsidiary enterprises and the service sector as well. Yet silk was a notoriously volatile commodity on which to stake the region's future—and a series of related fissures in Shimoina's development pathway compounded the danger.

Explicating the nature of Shimoina's new vulnerability requires that we revisit the 1921 census, viewing it this time not from a contemporary booster's vantage point but with an eye to the destructive undertow. The profits of the timber and silk booms may have been impressive at the time, but they exacted a high price in the long run. By degrading Shimoina's environment, simplifying its export base, eroding the valley's ability to feed its people, and creating a comparatively weak industrial core, commercial forestry and sericultural development together dealt a series of blows to the region's resilience, leaving Shimoina susceptible to serious depression when the silk market collapsed.

ENVIRONMENTAL DEGRADATION

The most visible way in which the developments of the early 1900s limited Shimoina's future options was by decimating its forest resources. Logging created a short-lived boom by liquidating the region's ecological capital, leaving little behind in the way of lasting employment or fixed investments. The forest economy that developed in the county's mountainous periphery was essentially extractive, generating no sustainable spin-off industries and in some areas leaving the regional environment severely degraded.

Gauging the extent of forest losses requires assessing the state of forest assets before the logging boom began. In fact, few woodlands in the southern Japanese Alps at the turn of the twentieth century could have been characterized as virgin forest. Most old-growth stands in the Tenryū watershed had been logged at least once before, nearly three hundred years ahead of the Meiji lumber boom. Yet since that early round of deforestation in the 1600s, aggressive efforts had been made to protect trees on Bakufu lands, and private operators too had begun systematic afforestation.[50] As a result, in the late nineteenth century as in the sixteenth, standing timber constituted one of Shimoina's primary stores of ecological capital. As late as 1876, forest cover is reported to have remained lush over two-thirds of Shimoina; the entire rim of the valley supported forest-dependent hardwood-craft artisans (*kijishi*), hunters, and lumberers, and only two small communities in the heart of the basin complained of hardships in obtaining fuel (map 33). All others (including Iida) reported that "while transport is inconvenient, fuel is abundant."[51]

Population growth and agricultural expansion changed the situation in the core rapidly. Already by the turn of the century, many village commons on or near the central terraces no longer supported trees at all. A prefectural survey conducted in 1900 found that nearly a quarter of Shimoina's private and village-owned "forests" (*min'yūrin*) were in fact grasslands (*kusayama*); more surprisingly, a different set of statistics published by the prefectural agricultural association three years later revealed that fully 80 percent of all so-called woodland (*rin'ya*) in the county was regularly grazed or mowed for fodder or fertilizer.[52] With the coming of commercial foresters, however, forests began to disappear even in the previously inaccessible mountainous periphery.

50. Tokoro 1973:6–11; Iioka 1977; T. Satō 1981:179; Totman 1989:63, 69–72, 151–52. On private forestry in the Tenryū valley, see Tsutsui 1974b; Shimada 1982.

51. This picture of early Meiji fuel forestry is culled from the first Nagano prefectural gazetteer (NCS).

52. The latter survey was conducted by the Nagano prefectural chapter of the Agricultural Cooperative Association, and published in its bulletin, *Nagano-ken nōkai hō*, nos. 47, 48, 50, and 51. The survey covered not only village woodlands but private and imperial forests as well. Startlingly, the proportion of effective grassland in Nagano Prefecture as a whole was reported to be even higher than that in Shimoina: 50 percent of all woodland in the prefecture was classified as regularly mowed for fertilizer (*hiryō to shite karitoru*), 30 percent as mowed for fodder (*magusa to shite karitoru*), 5 percent as grazed (*kachiku o hōboku*), and 10 percent as afforested (*jūmoku no uetsuke*). The remaining 4 percent was categorized simply as "other" (*sono ta*). In Shimoina County, the last category accounted for a higher percentage of all woodland (13 percent), including nearly all

The forest frontier proceeded rapidly through the county, beginning in the southwestern corner and sweeping north and east. The Kiso Range was the first to be denuded. The interlude of highly profitable cutting of prime evergreen species (including fir and hemlock as well as cypress and cedar) was brief; by the end of the Meiji era (1912), the best stands from Neba to Namiai had been felled. The first modern land-use map of the region, published in 1913 from a survey undertaken two years earlier, confirms that the woodlands of southwestern Shimoina already rivaled those near the agricultural core in the extent of forest degradation (map 34). Within another dozen years, heavily capitalized logging firms from Aichi and Shizuoka prefectures would push farther north to Chisato and Seinaiji villages, and soon the mature growth was gone there too.[53] Farther to the east, professional crews from Aichi Prefecture began cutting in the Nanbu uplands in 1897; output had peaked in most of this area by 1911. In contrast to the Kiso Range, moreover, few opportunities for related transport or milling employment were generated in the Nanbu uplands. Virtually all lumber from this area was dragged to the Tenryū River and floated downstream as uncut logs, to be processed along the wharves in the port of Kakezuka.[54]

But it was in the remote southeastern corner of the county, in the Akaishi Range, that deforestation proved most destructive. At the turn of the century, this area's Tōyama valley contained the largest expanse of prime forest remaining in the county. Title to these nominally village-owned woodlands passed from one speculator to another until 1895, when logging rights were purchased by the Ōji Paper Company, Japan's first modern paper manufacturer. Ōji had founded a pulp mill downstream in Shizuoka Prefecture six years earlier, largely at the instigation of Shibusawa Eiichi (Meiji Japan's foremost entrepreneur) and with capital contributed in part by Mitsui. With the surge in demand for newsprint after the Sino-Japanese War, the company expanded northward, acquiring the Tōyama forest reserves with their abundant fir and hemlock, the preferred species for pulp.[55] Local residents protested, insisting that village lands had been sold without their consent. But Ōji Paper proceeded for the next twenty-five years to clear-cut the entire Tōyama watershed, effectively denuding the steep Akaishi Range. The

(nine-tenths) of Shimoina's imperial forest, making it clear that "other" lands were the primary target of commercial logging. Calculated from figures reproduced in Nagano-ken 1971:419, 420.53. Miura 1988; S. Satō 1966:42; Seinaiji-mura 1982, 2:21, 279.

54. Miura 1988:194ff.; Shimoina-gun 1911.

55. Sawada 1981:178–82; Hirschmeier and Yui 1975:106, 235–36.

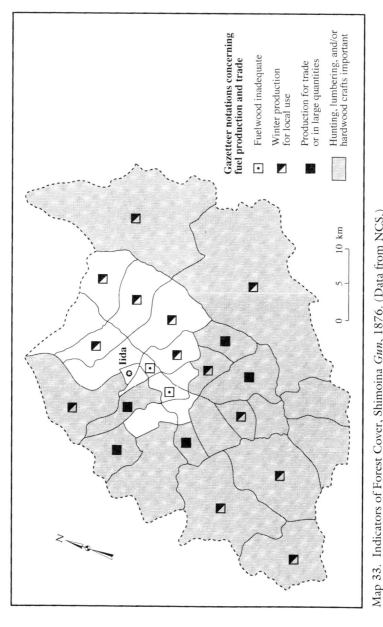

**Gazetteer notations concerning
fuel production and trade**

⊡   Fuelwood inadequate

◪   Winter production
for local use

■   Production for trade
or in large quantities

▦   Hunting, lumbering, and/or
hardwood crafts important

Iida

N

0   5   10 km

Map 33. Indicators of Forest Cover, Shimoina *Gun*, 1876. (Data from NCS.)

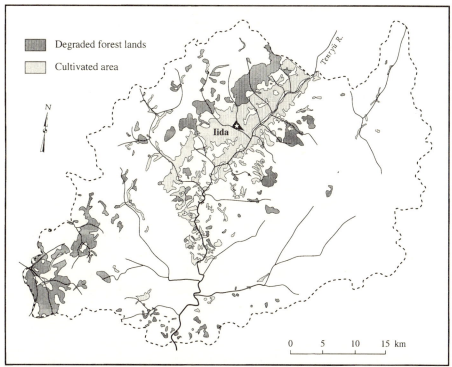

Map 34. Degraded Forest, Shimoina *Gun*, 1911. (Adapted from the Japanese government's 1:50,000 topographic map series of 1913, printed by Dainihon Teikoku
Rikuchi Sokuryōbu from data compiled in 1911.)

environmental costs were alarming. Within the first year of Ōji operations, the valley experienced the first of several massive erosion episodes;
as many as twenty to thirty years later, heavy rains could still set whole
hillsides of replanted forest growth sliding down the steep Tōyama
canyons.[56]

In short, the late Meiji logging frontier affected the various ranges of
the southern Japanese Alps in slightly different ways. The Kiso Range
saw heavy, but species-selective, cutting for locally milled timber; the
Nanbu uplands were also targeted for timber, but logs from there were
shipped downriver and milled elsewhere; and the southern Akaishi
Range was clear-cut for pulp. But despite these variations in form, the
lumber economy throughout the rugged uplands of Shimoina was an

56. Sawada 1981:183; Miura 1988:176; Minamishinano-mura Rōjin Kurabu
1979:1–6. Some of the logging was being done in national forests, and the state eventually insisted that the company undertake a modicum of forest-regeneration measures.

extractive and short-lived phenomenon; most of the profits accrued to investors from outside the region, and few multiplier effects were generated locally. Most of all, Shimoina's forests could not regenerate at a pace commensurate with their exploitation. When the big timber was gone, charcoalers would move in in their wake, and all but the southernmost fringe of the logged-over rim would be drawn into the orbit of the silk economy.

LOSS OF DIVERSITY

A second way in which developments from 1895 to 1920 weakened Shimoina's resilience was by diminishing the range of the region's linkages to the wider Japanese economy. A multiplicity of export goods and markets had been one of the hallmarks of Shimoina's protoindustrial development. But starting in the 1890s, one after another of the traditional commodities came on hard times, until the once robust craft production complexes headquartered in the Iida area had been reduced to a handful of marginal relics.

Several forces converged to erode the old industrial base, including changes in consumer demand, the advent of new technologies, government intervention, and competition for local resources. The traditional hairdress ties (*motoyui*) for which the Ina valley had been famous, for instance, simply lost their markets in the face of mechanization and changing tastes. As early as 1885, over a hundred local cordage artisans and papermakers left Iida permanently to settle in the Kantō region, where they began making other paper wares.[57] Demand for *motoyui* surged briefly in the 1890s, in a wave of cultural nationalism following the Sino-Japanese War, but the reprieve was short-lived; by 1898, interest in these tokens of the feudal era had resumed its long-term decline.[58] Other paper crafts fared better. Some *motoyui* makers were able to convert to producing decorative cordage (*mizuhiki*), whose popularity was growing; paper parasols also continued to command a regional market. Yet even within these surviving sectors, demand for handmade mulberry-bark paper declined sharply owing to the rise of large-scale pulp mills (including the one operated by Ōji Paper). Production of mulberry-bark paper in Shimoina had begun to slump as early as 1887; by 1925, factory-

57. Kiyomizu 1973:17.
58. Masaki 1978:412–15. Edward Seidensticker (1983:93) notes that only six years after the Restoration, a Tokyo newspaper reported that a third of all men in the capital sported short modern haircuts; a dozen years later, the estimate was 90 percent.

produced paper would account for over two-thirds of the total domestic market.[59] Local cotton processing was decimated for similar reasons. Beginning in 1896, newly established mills in neighboring prefectures began to flood the Iida market with inexpensive mechanically spun cotton thread. In the local hand-loom industry, homespun continued to be used only for the cruder sort of traditional footwear (*tabi*), which were still woven in the valley for sale in southern and central Nagano.[60]

While declines in the paper industries and cotton processing primarily affected settlements on the valley floor, the mountainous areas south and west of Iida were jolted by government intervention in the tobacco industry. Beginning in 1898, in the wake of the Sino-Japanese War, the state arrogated to itself the exclusive right to broker leaf tobacco. Where private dealers had previously paid cultivators eight yen for each bale, the government would pay no more than two yen, less than the local cost of production. Additional strictures were soon imposed on processing and growing, and finally cultivation was outlawed on steep slopes altogether. By 1908, tobacco had officially disappeared as a cash crop in the Ina valley.[61]

Other traditional enterprises, by contrast, were pushed aside, not by mechanization or new tax laws, but simply by the diversion of resources to more dynamic sectors of the new economy. In the case of lacquer crafts, commercial forestry was the culprit. The remaining lathe-turners living in the mountains south of Iida were finally displaced when the region was deforested. In 1892, the imperial woodlands to which they had formerly enjoyed access were divided into logging districts; by the 1910s, the *kijishi* had vanished from the valley. Some took up industrial work in the large cities along the coast; others turned to forestry or charcoal making. Those few who maintained their traditional livelihoods relocated out of the Ina valley, moving to Kiso's Urushihata ("lacquer field") village. By 1920, the once-flourishing lacquer industry employed fewer than ten persons in all of Shimoina.[62]

But it was not commercial forestry that displaced the majority of Shimoina's artisans; more often, the competition came from sericulture or filature. No longer could silk expansion be accommodated within the

59. Masaki 1978:413–15; Shinano Kyōikukai Shimoina Bukai 1934:176, 190.

60. Mukaiyama 1984:259–63. On the rise of modern cotton mills and associated economic changes in the former cotton-growing areas of Mikawa, see Kawaura 1960.

61. Seinaiji-mura 1982, 2:38–42; Shimoina Chiikishi Kenkyūkai 1982:67–68. The term "officially" is used advisedly, as illicit production is alleged to have continued in some areas; Murasawa 1963:22. One bale equals eight *kan*, or 32 kilograms.

62. Yamauchi 1980; NSTI:77.

existing economic fabric of rural production. During the first decades of the twentieth century, larger and larger shares of land and labor were diverted into silk production, pushing all the remaining protoindustries into a downward spiral.

Competition for land affected the county's ability to produce mulberry-bark paper and persimmons, both of which derived from orchard crops with which the silk mulberry directly competed. The large volume of persimmon wood appearing in a cargo manifest from the Ina road in 1896—some 5,000 *kan*, or nearly twenty metric tons, in one year alone—testifies to the rapidity with which persimmon orchards in this area were cut down to make way for silk mulberry after the Sino-Japanese War. In nearby Yutaka village, where paper mulberry and persimmons had been widely cultivated, silk mulberry replaced both crops everywhere during the early 1900s.[63] This phenomenon was evidently ubiquitous throughout the county. A census taken in 1911 listed only four Shimoina settlements as paper-mulberry producers; all were said to be decreasing their output at the time owing to the uprooting of the plantations to make room for silk mulberry.[64] By 1921, the entire remaining paper-mulberry acreage in the county amounted to a paltry 34 hectares, less than one two-hundredth the acreage planted to silk mulberry.[65]

In other instances, labor more than land constraints pushed traditional productive sectors into decline. Domestic cotton spinning, for instance, was rendered impractical by the wholesale exodus of young women to the filatures. As their reeling wages allowed factory hands' families to buy machine-spun thread, and eventually cloth and even ready-made clothing, the local demand for homespun declined in any case. The same shortage of female labor undermined the manufacture of a number of hand-woven specialty textiles and contributed to the decline of paper crafts, inasmuch as more and more women abandoned piecework to devote their time to sericulture and silk reeling.[66] Similarly, the shifting of men's labor from the tobacco fields to clearing new upland sites for silk-mulberry cultivation is said to have been one of the initial factors in the decline of tobacco production in Seinaiji village.[67]

63. Miura 1964; Miura 1988:195.
64. Shimoina-gun 1911.
65. NSTI:44, 47–49. Ishikawa Takeo and Takeuchi Atsuhiko (1986:19) allege another interesting factor in the demise of the persimmon orchards: reduced demand for dried persimmons in New Year's decorations.
66. Mukaiyama 1984:264.
67. Seinaiji-mura 1982, 2:15, 37.

In short, sixty years after the opening of the ports, Shimoina's Toku-gawa-era economic base had been significantly undermined. Some of the decline was driven by demand, as when the market for paper hair-dress ties collapsed with the demise of traditional Japanese hairstyles and machine-made cotton replaced homespun; in other cases, production changes were dictated by strictures on supply, as when loggers displaced lathe-turners and silk mulberry displaced other cash crops. Each of these changes represented a rational reallocation of resources, and none was to be mourned in itself. But their collective impact on the southern Japanese Alps was deleterious. Once the forest frontier had spent itself, the entire economic region centered on Iida would become dependent in one way or another on silk.

AN INSUFFICIENCY OF LOCAL STAPLES

Resources not only from former commercial enterprises but from the subsistence economy as well were increasingly diverted to sericulture. In the allocation of land, fertilizer, and labor, cocoon production increasingly gained the upper hand over food production. In fact, Shimoina in the 1910s was unique in the country in the extent to which sericulture there came, not merely to supplement subsistence farming, but to displace it as the primary occupation of the small-holding and tenant class.[68] Since this redirecting of resources coincided with significant population growth, the result was to render the county as a whole increasingly dependent on imported grains and pulses.

One of the clearest indices of this shift was the steady expansion of mulberry acreage at the expense of staple crops. In Shimoina overall, silk-mulberry acreage rose from 7 percent of total arable in 1884 to over 30 percent in 1912 and nearly 50 percent by 1921 (map 35).[69] Having already expanded into the terrain of Shimoina's former export crops during the last decade of the 1800s, after the turn of the century, the mulberry orchards began steadily encroaching on grain-producing fields as well. Dry-field grains and pulses, led by barley, soybeans, wheat, and millet, showed the sharpest downturns. Barley, the second most important staple in the region after rice, had been widely grown not only in dry fields but also as a winter crop in paddy fields along the banks of the Tenryū; 60 to 70 percent of all irrigated rice fields in the centrally located

68. Ōshima 1980:135. For a case study of one household in Matsuo village that exemplified this pattern, see Hirano 1980.
69. Masaki 1978:408; Shimoina Chiikishi Kenkyūkai 1982:61.

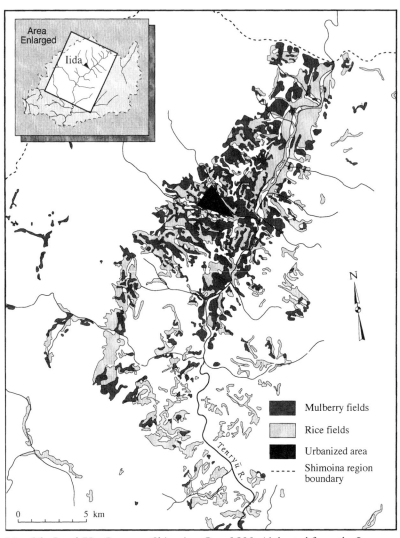

Map 35. Land-Use Patterns, Shimoina *Gun*, 1911. (Adapted from the Japanese government's 1:50,000 topographic map series of 1913, printed by Dainihon Teikoku Rikuchi Sokuryōbu from data compiled in 1911.)

village of Kamisato, for instance, were double-cropped with barley in the early Meiji period.[70] With the advance of the silk-mulberry orchards in the late Meiji period, however, barley began to be displaced from both its upland and its paddy niches. In the latter, it was partially replaced by mulberry nurseries, a highly localized new element in the wintertime landscape of the villages immediately surrounding Iida (table 21).[71]

In addition to taking large areas of land out of food production, farmers allocated a disproportionate quantity of fertilizer to their mulberry plantations. As early as 1892, fertilizer expenditures per hectare of mulberry in Shimoina were double the Nagano average.[72] Three decades later, when mulberry orchards accounted for 48 percent of total arable in the county, they absorbed 63 percent of all purchased fertilizers.[73] North of Iida, farmers specifically reserved the most expensive fish meal and chemical supplements for their mulberry orchards rather than their rice fields.[74]

At the same time that it was absorbing ever-larger shares of limited land and fertilizer, sericulture became an increasingly powerful magnet for household labor as well. A twelvefold increase in cocoon output between 1884 and 1912 owed little to the recruitment of new families; the total number of households participating in the sericultural economy appears to have risen by only about a third during this period. The decisive change was rather that already active households expanded their capacity, most importantly by beginning to follow up their spring silkworms with summer and fall broods. By 1902, summer and fall cocoons in the valley roughly matched the level of the once-dominant spring harvest.[75]

This increasing emphasis on mulberry cultivation and silkworm culture, with a corresponding decrease in food production, coincided with rapid population growth throughout Shimoina. In 1876, the county had

70. Kamisato-machi 1978:1115.

71. Between 1919 and 1921, Shimoina's nurseries produced some six million mulberry seedlings and cuttings. Chiba 1957:9; Shimoina-gun 1911; NSTI:48.

72. Local farmers spent 24 yen on fertilizers for each *tan* of silk mulberry, compared to 12 yen per *tan* in the prefecture overall. Partly as a result, mulberry productivity in the county was unusually high. Shiozawa 1982:7; see also Masaki 1978:406.

73. The total included 63 percent of all chemical nutrients, 54 percent of crushed bone, 63 percent of pressed soy cakes, 70 percent of marine products, and 30 percent of miscellaneous fertilizers. NSTI:34.

74. Toyooka-mura 1975:1047.

75. Masaki 1978:407–8. In Kanae village, for instance, while spring cocoon output increased roughly fivefold from 1882 to 1912, production of summer cocoons grew twentyfold. Kanae-machi 1969:559–60.

Table 21 *Changes in Crop Acreage, Shimoina County, 1885–1923 (in* chō)

	1885	1897	1912	1923
Rice	5,758	5,884	5,467	4,808
Barley	3,070	3,455	2,390	1,875
Soybeans	1,330	1,745	934	622
Wheat	1,127	1,143	469	199
Awa millet	322	544	151	112
Hie millet	749	379	162	58
Buckwheat	470	212	150	183
Sweet potato	50	48	55	40
White potato	221	292	129	202
TOTAL (all staples)	13,097	13,702	9,907	8,099
Silk mulberry	740	2,012	5,094	6,524

SOURCE: Ichikawa 1955:11.
NOTE: 1 *chō* = 0.992 hectare.

been home to under 100,000 persons; by 1920, the resident population was nearly 170,000, for a net growth of more than 75 percent in forty-five years. After the turn of the century, when Shimoina became the second-fastest-growing county in Nagano Prefecture, an increase of more than 14 percent a decade was sustained (see table 22). At roughly the time of this surge, the region lost its ability to provide sufficient staples for its population. During the fourth decade (1897–1907) of the Meiji era, the corner was turned; the Ina valley, once a net exporter of grain, started to import staple foodstuffs.

The shortfall could not be supplied from other agricultural regions within the country. Japan's rice production kept pace with domestic needs until near the end of the 1800s, but shortly before the turn of the century, demand began to exceed supply. As a result, the nation as a whole was forced to turn to overseas sources.[76] By 1915, Japan was importing 183,000 tons of rice annually; over the next twenty years, that figure increased by an order of magnitude. At the same time, domestic grain production became increasingly dependent on foreign fertilizer, with imports of soybean cake from Manchuria rising sharply after the Russo-Japanese War.[77]

76. From 1885 to 1920, Japan's population grew by 45 percent. Duus 1984:171, table 9, indicates the fluctuating yearly rice imports to Japan, by origin, from 1895 to 1912.

77. Francks 1984:74,79.

Table 22 *Population Growth in Shimoina County, 1876–1920*

Year	Population	Indexed to 1876 Base
1876	95,216	100
1883	105,574	111
1890	114,368	120
1900	127,874	134
1910	146,834	154
1920	168,064	177

SOURCE: The figure for 1876 represents the combined village totals recorded in NCS; the remaining data are from national census figures reproduced in Nagano-ken 1971:474.

Both of these developments had direct manifestations in the southern Japanese Alps. The first recorded instance of foreign rice being brought into Shimoina dates to 1897, just after the Sino-Japanese conflict. In that year, Iida merchants purchased 300 yen worth of Nanjing rice, already available on the open market in Nagoya, to dispel widespread resentment over rising grain prices in the region.[78] Shortly thereafter, starting in 1903, rice began to be imported into the county on a regular basis both from the coast of the Sea of Japan and from Korea. Six years later, when the Chūō Railroad began service to Tatsuno (the northernmost village in the Ina valley), rice led the list of Ina's imports. Dry-field grains, sweet potatoes, wheat flour, *sake*, fish, and fertilizer were on the first cargo lists too.[79] Such food and fertilizer imports—subsidized by the increased Japanese military and political influence on the Asian continent—were critical to maintaining competitive wages in early-twentieth-century Shimoina.[80]

WEAKNESSES IN THE INDUSTRIAL CORE

In addition to environmental degradation, a contracting export base, and the loss of self-sufficiency, the Ina valley would eventually suffer from weaknesses in the filatures themselves. Shimoina's boosters were rightly proud of the extent of local industrialization. As we have seen, reeling mills spread throughout the county during the

78. Hirasawa 1967:10.
79. Kobayashi 1959:62.
80. On the connection between grain imports and the increasing Japanese military presence in continental Asia after 1895, see Duus 1984.

1900s and 1910s, allowing Shimoina to retain a role as a thread producer through the early twentieth century. Yet a closer look at those plants reveals three telling weaknesses in the local industrial base: (1) most of their financing ultimately derived from sources outside the prefecture, (2) filature expansion simply did not keep pace with sericultural development, putting Shimoina increasingly in the disadvantaged role of raw-materials supplier to central Nagano, and (3) most of the filatures that *were* built in the region remained relatively small and undercapitalized. The resulting technology gap between the mills of the Ina valley and the rapidly expanding factories of the Suwa basin would prove harmful to Shimoina in the lean years to come.

The first cause for concern was the financial geography of the silk industry. In the decade leading up to 1920, filature firms consistently accounted for three-quarters of all bank loans extended in Nagano Prefecture. Most of those loans were secured through the roughly one hundred regional banks headquartered within the prefecture. Yet because of the sharp seasonal peak in cash demands (cocoons having to be bought as soon as they were harvested, at various points between June and early October), and because of the long turnaround time in the industry, local banks in turn had to look to central financiers to cover these huge loans. In fact, in 1923, more than half of all credit extended to the Nagano filaturists could be traced back to Tokyo-based banks. Since the risk involved in the trade meant that silk loans carried higher than normal rates of interest, a significant share of Shimoina's filature profits were repatriated to Tokyo financiers.[81]

The second cause for concern, the shifting balance between filature and sericulture, is summarized in table 23. During the four decades that spanned the turn of the century, local thread output may have risen from 5.8 to 8.9 percent of the prefectural total, but over the same period, sericultural output rose faster. Shimoina's share of Nagano cocoon output more than doubled, climbing from 7.2 percent of the prefectural total in 1882 to 16.4 percent of the much larger net prefectural output in 1919.

In short, despite its dramatic growth in reeling capacity, Shimoina's filatures failed to keep pace with the county's even more rapidly growing sericultural sector. While local thread output grew 17-fold during these years, cocoon output grew 22-fold. Clearly, Shimoina was exporting an ever-larger share of its cocoons before processing. Since the

81. Nagano-ken 1972:170.

Table 23 *Cocoon and Silk-Thread Production in Shimoina County as a Percentage of Prefectural Output, 1882 and 1919*

	Cocoons	Silk Thread
1882		
Shimoina output	6,847 *koku*	7,780 *kan*
Share of prefectural total	7.2%	5.8%
1919		
Shimoina output	151,706 *koku*	134,457 *kan*
Share of prefectural total	16.4%	8.9%

SOURCES: Nagano-ken 1971:155, 144; Nagano-ken 1972:105, 117.

greatest profits in the silk industry were to be earned in filature work, this put the county as a whole at a disadvantage relative to its more industrial neighbors.[82]

Shimoina was also disadvantaged in a third way, in that the industrialization that did occur was comparatively small-scale. To some extent this was generic to silk reeling. By international standards—and by comparison with Japan's own cotton-spinning mills—Meiji reeling plants were small concerns, employing on average only 52 persons in 1909. Even in 1921, when the average Nagano filature employed 146 operatives, numerous firms with no more than ten to thirty basins continued to function profitably.[83] In Shimoina, however, such small firms were particularly numerous. As late as 1904, over a quarter (28 percent) of Shimoina's silk-reeling work force continued to be employed in household workshops; a thousand treadle reels (*zaguriki*) remained in use, al-

82. Although available statistics do not directly state how many cocoons Shimoina exported in raw form, the scope of this trade may be teased out of existing data in at least two ways. A detailed county economic census for 1921 records that Shimoina produced 1,329,508 *kan* of cocoons that year, while local filatures are said to have purchased a combined 120,794 *koku* of the same to process into thread. Since one *koku* of cocoons before drying weighs 7.52 *kan*, this means that at most 908,371 *kan*, or 68 percent of the county's output, was used locally for commercial thread production. Except for the small amount that may have been reeled domestically (and thus remained unrecorded), the bulk of the remaining 421,137 *kan* of cocoons was presumably exported before reeling. This conclusion is supported by the alternative method of comparing the ratio of cocoons to thread produced in the county (approximately 11:1 by weight) with the actual proportions involved in converting cocoons into thread at the time (8:1). The result again indicates that roughly one-third of all cocoons produced in the region were exported for reeling elsewhere.

83. Sasaki 1978:125–30.

most all (98 percent) confined to domestic reeling operations with a single basin. Since each reeler typically stood behind her own basin (where her cocoons were immersed before reeling), the ratio of mills to basins shown in table 24 confirms the extremely small size of most filatures employing the older technology.

Such small firms were more vulnerable than their counterparts at the higher end of the size spectrum, and consolidation proceeded in ratchet fashion during crisis years. Immediately following the deflation of the early 1880s, for instance, bankruptcies and buyouts eliminated many small mills; a 2.5-fold increase in total capital investment during this period translated into a 4-fold increase in average capital per factory.[84] Moreover, the largest firms grew most dramatically. Whereas Shimoina's largest factory in 1874 employed only twenty women, fifteen years later the same firm had basins for nearly six hundred.[85]

Yet while a few large plants had appeared in the local landscape, Shimoina's filatures on the whole did not achieve par with the industrial powerhouses of Suwa. In 1921, when the average Suwa filature employed 146 workers, Shimoina's average was only 78.[86] And although Shimoina's largest mill at the time (the Katakura subsidiary) employed 617 workers, a fifth of all firms in the county remained minuscule operations with fewer than twenty basins each. These small producers held on tenaciously *as a class*, if not as individuals, throughout the period under consideration; but most would go under in the coming crash, when an oligopoly of giant firms took over the bulk of Japan's silk-reeling operations.[87]

A LOW-WAGE LABOR FORCE

A final index of the new economy's susceptibility was its exclusive reliance on low-paid labor. Keeping all laborers' compensation low—on the farm as well as in the factory—was key to the Japanese silk industry's international success. Yet this strategy also hindered the for-

84. Between 1883 and 1889, the number of factories in the county fell from 48 to 35, yet total fixed capital in the industry more than doubled (from 107,216 to 265,358 yen) and annual output more than tripled (from 2,880 to 10,329 *kan*). Masaki 1978:401. Takeuchi Toshimi (1976:79) alleges that even after the turn of the twentieth century, many ambitious farmers sold family land to build small filatures; most ended in bankruptcy.
85. The precise figure was 586. By the turn of the century, fewer than a third (31 percent) of the county's factories had 10,000 yen or more in equipment, accounting for over half (57 percent) of the total investment in the industry. Hirasawa 1952:152.
86. Nagano-ken 1972:104.
87. Sasaki 1978:125–30; Shindatsu 1985:135.

Table 24 *Comparison of Treadle-driven* (zaguri) *Filature Operations with Engine-powered Mills in Shimoina County, 1904*

No. of Employees	Treadle		Engine	
	Mills	*Basins*	*Mills*	*Basins*
1–9	1,055	1,150	1	8
10–49	3	28	55	1,531
50–99	—	—	12	726
100 or more	—	—	5	731
TOTAL	1,058	1,178	73	2,996
Average no. of basins per plant		1.1		41.0

SOURCE: Masaki 1978:405.

mation of a local consumer market capable of supporting future regional development. To understand the limitations of Shimoina's industrial transformation during this period, it is thus essential to recognize the extent to which local sericultural and reeling wages were depressed, and to identify the mechanisms by which this was accomplished.

Throughout Japan, the work force mobilized for sericulture, as for the filatures, was overwhelmingly female, predominantly young, and firmly embedded in a matrix of patriarchal household relations. Stephen Marsland describes how factory recruiters secured access to the labor of young women by invoking the feudal custom of *jochū minarai*, or "learning to be a woman": "By adhering to the traditional requirements (small payments to the household, addition to the dowry, some measure of supervision), factory owners could claim that they offered a form of *jochū minarai* where the girl learned a skill—the skills required to manufacture the factory's product. . . . [T]his custom opened up a potential source of labor that could be recruited over considerable distances, which otherwise would have been kept at home under the watchful eyes of concerned parents."[88] As this suggests, nineteenth-century Japan's silk industry developed wholly within the fabric of traditional family relationships (a sharp contrast with the silk districts around Canton, where high rates of male emigration are said to have allowed women with factory jobs to achieve a remarkable degree of freedom from conventional family obligations).[89] The virtual forced labor of young girls in

88. Marsland 1989:17–18.
89. For a fascinating study of the Canton case and the associated culture of "marriage resistance," see Topley 1975.

the Japanese textile mills, based on contracts whose central feature was advance payment to their parents, has been widely documented.[90] But women of all ages worked for similarly low compensation at home in the sericulture rooms.

The wages of Japan's labor force in the late nineteenth century were low across the board, but with a strongly gender-segregated labor market, women were typically paid one-half as much as men for a day's work.[91] While no quantitative information about compensation for female family members' sericultural labor exists, surviving statistics from the early twentieth century provide valuable information on temporary employees' wages. As had been true since at least the 1880s, sericultural workers in Shimoina were consistently among the lowest-paid in Nagano Prefecture. From 1906 to 1921, women's wages in sericulture in the Iida area were a full third lower than those pertaining in Matsumoto, Ueda, or Nagano City.[92] Moreover, income for silk workers was not only lower in southern Nagano than in other parts of the prefecture; it was also highly volatile. During the Matsukata deflation, for instance, when women's agricultural wages in Iida fell by 46 percent, local sericultural wages dropped fully 58 percent in a single year. And in the filatures as well, Iida's laborers earned less than those of central and northern Nagano. By virtually any comparative standard, then, wages for silk-related labor in Shimoina were low.[93]

Various mechanisms conspired to keep them low throughout the great expansion of the industry in the first decades of the twentieth century. The single most important such mechanism, in Iida as elsewhere, was the industry's ability to retain an almost entirely female labor force. Even in 1919, when silk prices reached their highest pitch and producers were desperate to expand output, only 9 percent of Nagano Prefecture's 96,349 mill operatives were men.[94] The incentive for maintaining a female work force was clear: women still commanded only one-half to two-thirds the level of compensation routinely paid to men in compa-

90. In English, see, for instance, Kidd 1978; Bernstein 1988; Hane 1982:172–205; Tsurumi 1984 and 1990. For an excellent introduction to the Japanese literature on female textile workers in this period, see Tsurumi 1986.

91. In Kumashiro village, for instance, women routinely earned only 50 percent as much as men for agricultural day labor; Toyooka-mura 1975:1070–71. Per diem, women's earnings from spinning and weaving cotton were roughly comparable to their agricultural wages. Takamori-machi 1972:477; Mukaiyama 1984:260ff.

92. Nagano-ken 1985:634. On patterns of sericultural wage labor, see Itō 1955.

93. Nagano-ken 1985:634; Matsukawa-machi 1965:802.

94. Nagano-ken 1972:105.

rable jobs.[95] Men's alternatives to silk-related work were almost uniformly higher paying;[96] by contrast, the other jobs open to women were even less remunerative than silk-related employment. The main alternative was farm work, where women's earnings averaged roughly 15 percent less than sericultural wages.[97] The few remaining jobs in the traditional craft industries were geographically restricted to a handful of hamlets in the old protoindustrial core; in the mountains, alternatives to filature work were rough and extremely low paying.[98]

What allowed the local silk industry to rely almost exclusively on female labor, and to keep local silk wages depressed, was continued in-migration from the surrounding countryside. By 1920, over a thousand Shimoina reelers, or 13 percent of the local filature work force, were officially residents of other areas. Fewer than half of these migrants had come to Shimoina from within Nagano Prefecture; the majority hailed from neighboring prefectures to the west and south (table 25). Like the infamous Nomugi Pass into Suwa, the steep Ōdaira and Aokuzure passes into Shimoina also became the site of annual pilgrimages for hundreds of young mill workers.[99]

Yet with the continued expansion of the industry, more powerful tactics than this were needed to hold down labor costs. The last important vehicle of wage suppression, in Shimoina and throughout Japan, was political (and sometimes violent) intervention in the marketplace, through the virtual outlawing of workers' organizations and the simultaneous fostering of an aggressive manufacturers' association. In order to cope with

95. Between 1906 and 1921, Shimoina's female sericultural workers' real earnings averaged 1.5 to 2.2 *shō* (2–3 kg) of rice a day (depending on skill level); men in the same line of work averaged 2.4 to 3.5 *shō* a day, or 60 percent more. Contemporary officials estimated that to cover all its food and nonfood needs, a family of five required income equivalent to the price of 4 *shō* of rice a day. Real wages are calculated from price and wage data in Nagano-ken 1985:634 and NSTI:62, 79–81; income and food requirements are discussed in Nagano-ken Achi-mura Rōjin Kurabu 1978:132; Achi-mura 1984:439; and Nagano-ken 1971:480.

96. Artisans and skilled men of all professions commanded two to three times the wages of male sericultural workers; even manual day labor paid nearly 50 percent more than silk-related work. Nagano-ken Achi-mura Rōjin Kurabu 1978:123.

97. Nagano-ken 1985:634. Calculated from figures for Iida, 1906 to 1921.

98. Production of paper cords and umbrellas continued to call for female labor at several stages. But work of this kind was limited to women in a few core families, where husbands and children also contributed to the intricate household division of labor that alone kept the industry viable. Shinano Kyōikukai Shimoina Bukai 1934:140–44, 164–68. In the mountainous hinterland, the only options were basketry, charcoal making, and transport work (which paid as little as one *shō* of rice per day). Seinaiji-mura 1982, 2:279.

99. Masaki 1978:404; Sawada 1981. A famous account dramatizing the plight of silk workers from Gifu, who were forced to cross the forbidding Nomugi Pass to reach the mills in the Suwa basin, is S. Yamamoto 1977.

a growing labor shortage (and accompanying problems of labor discipline) without bidding wages up, seven existing regional producers' organizations in Nagano had joined forces in 1901 to form the Silk-Manufacturers' League (Seishi Dōmei). Through this state-sanctioned cartel, filature operators throughout the prefecture tried to circumvent market forces to hold down both the wages of their female labor force and cocoon costs. While their efforts were not entirely successful, they did suppress wages and render it increasingly difficult for workers who ran away from an oppressive factory to find a willing employer elsewhere.[100]

The Meiji state, recognizing the importance of cheap labor for keeping Japanese floss globally competitive, supported the factory owners' efforts. The authorities not only sanctioned producer cartels but cooperated in political repression of labor organizers. The latter policy was codified in the Public Order Police Law of 1900, which included a provision banning all political activity by women. This punitive measure, which predated enactment of Japan's first protective women's and children's labor legislation by a quarter-century, effectively blocked women in the filatures from organizing to improve their miserable working conditions.[101]

This context of wage repression and limited opportunity for women casts the brothels of Iida in a different light. The new red-lantern district may have been seen by some as a symbol of Iida's prosperity; decorated women are much in evidence in photographs of Shimoina from the Taishō period, appearing at the head of festival processions, at the center of groups of springtime revelers, and in front of the inns and teahouses where they worked.[102] Yet while the mood of these scenes is gay, the unprecedented business enjoyed by Iida's bordellos during these years was not without its dark side. In years of hardship, local daughters were sold outright to the innkeepers of the district, as were a number of immigrants from impoverished villages along the coast of the Sea of Japan.[103] Moreover, much of the wealth dissipated there represented the hard-earned fruits of women's labor in sericultural rooms and reeling mills throughout the countryside. Some men would stay for ten or more

100. Tsurumi 1990:52–53, 74–75.
101. On attempts to unionize female textile workers, see Sievers 1986. The appalling working conditions in the early Japanese filatures do not need reciting here. For sympathetic accounts in English, see n. 90 above.
102. See, e.g., Kyōdo Shuppansha Henshūbu 1983.
103. It is reported that at the time the district opened its doors, a woman could be bought outright in Iida for ten to twenty yen. Shimoina Chiikishi Kenkyūkai 1982:39; Masaki 1978:303.

Table 25 *The Origin of Women Factory Workers Employed in Shimoina, 1920*

Place of Origin	Number	Percentage
Nagano Prefecture (by county)		
Shimoina	7,968	86.6
Kamiina	360	3.9
Nishi-Chikuma	193	2.1
Suwa	10	0.1
SUBTOTAL	8,535	92.7
Other (by prefecture)		
Gifu	300	3.3
Aichi	201	2.2
Shizuoka	190	2.1
Other	80	0.9
SUBTOTAL	671	7.3
TOTAL	9,206	100.0

SOURCE: NSTI:72–73.

nights in a row after selling their family's cocoons, and stories are told of the sons of wealthy farmers stealing from the family storehouse to finance a night of dissipation.[104] That their menfolk, on delivering the precious cocoon harvest to the warehouses, could choose to spend the returns to their labor in the company of prostitutes is one of the starkest manifestations of the gender division of power that underlay the regional economy of Shimoina in the early twentieth century.

Conclusion

The period from 1895 to 1920 brought radical change to the landscape of production in the southern Japanese Alps. As national tastes changed—and as cheap machine-made substitutes began to flood the market—demand for the area's traditional handicraft goods steadily contracted, depriving thousands of households of vital cash income. De-

104. On the high cost of spending an evening in the brothels, see Iwashima 1967:83; on how clients financed their visits, see Kobayashi 1959:20. For a firsthand account by a man who stayed for three months in a brothel between cocoon harvests, see Saga 1987:41–43.

clines were pronounced in every sector: lacquerware, paper goods, spun cotton, hand-woven specialty textiles, and even tobacco and dried persimmons. By 1920, Shimoina's protoindustrial economy survived only in fragments; while a few traditional manufactures continued at low levels, most were displaced altogether.

In their stead, more and more people had thrown their labor and their fortunes into sericulture, filature, and related enterprises, transforming silk from one sector among many to the vital core of the area's economy. In the process, the linkage between industry and the countryside underwent a dramatic reversal. During the first quarter-century of the trade, silk had been accommodated into an existing spatial and seasonal division of labor; after 1895, however, priorities were reversed, and the needs of silk increasingly came to dictate what went on in the rest of the landscape. By 1920, when sericulture and filature together accounted for fully two-thirds of Shimoina's total regional product, the entire productive landscape of the Japanese Alps had been reoriented around silk.

As much as its traditional industries declined, Shimoina did not simply devolve into a monocropped, single-commodity economy. Erosion of traditional activities was offset by the growth of new ones, giving rise to new connections between industry and rural production. Villages in the region's periphery discovered a variety of backward or upstream linkages, since three woodland products—lime ash, firewood, and charcoal—proved vital to the new production complex. Only improved roads and the advent of a massive silk industry made transportation of these bulky materials over long distances to the region's core commercially feasible. Meanwhile, areas near the primary silk centers established comparable downstream linkages, taking advantage of the copious by-products of filature work, while scores of new businesses sprang up in the Iida area to provide services that were suddenly in great demand. Some catered to women who were now too busy to perform certain time-consuming routines of food preparation. It is no coincidence that a regional (and in fact national) market arose during these years for various "fast foods": freeze-dried tofu, prepared miso, and even crackers and sweets. Others, like restaurants and bordellos, catered to men who left the cocoon warehouses flush with cash.

The long-term effect of these developments was to redraw the economic map of Shimoina at every level. As silk replaced the heterogeneous complex of commercial putting-out industries that preceded it, Iida's hinterland was reorganized and assigned new functions. As we have seen, the steady growth of sericulture and filature output did not result in a massive simplification of Shimoina's productive complex; in place of the declining protoindustrial diversity, a new kind of diversity arose, based on the production of inputs for, and spin-offs from, the increasingly dominant silk industry. Yet despite the number and variety of these industries, the region's export base had lost both breadth and adaptability. Silk had come to account for the great bulk of Shimoina's extraregional trade, and its fortunes affected virtually every sector of the new regional economy.

As a result, for all the interdependence exhibited within its backward and forward linkages and service trades, Iida failed to generate the critical momentum needed to sustain regional economic health. Diversity in the service of a single-commodity export economy simply cannot have the desired effect; as the geographer Harold Brookfield observes, even "an export industry with large backward and forward linkage, or 'multiplier' effects, will not generate [self-sustaining] economic transformation if it remains wholly related to a national or regional export."[105] The upshot of this process, in other words, was the remaking of Shimoina as a whole into a periphery within a global division of labor. Where a semiautarkic, well-rounded regional economy had been, a very different entity stood by 1920: focused around a single industry, dependent on imported staples, and at the mercy of a capricious international market.

The inevitable downturn began in the early twenties. By the 1930s, when the American depression wiped out what was left of a once-widespread consumer market for silk, the cost of Shimoina's dependency became starkly evident. Business and political leaders joined the county's swollen population of working men and women in a desperate search for alternative means of livelihood. But no ready substitute for silk was to be found. Although numerous families turned to charcoaling in the hills or attempted to revive the old craft industries, intense competition in both made remuneration abysmal.[106] Many had no recourse but to leave. The population growth that had been sustained for half a century slowed to a crawl, with peripheral villages in particular experiencing the start of a long-term decline. In one dramatic index of local hardship, southern Shimoina's Yasuoka village sent over eleven hundred migrants to Manchuria. Nor was Yasuoka a special case. During the later 1930s, the population of Shimoina as a whole experienced an absolute decline

105. Brookfield 1975:96. Sidney Pollard (1981:106) similarly stresses the importance of "interdependent diversity," seeing in it the origins of that "critical interactive mass" needed for a progressive industry to maintain its competitive momentum.

106. Namiai-mura 1984:1513; Ōshika-mura 1984:229; Seinaiji-mura 1982, 2:21.

for the first time in half a century of record keeping; by 1945, on the eve of the postwar repatriations, the county was home to only 9 percent more people than it had housed in 1920.[107]

Yet even in its new incarnation as a regional periphery, Shimoina was greatly privileged as part of an emerging imperialist core. The food and fertilizer imports that had sustained the valley's growing population during its boom years were not of local provenance; they were extracted from Korean and Chinese farmers in the shadow of an aggressive Japanese military presence on the Asian mainland. And when the local economy collapsed, desperate villagers were able to escape to the colonies as settlers or soldiers. This emigration, which spoke eloquently of hardship at the local level, was also indicative of Japan's military might in the western Pacific.

In short, the reshaping of Shimoina's landscape cannot be read simply as the industrialization of a protoindustrial economy. Rather, that process must be understood as an integral part of a global capitalist transformation. In the last analysis, the development of the silk industry played a highly contradictory role in regard to Shimoina's space-economy: while it turned the valley into a clearly subordinate periphery of the Japanese state, that very subordination was critical to the contemporary Japanese state's ability to escape becoming a periphery of the global capitalist system. It is its role in this larger drama, I believe, that gives Shimoina's successful-yet-perilous mobilization for silk production a claim on the historical imagination.

107. On Yasuoka's "branch village" (*bunson*) in Manchuria, see Kobayashi Kōji 1977. County population figures from Nagano-ken 1985:1730.

CHAPTER EIGHT

Regional Inversions

The Shifting Matrix of Production, Power, and Place

As told here, the history of Shimoina's development from 1750 to 1920 is a study in regional metamorphosis. Over the course of the Tokugawa period, the Ina valley was slowly knit into a cohesive economic region, only to be unraveled and reworked into a very different fabric when the Japanese countryside was incorporated into a globalizing economy and a centralizing state. Yet as Derek Gregory insists, " 'integration' is not a morally neutral term."[1] As Shimoina was subsumed, it was also subordinated, reconfigured as a periphery in this larger politico-economic universe. While many in the Ina valley prospered temporarily from the transition, by 1920 the region as a whole had become dangerously dependent on a single, distant commodity market. As we have seen, that dependency would have harsh repercussions for the southern Japanese Alps in succeeding decades.

The shifting contours of production, finance, and exchange discernible in the landscape of Shimoina during these years both reflected and effected this realignment. Attempting to map out and make sense of those shifts calls our attention to a series of linkages: between the early modern and the modern, environment and politics, geography and history. Since each pair of terms also signals a significant disjuncture in much academic practice, it may be worth revisiting three conceptual strategies by which the present analysis has attempted to articulate these linkages.

1. Gregory 1988:51.

My first starting point has been an insistence that social life is spatially patterned. History "takes place" in a spatial as well as a temporal dimension; by the same token, places are not merely geographically given, but historically constituted as well. This truism can be formulated in less abstract terms with regard to the economic and political institutions whose intersection most concerns us here. Both the economy and the state may be conceived as sociospatial networks of power; each, however, obeyed fundamentally different spatial logics. In 1750 as in 1920, economic power operated through fluid and essentially linear circuits of production and exchange: always anchored in (and continually reshaping) particular physical environments, they nonetheless united producers, consumers, and financiers who might be separated by great distances. Political power, by contrast, was typically exerted over discrete, contiguous areas (albeit of widely varying sizes), forming a patchwork of more or less stable blocks of terrain. The clash between these two different sociospatial orders has been an essential line of interest in the drama traced here.

A second premise concerns temporal process. As we have seen, relatively stable periods, marked by gradual evolution, alternate with phases of more dramatic or revolutionary restructuring. The present study embraces one such turning point, when the relative equilibrium of the Tokugawa period gave way to rapid structural change as Japan headed into its industrial and imperial age. This in turn was paralleled in economic life by a new round of accumulation, marked by novel arrangements of circulation and dictating a new locational logic of production. To borrow from the lexicon of evolutionary biology, the historical vision articulated here is one of punctuated equilibrium.

The last touchstone for this project is an appreciation of the nodal region as a framework within which the interaction between these synchronic structures and diachronic processes may be apprehended. The region may be one of the hoariest concepts in the geographical literature, but it remains indispensable for historical geographic research. In broadest terms, the kind of region that has been of interest here is a subnational or mesolevel matrix in the sociospatial order: a more or less tightly structured, more or less clearly bounded concatenation of people and resources, drawn together through a complementarity of economic roles, and both reproduced and transformed by the interaction of competing power networks that pervade its territory. The importance of recognizing these formations when thinking about Japanese development is twofold: on the one hand, they represent an important crucible within which the upheavals of the day were played out and experienced; on the other hand, they constituted social building blocks that both enabled those transformations to take place and, at the same time, helped to determine the forms that the new political and economic order would take. In both senses, regions mediated between the macro and the micro, between national and global forces and the local communities and households in which the people of Japan worked and lived during these convulsive years.

Triangulating from these conceptual compass points, it is possible to take a new reading of Japan's modern political and economic revolutions. The changing geography of the southern Japanese Alps can be made to serve in effect as a prism through which those revolutions are refracted, allowing us both to glimpse the sociospatial contours of the Tokugawa and the Taishō regimes and to discern the often-jarring processes by which the one became the other. The fact that the rest of Japan did not undergo a metamorphosis identical to that of Shimoina does not lessen the value of such a study; to paraphrase Andrew Sayer, even a "quite atypical" locality may be used as a vantage point to illuminate a much wider canvas provided it is "strategically situated . . . at the intersection of the key structures in contention or transformation."[2] The present study is animated by a firm belief that Shimoina was so located, and that even the highly particular vagaries of its passage to modernity have the potential to illuminate the wider transformation taking place in Japan and, indeed, the process of regional formation more generally.

The following ruminations on Shimoina's regional development accordingly interweave findings particular to the southern Japanese Alps with more generic observations. It may be useful to begin by enumerating the discontinuities that justify labeling this transition a "regional inversion." From its internal configuration to its linkages with the wider economy, the spatial order of the Ina valley economy can be shown in more ways than one to have been metaphorically turned inside out, exemplifying what Derek Gregory aptly terms the "fluidity and precariousness of regional structures" during the Industrial Revolution.[3] Yet beneath these highly visible changes lay an unchanging substratum that calls for explication as well. In the last analysis, the constants of Shimoina's regional development across these 170 years are as revealing as the more visible changes in the land.

2. Sayer 1989.
3. Gregory 1988:54.

A Study in Contrasts:
Regional Development under the
Tokugawa and Taishō Regimes

THE SHIFTING CONTEXT

In 1750 as in 1920, the possibilities of local development regimes were largely fixed by a trio of forces emanating from beyond Shimoina's borders. The first important component of this force field was the geopolitical configuration in which the region was lodged; second was what might be called the technology of exchange governing its integration with the wider economy; and third, the relations of commodity production in the leading export sectors.[4] Between these two dates, dramatic changes in all three of these vectors would combine to elicit sharply contrasting patterns of regional development.

Since polities tend to have easily mapped structures—and an incentive to generate and preserve cartographic records—political context is the feature most readily specified in geographical terms. In the southern Japanese Alps, a glance at the early modern and modern political maps shows contrasts on every scale. The process of centralization lying between the two is a familiar story in the literature of modern political history, but its variable implications for long-term development in regions *within* the incorporated provinces have seldom been systematically drawn out.

For Shimoina, the shift from a landscape of fragmented imperium to one of centralized administration had one powerful implication: it undermined the fiscal basis of local self-rule. The old regime had left the taxation and governance of the Ina valley to half a dozen different-sized fiefdoms, scattered small temples and shrines, and an intendant who supervised the extensive shogunal lands in the mountains. These local authorities, while varying greatly in the extent of the terrain they commanded, were essentially equals in controlling the most significant revenues generated from the Shimoina landscape; none was subject to oversight from its neighbors. Beneath these local authorities, the region's 160-odd villages maintained extensive powers of self-regulation as well,

---

4. This formulation dovetails with the more generic observation of Derek Gregory, who sees regional production in the Industrial Revolution as based on two new tendencies: increasing ties to the world economy, and the increasing centralization of knowledge/power at the level of the state capital over its hinterland. Gregory 1988:54–55.

albeit with mandated tax liabilities to their respective lords; above those lords was only the Bakufu government, whose interventions were limited essentially to peacekeeping and adjudication. Despite the formal existence of intermediate aggregations, neither Shinano province nor its constituent districts (*gun*) had any meaningful role in the geography of power.

Sweeping away this markedly decentralized arrangement, the new Meiji state forged a single, nested hierarchy of standardized administrative units. The effect of comprehensive centralization of government revenues, and of the amalgamation of lower units into a uniform hierarchy, was to consolidate resources at the top of the national pyramid. The eight-score Tokugawa villages in what was now Shimoina County were reduced to the status of hamlets and forced to amalgamate into fewer than forty supervillages, whose orders on numerous matters were handed down from the county offices in Iida.[5] Shimoina in turn was subsumed into the newly empowered Nagano Prefecture, whose governor was for many years appointed directly from Tokyo. Since these dramatic changes in local administration paralleled a centralization of tax collection and disbursement, the new apparatus of local government became largely an instrument of top-down rule.

This consolidation of the local government pyramid came in tandem with a very different set of changes at the supranational level. The celebrated arrival of Commodore Perry, and Japan's subsequent immersion in the perilous world of late-nineteenth-century imperialism, altered the wider geopolitical context of the southern Japanese Alps in two ways. On the one hand, it expanded the sheer size of Shimoina's world by an order of magnitude. No longer were the region's political preoccupations (or economic ties) limited to the Japanese archipelago; they soon extended from one shore of the Pacific Ocean to the other. At the same time, however, the highest level of Shimoina's relevant political universe was converted from a stable order, revolving around Edo and Kyoto, into a world with multiple, contending centers. Not only did Tokyo enter the fray as a distinctly secondary power, but no single city played the role of capital in the new world order, and no ruling authority was able to enforce the status quo as the Bakufu had done for 250 years. Thus, while its local frame of reference was centralized and stabilized, the larger

5. On the government's attempt to transfer loyalties from the hamlet to the newly consolidated towns and villages (by transferring hamlet lands to village and town control, merging shrines, integrating citizen groups into the administrative structure, and calling for village and town development plans), see Pyle 1973.

(now international) framework within which Shimoina found itself was radically *de*centralized and *de*stabilized. Both changes would have profound implications for local development.

In addition to geopolitics, a second determinant of local development that changed dramatically was the technology of exchange. Despite the Ina valley's function as a corridor for Tokugawa overland trade, Shimoina residents in the eighteenth century—like their counterparts in all landlocked basins of Japan—had inhabited a largely self-contained region. With cargo flows across the surrounding mountains limited to what a packhorse could carry, the Ina valley had perforce been self-supporting in such fundamental resources as food, fuel, and fertilizer; exchange with distant regions, as significant as it may have been for local commerce, was confined to specialty products and lightweight luxury goods. But in the later nineteenth century, the barriers that had produced and protected this semiautarkic economy were all but broken down. Riverboats, wheeled vehicles, and aerial cables multiplied both the efficiency and the capacity of the transport system several times over. At the same time, the mobility of labor and capital across Japan was greatly enhanced as well, encouraging unprecedented movement, not only of things, but of money and people too. The net result was a significant loosening of the bonds tying regional development to local resources.

Nor were innovations within the valley the only ones to affect Shimoina's connections with the outside world. Although they did not directly penetrate the county, railroads and steamships brought about revolutionary breakthroughs at the outer edges of Shimoina's world during these years. The effect was to increase the Ina valley's permeability in several ways at once. Not only could more and larger shipments be hauled across its borders, but the spatial reach of the region's circuits of exchange was massively extended at the same time. No longer dependent on local resources, the region was also no longer dependent on national markets. If the effective end points of local commodity trade in 1750 had been Edo and Osaka, by 1920 it reached as far as Harbin and New York. Iida's sphere of exchange had expanded to embrace the entire North Pacific.

Compounding these revolutions in geopolitics and transportation were shifts in a third parameter of regional development, the relations of production in the export sector. Under the protoindustrial regime of the Tokugawa period, most of Shimoina's commodity producers were also tillers of the soil, peasants who engaged in craft production on a seasonal, by-employment basis during slack periods in the agricultural calendar. The geographical implication was one of dispersed and often atomized production, as materials were processed within peasant households and artisanal workshops in Iida and surrounding villages. The financial infrastructure supporting these complexes was more centralized, but both small in scale and local in origin, the main providers of capital for commercial production in Edo-period Shimoina being a few score wholesalers residing in the region's largest castle town and its rural satellites. Most of these regional merchant households were not rich by national standards, but since the turnaround time in their operations was relatively rapid, they were able to control considerable production networks with a small pool of constantly circulating capital.[6]

By the early Taishō period, these arrangements no longer characterized the county's main export sector. Although traditional outwork relationships may have obtained well into the twentieth century in the smallest domestic reeling works (as they certainly did in charcoal making and the remaining craft industries), the bulk of the filature industry had given rise to markedly different social and spatial relations of production. In 1920, the majority of Shimoina's silk floss was produced in modern factories, where ten to a hundred or more wage-earning operatives were employed for much of the year. Meanwhile, mechanization had multiplied the scale of capital requirements, and dependence on international shipping had simultaneously greatly lengthened the lag-time before expenses could be recuperated. This stretching out of the scale and timing of capital requirements gradually forced individual local merchant houses either to give way to larger corporations or to rely heavily on national banks, which by 1920 supplied the bulk of funds for the area's export-oriented industries. In sum, the relations of production in the export sector had undergone a series of related transitions: from piecework to wage work, from dispersed domestic workshops to larger mechanized factories, and from small-scale mercantile finance to larger-scale industrial and finance capital. In tandem with the concurrent revolutions in geopolitics and communication, these novel arrangements in the export sector would lay the groundwork for a fundamental transformation of the Shimoina landscape.

6. For a suggestive and detailed treatment of the critical issue of turnover time for capital in a British textile region, see Hudson 1989b.

THE CHANGING FULCRUM OF
REGIONAL DEVELOPMENT

As compelling as they may have been, geopolitical realignments and technological innovations did not directly induce
new forms of regional production and exchange. Global and national
forces were mediated through—and resisted by—local people; the Ina
valley's own inhabitants were the direct creators of its Tokugawa and
Taishō economies. Accordingly, an essential second step in delineating
the contrast between these two regional development regimes is to
identify the primary alliances through which local resources were mobilized in each.

The crucial agent of Shimoina's production as a protoindustrial region during the Tokugawa period, I have argued, was the alliance of
ton'ya and *chūma* leaders. Castle-town wholesalers and rural packhorse
drivers existed in mutual symbiosis; the packhorse network helped foster local commerce, advancing the circulation of capital as well as cargo,
while merchants reciprocated both by coordinating production and exchange and by offering decisive help in the battle to legitimize the private carriers. In time, a third group would muscle its way into this alliance as well: the rural notables, or *gōnō* class, through whose hands
agricultural rents began flowing into the spheres of craft production and
circulation. Through the buying-up and putting-out activities of both
urban and rural *ton'ya*, effected through their mobile brokers and agents,
each of the half-dozen local export industries extended commercial ten-
drils into the surrounding countryside, seeking raw materials and labor
for the diffuse processing characteristic of protoindustrial production. It
was these routine exchanges that drew Iida and its rural hinterland together into an integral economic region.

Merchants, landlords, and packhorse drivers did not effect this ac-
complishment alone; important support was provided by two distinct
levels of government. Local domain lords were the most actively involved, initiating, protecting, and even recruiting their retainers for the
Ina valley's first protoindustries. In contrast to these positive interventions by local rulers, the central government in Edo was relatively passive. Yet in its more distant way, the Bakufu too played an important
part in the protoindustrialization of the Japanese Alps. For one thing,
through its international policies, the Bakufu indirectly (and perhaps inadvertently) stimulated the Japanese economy by slashing silk imports
from China in the 1680s. Although the stimulus was selective, it had an
immediate impact on Shimoina. The resulting spurt in demand for do-

mestic silk floss prompted local silkworm raisers to begin producing thread for Kyoto, providing a classic instance of import substitution. At the same time, the shogunate effectively maintained the peace at home, adjudicating disputes and circumscribing the predations of local rulers. Benefits clearly accrued to Japanese merchants from the enduring Pax Tokugawa and the parcelized imperium that resulted. The Bakufu's vigilance over the domains not only prevented destructive wars; it also curtailed local lords' ability to dominate or destroy the fragile economic reticula that extended across their borders.

The shogunate's already passive support for Tokugawa commercialization was certainly not unalloyed; protection was offset by sumptuary legislation and a barrage of invective against the evils of commerce. Dependent on land taxes for the bulk of its revenues, the Tokugawa regime was compelled both morally and fiscally to defend the integrity of its agrarian base. The result was a conservative regulatory environment, where Bakufu reformers attempted repeatedly to limit the latitude of Japanese merchants—and regularize their activities—so as to prevent commodity production and consumption from interfering unduly with agricultural output. But I would suggest that even these restrictions on trade were not altogether a bad thing for the Tokugawa commercial economy. Edo officials may have hampered Japan's protocapitalism and even deformed it in particular ways, but they also contributed tangibly to the survival of the regime's social and ecological foundations by protecting peasants and their lands against the most extreme forms of financially motivated depredation.

If regional production under the Tokugawa regime may be characterized as driven by a ton'ya-chūma alliance, aided by active patronage from local rulers and passive support from Edo, a very different configuration of actors was responsible for promoting regional development in the Meiji and Taishō years. On the one hand, the propertied classes were considerably freer after the Restoration. Land had become real estate, to be bought, sold, divided, and converted to new uses as its owners saw fit; likewise, those in control of more liquid forms of capital could now invest legally in a wide range of enterprises, unhampered by the feudal era's spatial and social restrictions on commerce. On the other hand, with Tokyo assuming primary power over revenue collection and dispersal, local rulers were replaced by appointed administrators, whose role was more to enforce the mandates of the center than to issue their own orders for resource allocation. Finally, the philosophy of the new state represented an about-face from the agrarian mercantilism of its Tokugawa predecessor. The Meiji leadership, galvanized by an urgent sense of impending national crisis, fashioned a capitalist development state, driven not to preserve the status quo but to revolutionize it as quickly as possible.

While leaving considerable responsibility for resource disposition in the hands of millions of private property owners, Tokyo managed to pursue its goals by deliberately setting the parameters within which private decisions would be made. After a decade of experimentation, the Meiji government disentangled itself from direct oversight of industrial enterprise, but the regime continued through other means to make investment in export-oriented production an attractive venture for private parties. In the case of silk, a coordinated series of policies was aimed at both farmers and filaturists. By establishing agricultural experiment stations capable of developing improved varieties of mulberry and silkworms, by disseminating information through a national network of agricultural cooperatives, by screening silkworm eggs for disease, and in a myriad other ways, the government was able to boost the attractions (and reduce the risks) of sericulture for Japan's farmers. And by setting standards for the grading of silk floss, establishing the Yokohama specie bank, helping set up contacts in the United States, and, not least, facilitating the creation of producer cartels, it also enhanced returns to filature owners. Moreover, the central government also took it upon itself to build a modern communications infrastructure for industry, including telegraph, rail, and shipping lines.[7] Nor was Tokyo the only locus of public support for export-oriented industry. Prefectural officials played an important part as well, investing in technical training and local infrastructure and staging competitive industrial expositions.[8]

The floundering Chinese state proved unable to perform any of these crucial supporting roles. As Lillian Li concludes from her extensive study of the early Chinese silk industry, "the problem was that there were no steps taken to overcome risk and uncertainty either by collective action from below or by government action from above." With its officials spread much more thinly than their Japanese counterparts in any case, the Q'ing government in the late nineteenth century "had not even begun to conceive of a protectionist or nationalist policy."[9] This lapse on

7. For a succinct survey of Japanese government efforts to aid the silk industry, see Eng 1978:232–41; on the scale and importance of shipping subsidies maintained by the Meiji government, see Wray 1986:254ff.

8. Nagano-ken 1979:146–47. For a careful study of prefectural support for sericulture in an area farther from the nation's filature core, see Kawahara n.d.

9. Li 1981:205, 200. For a parallel argument concerning the U.S. and Chilean copper industries in the late nineteenth century, alleging that the former prospered owing to the "extent and quality of government intervention," whereas the latter floundered because it failed to win the state as an ally, see Culver and Reinhart 1989.

the part of a potentially formidable competitor gave the Japanese a compelling advantage in the Asian silk trade, helping to set the stage for massive transformations in Tokyo's industrializing hinterland.

But if actions at the national and prefectural levels set the stage for a new regional production, the content and quality of Shimoina's performance also depended heavily on the local cast. The Ina valley could not have effected its rapid shift to a silk-based economy without a purposeful crew of indigenous industrialists and boosters, who soon succeeded the *ton'ya-chūma* alliance in forging a new kind of regional development alliance for the modern era. Spearheading the new coalition were village leaders of wealthy peasant (*gōnō*) descent. Heirs to the valley's proud nativist tradition, these central Shimoina landlords had also inherited material wealth and business experience, accumulated during two centuries of protoindustrial development. Moreover—and in pointed contrast to speculative dealers like Tanaka Heihachi, whose ventures took him ever farther from home—they had their feet planted firmly on Ina valley soil. Animated by a potent combination of community spirit and enlightened self-interest, it was these local notables who led the way in both sericulture and filature development.

Outstanding among this group was Hasegawa Hanshichi, founder of the first modern filature in the county. Hasegawa proved adept at navigating through the hazardous economic environment of the early Meiji era, taking calculated risks in the marketplace but also securing political support when his investments were threatened. Son of a former village headman, and friend to one of the region's deposed overlords, he parlayed his political connections into a sizable government loan during the Matsukata deflation. But few of Hasegawa's contemporaries could activate personal networks as formidable as his. For the silk industry in the county as a whole, a crucial partner of private investment was the local booster association, where landlords were joined by merchants, bankers, journalists, teachers, students, and politicians. What brought these coalitions together were concrete proposals for regional development, most conspicuously the opportunity to secure a national railroad line for the Ina valley. In this effort, as in others like it, the boosters' role was a dual one: to mobilize community support for a project (including pledges of money and labor), and to represent the region in deliberative bodies that distributed public works funds. Here elected officials played a valuable role, joining the concerns of their constituents with the interests of their backers as they argued the region's case before national and prefectural assemblies.

In short, a revolution in Shimoina's geopolitical and technological context had been matched by a revolution in the social alliances that mobilized local resources. As the prerogative of patronage shifted from the provincial daimyo to the central government, and as the conservative Bakufu gave way to an aggressively developmentalist state, the *ton'ya-chūma* alliance yielded to local booster associations as the primary mediator of regional production. Nor was Shimoina unique in this regard. Similar alliances of private capitalists and regional lobbyists arose throughout Japan, helping to shape a new local-center relationship characterized, in Gary Allinson's words, by "responsive dependence." And precisely because every local lobby had eager counterparts across the country, the obverse of responsive dependence was fierce interregional competition: a variety of social conflict that served, whether by design or by default, to shore up the conservative social order. In the new Japan, as Allinson notes, the focus of political struggle would be "not so much a resolution of conflict within communities, but more a contest for scarce resources *among* communities."[10] Such pervasive regional rivalry would form an enduring backdrop to Shimoina's economic transformation in the prewar years.

INVERTING THE ECONOMIC AND SOCIAL LANDSCAPE

As Ina valley peasants, artisans, property owners, and politicians responded to the changed conditions in which they found themselves after 1868, members of every class experimented with new deployments of the labor, capital, and material resources at their command. From a geographical vantage point, the collective impact of those myriad innovations was to invert the Ina valley landscape. That inversion may be read in a trio of related changes: a radically expanded scale of territorial specialization, a fundamental reorientation of regional trade, and a new map of social differentiation.

The enlargement in the effective scale of regional economic specialization constituted one of the most significant changes affecting Shimoina's geography. With innovations in transport and a steady expansion of world trade, the intricate landscape of local specialties increasingly

---

10. Allinson 1975:59; emphasis in the original. As Gary Allinson later elaborates, "This kind of competition characterizes politics of the pork barrel in many societies. But in prewar Japan it was especially pronounced. Legal codes and financial dependence kept communities subservient to the state. Under such conditions, . . . political struggle was a competition among communities for the scarce resources distributed by the state" (1975:77). Similar observations have been made concerning the contemporary United States, where again "competition among localities rather than conflict within them" has been identified as the distinctive form of local politics (Cox and Mair 1988:307).

gave way to much cruder, supraregional divisions of labor. While micro-geography still mattered in the production of goods for local markets, the range of the Ina valley's export commodities contracted dramatically. In place of the half-dozen protoindustrial wares produced for sale to cities outside the valley during the Tokugawa period, by the early Taishō years Shimoina's exports were tilted heavily toward two categories of goods: forest products and silk. Moreover, within these sectors, the Ina valley's role was defined increasingly as one of raw-material supplier to more advanced industrial centers elsewhere. Even silk floss, the area's most highly processed industrial product, was woven and made into garments elsewhere (predominantly in the United States). In a word, the expanding scale of territorial specialization in the new North Pacific economy meant a proportionally larger but substantively narrower export base for Shimoina, and a diminution of its role in finishing the materials it produced.

A second geographical index of the regional inversion may be observed in the shifting focus of trade. The 1763 trade surveys indicate that Iida in the Tokugawa period was tied primarily to greater Nōbi; the Ina valley secured most of its long-distance imports from, and shipped half of its local specialty goods to, an area stretching along the Tōkai coastline from Yoshida to Nagoya. Interior Shinano and Edo, by contrast, constituted distinctly secondary poles in the region's economic field, absorbing fewer of the area's protoindustrial manufactures and supplying a smaller volume of specialty goods in return. Moreover, mirroring this southward bias in long-distance trade was a comparable but smaller-scale tilt within the regional core. It was the villages straddling the main trade routes on the approach to Iida from the south that grew to be the most prosperous during the Tokugawa period; even the fertile area north of the castle town was a hinterland of sorts, referred to as the *oku*, or interior.

By 1920, however, the situation was reversed: Iida's economic orientation on both scales had swung decisively to the north. Suwa and Yokohama, rather than Nagoya or the Tōkai ports, now absorbed the bulk of Iida's export goods, and the villages north of the former castle town had displaced their southern neighbors as the center of rural prosperity in the region. As the Ina road lost its role in the through-trade, converting the valley from a corridor to a cul-de-sac, this northward orientation became even more pronounced. While Nagoya's reach as a regional wholesaling and shipping center steadily expanded, the configuration of the new railroad network assured that Iida's relations even with its southern neighbors would be routed through the north. By 1910, Iida had begun importing virtually all of its supplies from the railhead at Tatsuno in the upper Ina valley, converting the extensive terrain south of the former castle town into Shimoina's most remote hinterland.[11]

At the same time, however, another change occurred: the southernmost fringe of the county split off completely from Iida's economic sphere. Whereas nearly the entire area later encompassed within the boundaries of Shimoina *gun* appears to have been integrated with one or another of the protoindustrial crafts centered in Iida by the late Tokugawa period (as indicated in map 13), the same area in early Taishō had been split into two distinct commodity-supplying zones (maps 30 and 32). The fuel-producing uplands nearest the valley core were drawn into the orbit of the silk economy centered in Iida and Suwa, but Shimoina's southern forests had been essentially captured by the booming Tōkai coast, as first timber and later charcoal were extracted in large quantities for shipment toward the rapidly industrializing Pacific littoral.[12] In at least a limited sense, then, the twentieth century saw the sundering of Shimoina as an integrated economic region into two distinct peripheries, captive to larger and more dynamic economic centers on either side.

This economic cleavage was in turn reflected in the region's new social map, affecting spatial variations in population growth, sex ratios, and class composition. The overall direction of the change was somewhat ambiguous, but the crosscurrents that created it are readily discerned. On the one hand, the extension of a transport web throughout the county and the diminution of through-trade along the Ina road lessened the significance of the trade corridor as a differentiating device, eroding one axis that had underlain the fourfold settlement typology of the Tokugawa period. At the same time, the powerful centralizing tendencies at work in both the political and the economic world led to convergence across the region in some features, while sharpening the social differences between core and hinterland in other ways. Although the ineluctable loss of local detail (owing to the consolidation of Tokugawa

11. Iida's relationship with the Nōbi plain was not eclipsed altogether; on the contrary, new railroad and harbor facilities significantly elevated Nagoya's economic role within central Japan. In fact, according to Nakajima Katsumi (1971:23), as much as 80 percent of Iida's imports in the Taishō period—including those brought in through the railhead at Tatsuno—originated in Nagoya. This does not, however, alter the fundamental argument here. Regardless of their original port of entry, the routing of Shimoina's imports through Tatsuno station reoriented the region decisively to the north. On the gradual shift of Iida's major import route toward successively northward passes as the railroad pushed north through the Kiso valley, see Ōsawa 1966:22.

12. After the departure of the Ōji Paper Company, even the bulk of the charcoal from the Tōyama valley was shipped not to Iida but to Shizuoka Prefecture. Minamishinano-mura Rōjin Kurabu 1979:1–6.

villages into the larger supervillages of the Meiji period) makes generalization perilous, I would suggest that the main effect of these changes was to smooth out the microgeographical perturbations of the early modern social landscape.

Demographically, the Tokugawa-to-Taishō transformation was characterized by three interlocking trends, which were expressed in virtually every village across the county. For one, villages throughout Shimoina began attracting substantial numbers of interregional migrants. At the same time, average household size throughout the county began to rise (a trend that had characterized only the protoindustrial satellites during the Tokugawa period), confirming other evidence that suggests a rising rate of natural increase. This combination of immigration and rising birthrates had a profound impact on the county's overall population regime, changing it from one of highly localized demographic increases, offset by comparable areas of decline, to one of sustained if uneven growth throughout the county for over half a century.

If overall population trends suggest a growing uniformity across the county, however, the sharp rise in sex ratio differentials provides one index of persistent core-periphery distinctions. The cause was simple. With the concentration of sericultural and filature work on the valley floor, women came to outnumber men in all of Shimoina's centrally located villages, while the rise of logging camps and charcoaling work created a reverse trend in outlying communities. A mild tendency in this direction had characterized Shimoina's population during the Tokugawa period as well, but as is clear from map 29, by 1921 the split was much more sharply defined.

Related to these demographic shifts were equally important realignments in the social landscape of class. From 1750 to 1920, the configuration of workers and property owners in the leading export sector was altered in fundamental ways. The supplanting of the Tokugawa craft complexes by the Taishō filatures entailed crucial, if in some ways contradictory, transformations in the sociospatial relations of labor and capital. For those employed in the export sector, increasing concentration in large workshops went hand in hand with new forms of segregation and labor discipline.

In the protoindustrial era, Shimoina's commodity producers had been scattered throughout a dozen Iida-area villages, working either within their own households or in small artisanal workshops. The latter, which represented the largest congregations of workers, were internally stratified by seniority, with journeymen and apprentices serving a master crafts-

man who combined skilled labor, investment capital, and managerial re-
sponsibilities. Organized into self-governing guilds, these artisans had
proven capable both of negotiating with merchants and of mobilizing
large-scale protests when provoked. As the paper industry declined, however, many skilled artisans emigrated from the region; those who remained began to take on other by-employments, including sericulture
and tenant farming. In short order, the proud professional artisans who
had constituted the labor aristocracy of protoindustrial Shimoina were
transformed into domestic outworkers.[13] Meanwhile, as the complex
reticula of craft production began to contract, reeling mills steadily arose
in their place. Here the social and spatial lines were drawn very differently: by 1920, most of Shimoina's industrial labor force worked, not in
isolation at home or in hierarchically stratified workshops, but in groups
of tens or even hundreds, all engaged in the same kinds of work and subject to constant oversight from their employers. The spatial convergence
of this essentially homogeneous work force in Shimoina's factories coincided with the diffusion of socialist ideology and the formation of proletarian parties like the Aikoku Seirisha, creating the potential for large-
scale working-class activism.[14]

Yet even as they came into contact with one another in the factories,
the region's industrial workers were simultaneously segregated by gender, and subjected to new forms of labor discipline. In the case of the
silk mills, this discipline was particularly harsh, effectively preempting
the emergence of any widespread class-based mobilization. The Japanese filatures primarily recruited rural girls and young women, binding
them to serve out fixed-term contracts in exchange for cash advances
paid to their parents (or husbands). As subordinate members of their
natal (or marital) households, these workers were constrained by powerful claims of familial loyalty and filial obligation. Such claims were both
manipulated and compounded by factory owners and the state, for
whom securing a skilled and compliant work force in the textile industry was a grave national priority. Japan's female textile workers were thus
not only abused at the workplace; they were also denied the right to vote
or strike, and even barred from participating in public rallies or joining
political associations of any kind. Finally, compounding this truncation

13. Shinano Kyōikukai Shimoina Bukai 1934:114.
14. Although the state began its campaign against the labor movement as early as
1900 with the passing of the Public Order Police Law, nationwide socialist party activity
continued until the 1930s. On the evolution of labor activism and popular politics under
"imperial democracy" in early-twentieth-century Tokyo, see Gordon 1991; on conflict
between police and socialists in Nagano Prefecture, see Nagano-ken 1971:380ff.

of their political rights was another feature of the new work force in export production: their less-than-permanent involvement in industrial work. The demise of the Iida-area paper crafts meant the waning of a core of relatively privileged full-time artisans, and their supplanting by a mass of essentially temporary workers, most of whom would leave the filatures after only a few years. While some of Japan's female mill workers overcame all of these obstacles and found creative modes of labor activism, silk reelers did so less often than their sisters in the cotton mills; filatures tended to be smaller, more rural, and less highly mechanized than spinning mills, and filature operatives' contracts tended to be of shorter duration than those in the cotton industry.[15]

The demise of the Ina valley's sometimes militant protoindustrial work force and its replacement by a young filature work force thus had profound implications for the political complexion of the Ina valley. As Shimoina's most advanced industrial sector shifted to disenfranchised female factory workers, a county known for having one of the highest protest rates in the country was transformed into one of Japan's most quiescent regions. Following a brief surge of political activity, which climaxed in the Iida Incident and its forceful suppression, both class-based political organization and public revolt all but disappeared from the county. In fact, the last major uprising in Shimoina was a rice riot in early September 1897, one of forty-four such incidents around the country in that year, during a brief recession following the Sino-Japanese War. While a wide variety of laborers were among the two thousand protesters who stormed a mill in Iida that fall, the crowd was led by an angry group of cordage artisans from Ajima, who needed rice paste for their paper crafts. The price of rice in Iida had soared in the previous months— not owing to a poor harvest, but because a local merchant was buying up grain supplies to send to a lumber camp in the Tōyama valley. After this incident, however, Shimoina turned remarkably quiet, even in 1918, when rice riots swept most of Japan. Far from prompting increased social unrest, then, the displacement of dispersed outwork by factory production coincided with a reduction in the incidence of popular disturbances in Shimoina.[16]

15. For forceful arguments against the view that Japan's female industrial workers constituted an uncommitted and passive group, see Tsurumi 1984 and Molony 1991. Barbara Molony concentrates exclusively on cotton mills, however, acknowledging the additional obstacles to activism in the filatures (p. 219 n. 9).

16. While Iida's calm in 1918 was undoubtedly owing in part to early relief efforts organized by police, administrators, and area merchants (Shimoina Chiikishi Kenkyūkai 1982:95–96), the significance of the region's social transformation should not be over-

As the Ina valley's industrial work force underwent this simultaneous convergence and subjugation, the ranks of the county's elite went through a contradictory transformation of their own. On the one hand, the area's wealth and power were slowly and fitfully consolidated in fewer hands. One early means by which this was effected was the abolition of formal barriers between warrior, peasant, artisan, and merchant. The end of the Confucian status system meant that individual persons could legally exercise different kinds of power, permitting a fusion of landholding, commercial investment, and political leadership in a handful of families. This multiplicity of roles certainly had a precedent in the village notables (*gōnō*) of the Edo period, who operated simultaneously as landlords, traders, and protoindustrial merchant-producers. The crucial difference with Meiji was one of scale: persons holding formal political as well as economic power now began to appear at the highest levels of both business and government. The Matsukata deflation and subsequent economic crises strengthened this trend. As mid-level landlords lost their assets in bankruptcy proceedings, the ranks of the emergent capitalist class were thinned, contributing to the concentration of the county's wealth in fewer hands. Finally, the Shimoina elite was further narrowed by the coalescence of commercial functions in a single spatial core. The lifting of feudal trade restrictions on the rural satellites may have curtailed monopoly privileges for Iida's merchants, but its designation as county seat and site of the 117th National Bank lent Iida undisputed primacy as the political and financial center of the region. No longer would the Ina valley's commercial elite be riven by political rivalries at the core; the days of a spatially divided merchantry were over.

Yet despite these consolidations, Shimoina's local notables now inhabited a much wider and more competitive economic world, one that would expose them to a number of insults from beyond the county's borders. The Meiji land-tax reform, for instance, sharply raised the obligations of landlords throughout the valley, cutting substantially into one important source of local revenues. Meanwhile, as the new legal framework made capital more mobile, corporations based in Suwa, Tokyo, and Nagoya began to corner some of the most lucrative new en-

---

looked. For confirmation of Shimoina's high rates of protest during the Tokugawa period, see the detailed spatial and temporal analysis in Fujimori 1960 and Hirasawa 1972a. On the role of cordage artisans in the 1897 riot in Iida, see Hirasawa 1960b and Hirasawa 1963; on regional differentiation across Japan in the 1918 riots, see Lewis 1990. Note that Lewis's national map of riot outbreaks in 1918 shows no activity in southern Nagano.

terprises in the county, elbowing the local elite aside. By 1920, Shi-moina's major export industries were no longer financed solely by local capital; nor were they managed exclusively for the benefit of local power holders. In fact, during the same years when outsiders purchased the majority of stock in the valley's private railroad company, the most spectacular fortune ever accumulated by an Ina valley native was in-vested elsewhere. After his death in 1884, descendants of Shimoina's wealthiest native son—the adventurous thread dealer Tanaka Hei-hachi—invested their inheritances in Hokkaido coal and railroads, be-coming partners as well in a variety of national shipping, lumbering, utilities, and banking enterprises.[17]

Like the influx of investment funds into Shimoina from Tokyo and elsewhere, this exodus of profits from the region signals the growing mobility of capital in the new Japanese state. In most cases, the out-come of that mobility was increasing dependence for Shimoina. The bulk of the local timber, for instance, was harvested by firms headquar-tered in neighboring prefectures; in filature as well, important profits accrued to outsiders, both those who extended credit to the local banks and those who bought up an increasing share of local cocoons for reel-ing elsewhere. The story was similar in other sectors as well. Even con-struction work was dominated by outside companies; only one of the top four local contractors for road, bridge, and railroad work in the early twentieth century was a locally based firm.[18] In short, those sur-vivors of the local consolidation process who chose to remain in the Ina valley were less and less able to muster the resources to finance indus-trialization and infrastructural development. They may have been big-ger fish, but not by the standards of the greatly enlarged pond in which they now found themselves.

As this overview suggests, Shimoina in the Meiji and Taishō years was in essential ways turned inside out for silk and timber. The region's econ-omy was decoupled from the circulation services and protoindustries that had driven early Tokugawa development and harnessed instead to distant commodity markets, which ultimately laid claim to the vast bulk of local resources. To speak of this change as an inversion is more than a loose figure of speech. Not only was the content of Shimoina's econ-omy radically transformed, but the very compass of local identity was si-multaneously redirected. By the early twentieth century, the vertically

17. Kobayashi 1967:66ff.
18. Iwashima 1967:54–55. Of the other three firms, two were from Tokyo and one from Mito.

integrated, autonomous economic region centered on Iida had been transfigured beyond recognition. That the primary markets driving southern Shinano's economy by 1920 were located entirely beyond the pale of its previous circulation network represents a simple, powerful index of the scope of change in the intervening years. Yet equally important, although more elusive, was the transformation of the domestic matrix through which the county's economic life was ordered and controlled. Local workers largely lost their political voice, and local capitalists lost their monopoly over profitable investment opportunities, even as the area centered on Iida lost its integrity as a territorial unit of production. As new and more powerful mechanisms of integration were introduced in the Meiji era, the Ina valley was effectively subordinated to the embrace of national and international institutions. After the turn of the century, Shimoina was firmly positioned in a new and much more tightly centralized universe, cementing the inversion of the area's role in Japan's space-economy.

Legacies and Linkages: The Geographical Foundations of the New Order

To this point, Shimoina's history from the Tokugawa to the Taishō era has been told in terms that emphasize its transformation. The autonomous, small-scale region centered on Iida proved to be a transitional construct, corresponding to a specific historical period: one characterized by animal transport, merchant capital, and a political order of parcelized imperium. As those conditions gave way, so too did the region they had produced. Yet despite the pervasiveness of change, the earlier landscape did not just melt away; the early modern period was not simply erased. Rather, important elements of the regional formation developed during the Tokugawa years carried over into, and helped shape, the new political economy. Some carryovers from the old to the new landscape were tangible assets. Other continuities were more abstract, residing in features of regional production that might be common to a wide area of Japan or to the processes of state formation and industrial revolution more generally. Excavating these more enduring layers beneath the transformed landscape is essential for grasping the true significance of the historical dislocations sketched above.

THE PHYSICAL LANDSCAPE AND
FEATURES OF DEVELOPMENT

Of the numerous threads connecting the sociospatial or-
der of 1750 to that of 1920, physical entities that survived the transition are the easiest to identify. The southern Japanese Alps may have experienced an economic revolution, but the new landscape had by no
means obliterated the old. Infrastructural investments made during the
Tokugawa era constituted a sort of "second nature," augmenting the
physical geography of the region with social investments that endured,
in many respects, into the current century.[19] The city of Iida, for instance,
remained throughout this process the largest urban fixture in the region.
Even the internal spatial order of the castle town was replicated in the
new county seat: the merchant wards of the former became the commercial district of the latter, while the daimyo's compound came to
house the new order's public buildings (county offices, an elementary
school, and a library fitting easily into the space once inhabited by the
local lord and his highest retainers). In rural areas, too, the resemblance
between the old and new maps was striking. Villages may have acquired
new boundaries in the process of centralization, but the old communities survived as hamlets of the new, retaining their Tokugawa names and
many of their residents as well. Roads were widened, but wagons and
carriages in 1920 traveled essentially the same routes that the packhorses
had traversed in 1750; bridges spanned the Tenryū, but generally at the
sites of the old ferry crossings. And whatever the nature of the new crops
being grown in Shimoina soil, the canals, dikes, and field boundaries
that constituted the physical infrastructure of its agricultural system had
undergone only gradual, evolutionary accretion. In short, from county
seat to countryside, many of the material constituents of the Taishō landscape had deep roots in the early modern order.

A different kind of survival from Shimoina's Tokugawa geography
was to be found in vestigial protoindustrial enterprises. Only fragments
of the Ina valley craft complex outlasted the period of massive conversion to silk; tobacco processing, cotton spinning, and lathe- and lac-

---

19. The concept of "second nature" is elaborated by William Cronon (1991:56): "A
kind of 'second nature,' designed by people and 'improved' toward human ends, gradually emerged atop the original landscape that nature—'first nature'—had created as such
an inconvenient jumble. Despite the subtly different logic that lay behind each, the geography of second nature was in its own way as compelling as the geography of first nature." For a philosophical critique of the concept, see N. Smith 1984:19–20.

querware production faded entirely from the scene. But the advanced paper crafts and certain textile arts hung on in their traditional centers, sharing turf with the mulberry fields and filatures. Although profit margins and wage levels were low, both the technical know-how and the commercial organization required to produce decorative cordage and specialty fabrics were still in place when the silk bubble burst in the 1920s. When local politicians, businessmen, and laborers alike cast about over the next decade for alternative job-creating enterprises, reviving these and other traditional crafts became the subject of earnest debate. It was in this context that the Shimoina Educational Association undertook a comprehensive historico-geographical investigation of the traditional craft complexes, producing an invaluable study of their origins, spatial structures, technologies, markets, and potential for expansion. While most ultimately proved incapable of resuscitation, the lingering embers of the paper-cordage industry were already being fanned back to life, sustaining a regional specialty that even today supplies livelihoods for scores of Iida-area residents. In this sense, the retention of earlier economic assemblages provided another important link with the past: one that diversified local residents' options during the next round of structural change. Here again, Shimoina exemplified a more general development pattern. As Pat Hudson contends, the persistence of such "traditional sectors" (which she characterizes as "labor using" rather than "labor saving") into the era of factory development is a widespread, if not universal, feature of early industrialized regions, contributing to the economic dualism manifest in most First World countries.[20]

In addition to these material legacies of Shimoina's Tokugawa geography, there was a more abstract set of continuities: characteristic development patterns common to much of Japan during these years. While not all of these patterns would last into the postwar period, each may be identified as a distinguishing feature of Japanese development in the period under consideration. One such characteristic was the embedding of commodity production in an agrarian matrix. That craft development in the Edo period should have had a markedly agricultural character comes as no surprise; roughly nine-tenths of the Ina valley's residents were first and foremost tillers of the land, and regional self-sufficiency in food and raw materials was mandated by the severe constraints on transport. For the first quarter-century of the international

20. Hudson 1989a:8.

silk trade, as we have seen, these constraints remained intractable, forcing the expanding sericulture and filature sector to adapt to the strict requirements of a largely unchanged agricultural regime. But even after transport innovations had been pushed through and dramatic changes in production were under way, the silk economy remained profoundly rooted in the agricultural landscape.

The range of agro-industrial linkages identifiable during Nagano's heyday as a silk kingdom was broad. Cocoons, the silk mills' primary raw material, came directly off the farms where mulberry was grown and silkworms were raised. Since their inputs were strictly organic, the filatures' waste products were likewise cycled back into the agricultural system from which they had come, as either fish food or fertilizer. Socially, too, silk was inextricably embedded in the rural economy. The county's reeling plants were built primarily in farming villages, and staffed primarily by women and girls from farm households. Even their owners were often landlords; in Shimoina as across Japan, it was more often landowners than urban merchants who played the leading role in promoting both sericulture and filature.[21]

This pattern of agrarian-based development has been identified as a distinguishing feature of early Japanese industrialization. As numerous scholars have observed, its effect—particularly in the silk regions—was to boost the attractiveness and viability of agriculture, leading to an unusual situation in which industrialization served, not to uproot rural producers, but to keep them on the land. Saitō Osamu, for instance, concludes that "commercial cropping and sericulture acted as a brake on the dissolution of the peasantry"; it was not the silk industry's development but its later "disappearance [that] became an incentive for peasants to leave farming."[22] Yet the relationship could also be construed differently. As Thomas Dublin notes (in an article on nineteenth-century New Hampshire), the availability of industrial outwork "played a contradictory role—for women, and for the countryside itself. On the one hand, it allowed farmers' daughters to remain at home as contributors to a traditional family economy; yet it also drew them into broader economic networks that steadily undermined the semi-subsistence practices of the rural order."[23]

In time, of course, these agro-industrial linkages would be loosened. In Japan as in New England, factories would gravitate to expanding met-

21. See, e.g., Shindatsu 1985:134–35; Unno 1985.
22. Saitō 1986:417.
23. Dublin 1985:65.

ropolitan belts, the industrial labor force would break free of its rural roots, and fossil fuels and other imports—rather than the domestic rural economy—would come to supply the bulk of raw materials for industrial production. But until 1920, the industrial economy of Shimoina remained essentially what it had been throughout the previous two centuries: a processing arm of local agriculture. As a result, the Ina valley (like other silk regions) did *not* experience the "breakdown" in agro-industrial linkages that Penelope Francks identifies as a central feature of Japanese economic development during the first decades of the twentieth century.[24] While rural households elsewhere in Japan may simultaneously have been deprived of both by-employment income and the organic wastes they had received from traditional industries, as she suggests, residents of sericultural regions like Nagano were an exception to that rule. Given that silk accounted for between a quarter and a third of Japan's exports for half a century (from the 1870s through the early 1920s), such exceptions must have been relatively commonplace.[25]

A second supraregional continuity across the Tokugawa/Taishō divide was the comparative labor intensiveness of production. In early modern Japan, agricultural productivity increases had been secured less by extensive growth than by intensification, substituting human energy for that of draft animals as well as increasing total energy expended per hectare. This pattern of development perpetuated what Thomas Smith has called the "handicraft character" of Japanese agriculture, contributing as well to what Hayami Akira identifies as Japan's "industrious revolution."[26] Artisanal craft work and the associated rural by-employments appear, not surprisingly, to have been characterized by similar levels of labor intensiveness. Wind, water, and draft power played almost no role in early Japanese commodity production; Shimoina's paper cordage, umbrellas, lacquerware, and textiles alike required lengthy and intricate handwork procedures, performed for the most part by skilled practitioners wielding simple hand tools. The late-nineteenth-century revolution would bring nonhuman power sources and machine technology to bear in the local textile sector for the first time, transforming a protoindustrial craft into a recognizably modern industry.

But the introduction of novel mechanisms and techniques did not alter the fundamental factor relations in the Japanese economy. In com-

24. Francks 1984:85.
25. On the composition of Japan's prewar export trade, see Mizoguchi Toshiyuki 1989:13.
26. T. Smith 1959; Hayami 1989. This issue is addressed in chapter 4, n. 6 above.

parison with the industrialized West, capital in Japan remained expensive relative to labor. Indeed, the first round of industrialization in the textile industry was primarily impelled by concerns of quality rather than by the imperative of raising worker productivity. But factor ratios in the economy were not the only force perpetuating labor intensiveness in the filatures. Further capital improvements in the silk mills appear to have been held back as well by deliberate government policy: Shindatsu Haruki has shown that substantial proceeds from the sale of silk floss were diverted toward the purchase of ginned cotton, machinery, and petroleum for the growing cotton textile industry, rather than being returned to boost productivity within the silk sector.[27] As a result, Japanese reeling—like Japanese agriculture—retained its handicraft character throughout this period. Despite the rise of factories, a labor-intensive regime of production was thus an important continuity in the Ina valley's regional economy from the Tokugawa to the Taishō era.

This continuity underscores the importance of Abe Takeshi and Saitō Osamu's observation that the move "from putting-out to the factory" in fact masks two independent shifts: the spatial aggregation of laborers in large workshops on the one hand and the introduction of machine technologies on the other. Shimoina's experience confirms a widespread view that although the first shift occurred fairly early in the Japanese filature industry, the latter was delayed; reeling remained incompletely mechanized in that it continued to demand skilled operatives, capable of deftly removing and combining the fibers from several cocoons to feed a thread of consistent thickness onto the reels. Indeed, many scholars contend that as late as World War II, the Japanese reeling industry was still technically in the stage of "manufacture" (a term used by Marx to describe an early stage of production based on a large-scale division of labor, and involving a distinct separation between capital and workers, but where the production process is carried out primarily by hand rather than by machine). Although Shimoina had moved to the factory system, it had not entirely entered the age of "machinofacture."[28]

If the result of sericulture's agricultural embeddedness was to keep large numbers of Shimoina farmers on the land, the effect of labor-intensive industrialization was to perpetuate the dependence of its in-

27. Shindatsu 1985:128–29.
28. Abe and Saitō 1988:147; Yagi 1980. For more on this distinction, see Hoston 1986:105; Mantoux 1961:35–41. For a study of the complicated geographical realignments involved in the transition from manufacture to "machinofacture" in a British textile industry, see Gregory 1982.

dustries on a low-wage work force. This in turn had several repercussions for regional development. For one, since Japan as a whole had staked its claim to a place in the new international division of labor on the basis of a low-wage economy, the state was pressured to assume a coercive relationship to labor. The result was a mode of industrial regulation akin to so-called bloody Taylorism, where limited segments of a particular industrial process are relocated from the core to the periphery, "in social formations with very strong rates of exploitation (in wages, duration and intensiveness of labor etc.), the products being mainly re-exported to the core."[29] (The implicit contrast is the system known as Fordism, where industrial wages are high enough to allow domestic workers to absorb the products of industry rather than export them, and where violence is not a habitual characteristic of state-labor relations.)[30] As defined by Alain Lipietz, "bloody Taylorism" has two distinguishing features. First, the production process is brought within the purview of managers in a controlled factory setting, yet remains relatively unmechanized, often mobilizing a largely female work force. Secondly, it is bloody in that to the traditional oppression of women, it adds "all the modern arms of anti-worker oppression (managed unionization, absence of social rights, imprisonment and torture of opposition)." Although Lipietz is writing here about electronics and textile manufacturing in postwar Singapore and Hong Kong, his observations have equal salience for the prewar Japanese silk industry.[31] In 1920s Shimoina, as in many of today's Asian NICs, the rise of an export-oriented manufacturing base occurred within the embrace of a strong-armed, authoritarian government: one that became formally democratic, but remained antagonistic to broad-based popular participation.[32]

A third and final feature that characterized Shimoina's regional development regime from the seventeenth to the twentieth century was

29. Taylorism is a mode of scientific management in the workplace, based on the time-and-motion studies pioneered by Frederick W. Taylor, which involves the step-by-step control of the labor process by managers and the deskilling of that process through minute subdivision. On the rise of Taylor-style scientific management in American and British factories during the early twentieth century, see Urry 1986.

30. The so-called Fordist regime, which dominated in the wealthy industrialized countries in the early postwar era, is named for Henry Ford, who made the unprecedented move of paying assembly-line workers enough to enable them to buy the company's automobiles. For more on Fordism (and its limits), see Schoenberger 1989 and Lipietz 1986.

31. Lipietz 1986:31–32. For a subtle analysis of discord within the Japanese bureaucracy over how heavily the prewar state should rely on repression to cope with the rising industrial labor force, see Garon 1987.

32. On the failure of the democratic movement in prewar Japan, see Gordon 1991.

the deep involvement of political powers in matters of economic policy. Laissez-faire has never truly applied to Japan; evidently, it has rarely appealed even to the country's reform-minded critics.[33] To be sure, the nature and form of governmental intervention evolved significantly, from the local patronage and central arbitration of the Tokugawa period to large-scale infrastructural support and aggressive promotion of technological transfer in the late nineteenth and twentieth centuries. Yet throughout these mutations, significant state intervention in the economy remained a constant in Shimoina's as in Japan's history during these years (even if the contrasting idioms in which this interdependence is discussed obscure the continuity).[34] Considered in tandem with the enduring agricultural character and labor intensity of industrial production, this persistence of close ties between the political and economic orders suggests the need to recognize the permeability of temporal as well as spatial boundaries: despite the radical discontinuities between them, Japan's "Early modern" (1600–1868) and "early Modern" (1868–1925) periods shared many fundamental characteristics.

CONTINUITIES IN REGIONAL STRUCTURE AND PROCESS: TOWARD A HISTORICAL GEOGRAPHY OF POWER

Through these physical and developmental legacies, Shimoina's protoindustrial geography lived on in the Taishō landscape in structural ways as well. If particular communities and production complexes inherited from the past provided building blocks that could be cobbled into a new regional architecture, the less visible commercial relationships elaborated during the Tokugawa period provided an essential mortar. In fact, social organization may represent the most impor-

33. As Germaine Hoston (1992) documents, from the Meiji and Taishō through the early Shōwa periods, very few would-be reformers of the Japanese political system argued that the scope of government intervention in social and economic life should be reduced. A few union activists opposed social policy on the grounds that it made workers overly dependent on the government, but most progressive Japanese thinkers advocated statist rather than individualist versions of liberalism.

34. Tokugawa merchants' close alliance with their political overlords is traditionally disparaged as a weakness (dependence on government patronage), and is often identified as a major obstacle to the emergence of an indigenous Japanese capitalism. On the prevalence of this idea among Japanese Marxists, see Hoston 1986. The modern incarnation of a strong government-business relationship in Japan is assessed in quite different terms, however. From Ezra Vogel (1980) to Chalmers Johnson (1982) and Karel van Wolferen (1989), Japan's admirers and critics alike identify the interdependence of state and merchant as one of the signal strengths of postwar Japanese capitalism.

tant legacy bequeathed by Japan's protoindustrial era to the Meiji and Taishō regimes. The development of a national communications network, the extension of circuits for the exchange of credit and commodities, and the emergence of an incipient spatial hierarchy of power were crucial, I would suggest, to the Meiji oligarchs' later success at political centralization and economic mobilization. When it came time to create a new administrative schema, officials could be recruited from existing positions of authority, just as local administrative units could often be transposed and reorganized rather than invented from scratch. As Prasenjit Duara has pointed out, the Meiji regime proved singularly willing and able to take advantage of this inherited geography of power.[35] Likewise, two hundred and fifty years of economic development under the Pax Tokugawa had generated extensive mercantile and credit networks that were also assimilable. Just as the *ton'ya-chūma* alliance had produced Shimoina as a region, so routine patterns of commodity exchange had forged the Japanese archipelago as a whole into an increasingly integrated economic entity.

Not only in Japan did mercantile relationships such as these give protocapitalist enterprise the power to call economic regions into being. Similar networks were established by early modern merchants and shippers in many parts of the globe, uniting ever wider units of territory in regular patterns of economic exchange.[36] Frank Perlin is emphatic about the importance of this phenomenon as an apparatus of regional integration in South Asia—one that radically changed the circuitry of economic life, if not the look of the land.

> The lack of durable capital investments, and a high mobility of commercial and financial capital, has meant that we lack the more obvious kind of visible evidence for changes. . . . [But] what we need to observe is that economic development in the period before industrialization was mainly characterized by changes in the size and *organization* of circulating capital, and in its increasing control over large quantities of labour extensively dispersed through space in households and large workshops.[37]

I would extend Perlin's important insight in two directions: first, by noting that changes occur not only in the scale and organization of cir-

35. Duara contrasts the Meiji oligarchs, who co-opted traditional power structures in the process of state building, with their Chinese counterparts, who attacked and ignored them (1988:247).

36. An emphasis on circulating capital appeared in the 1980s as a new trend in the protoindustrial literature; see Berg, Hudson, and Sonenscher 1983:6; Toby 1991.

37. Perlin 1983:91; emphasis in the original. For a thoughtful appreciation of Perlin's argument, suggesting that it could usefully be applied to agriculture as well, see Eley

culating capital but in the movement of people and commodities as well, and second, by emphasizing that this process *produces regions* of a fundamentally new kind. This formulation should be of some utility for advancing comparative inquiries into early industrial history, suggesting a way to transcend the stale debate over whether the notion of protoindustry is inherently teleological. Despite early hypotheses suggesting that domestic outwork was essentially an embryonic stage of factory production, it has by now been amply demonstrated that many protoindustrial regions did not go on to become industrial centers; there was nothing intrinsic to protoindustrial production that ensured its hearths would also prove fertile grounds for the rise of factories.[38] Yet even in the absence of concrete examples, an informed geographic perspective suggests the fallacy of conceptualizing protoindustrial places as incubators for more "advanced" types of industry. With changes in technology and social organization from one regime of accumulation to the next, the logic of industrial location typically changes in fundamental ways; a site amenable to extensive domestic weaving, for instance, might prove wholly inimical to large-scale factories dependent on water-powered machinery. Given the episodic nature of economic restructuring, our efforts ought to be directed, not toward predicting the emergence of mechanized industry from a map of protoindustrial regions, but, in the words of Alan Warde, toward "specify[ing] what might be called the transformational rules between the logics of location in each layer."[39]

This geographic perspective suggests a promising new approach to the issue of continuities across the Tokugawa/Meiji divide. Rather than focusing solely on developments in the realm of production, I believe we ought instead to follow Perlin's lead in seeking the legacy of protoindustrialization in the less visible commercial networks that it engendered—networks that could as easily be directed toward the peripheralization of a region as toward its industrialization.[40] The essential point here is that consolidating and redirecting

---

1984. Hahn 1985, a study of yeoman farmers in upcountry Georgia, demonstrates the salience of Geoff Eley's suggestion (see esp. Hahn 1985:195).

38. Franklin Mendels (1972) explicitly defined protoindustrialization as "the first phase of the industrialization process"; Gutmann (1988:93–95) elaborates by suggesting four specific ways in which protoindustry "made the later growth of factory industry particularly easy and nearly inevitable." The notion that factories *necessarily* followed in the wake of protoindustrialization has, however, been forcefully criticized by Berg, Hudson, and Sonenscher 1983; Perlin 1985; and Pollard 1981:76–78, among others.

39. Warde 1985:198.

40. For a case in point, where protoindustry in one era facilitated peripheralization in the next, see Short 1989.

an existing set of sociospatial relationships poses much less of a challenge than creating them from the ground up. Once in place, regional ties between producers, consumers, merchants, and shippers made larger-scale operations possible; who would rule those operations, however, and to what ends, was a separate matter. In South Asia, advances in capital circulation during the sixteenth and seventeenth centuries were ultimately turned against the peoples who had developed them, facilitating colonial rule and, by some accounts, a century or more of *de*industrialization. The indigenous infrastructures of commercial production and exchange in India proved susceptible to capture by the British, who could readily redeploy them for the benefit of their own power and pocket. Clearly, it is not as a basis for industrialization, but as a basis for *incorporation into larger networks of power*, that the significance of these early modern developments should be sought.[41]

This finding has an immediate bearing on the present project. The notion that the protoindustrial economy helped erect a sociospatial scaffold on which the modern national and international order could be built points a way across the multiple cleavages that run through the terrain of this study. By linking the early modern and the modern, the local and the global, economy and polity, geography and history, it allows us to negotiate a passage between the production of an integral economic complex in the Ina valley from 1750 to 1860 and the process by which Japan emerged as an imperial power in East Asia in subsequent decades. The bridge between the two is none other than the regional inversion: the subordination of scores of overlapping but quasi-autonomous economic regions, like Shimoina, to a single national center. For in an expanded frame of reference, the making of regional peripheries *was* the making of a national core. Tokyo's emergence as a locus of power in the North Pacific was predicated in large part on the speed and thoroughness with which it accomplished this geographical revolution—a speed and thoroughness in turn predicated on the previous establishment of extensive protoindustrial circuitry.

41. Frank Perlin (1983) goes so far as to argue that, owing to brokers' ability to depress wages below market levels, the whole protocapitalist system had a built-in tendency to technical stagnation. In fact, two illuminating studies of early textile outwork in southeastern India (Brennig 1990; Arasaratnam 1990) suggest that merchants' power was often far from absolute. Yet these studies also confirm that, even where brokers exerted relatively weak control over the production process, the elaboration of their commercial circuitry laid the groundwork for subsequent incorporation by the British.

I would thus contend that imperial Japan had its roots in a particular spatial formation: a dense, polycentric commercial reticulum, woven during two and a half centuries behind a barrier of effective economic protection. When Perry's arrival shattered the walls to which that web was anchored, the single most important action of the Bakufu was to confine the breach spatially, interposing its own agents between the provinces and the in-rushing world. By funneling all contact between foreign traders and domestic producers through a handful of carefully patrolled treaty ports, Edo not only defended the Japanese countryside from the immediate threat of invasion; it also inaugurated a new economic role for the Kantō, laying the foundation for its emergence as the focal point of a truly national economy. The conversion of Edo to Tokyo shortly thereafter—reconfiguring the headquarters of a federation of daimyo into the command post of a unified administration—gave the new rulers unprecedented power to mobilize the resources of the archipelago. Through the confluence of these political and economic processes, both of which involved irreducibly spatial strategies, the modern Japanese state was born. In contrast to India, here it would be an indigenous junta that seized the existing economic circuitry and turned it to its own ends, re-creating the Japanese provinces as peripheries of Tokyo rather than London or Washington. But the geographical revolution would not stop there. The same processes by which power was consolidated at home launched Tokyo on its imperial project abroad, prefiguring Japan's emergence as a truly global power in the years to come.

Epilogue

Iida in the 1990s is an unremarkable provincial city of 100,000, unimposing in aspect and largely unknown beyond a handful of neighboring counties. The most impressive building in town, a branch office of the Nagano Prefectural Administration, perches atop the same promontory where the local daimyo once held court. The view from this persistent center of power takes in a drab urban sprawl, commingling near the Tenryū's banks with scattered paddy fields and giving way on the higher terraces to a precise grid of laboriously tended fruit trees. In the distance rise the Japanese Alps, forested once again, this time in neat coniferous rows.

Increasing incorporation into the growing Japanese economy has been a mixed blessing for the people of the Ina valley. Ironically, it was just after the silk market began its decline, in 1923, that the Ina Electric Railroad—a private line designed to connect Shimoina's filatures with the Suwa basin—finally reached Iida. As cocoon prices continued to tumble through the twenties and thirties, the rail link helped area farmers find a new livelihood in growing fruit for the Tokyo market; by the first years of the Shōwa era (1925–89), mulberry trees were already being uprooted to make way for the peach, pear, cherry, and apple orchards that blanket the Ina valley terraces today. Half a century later, the railroad was superseded by the Chūō Expressway, a gleaming four-lane toll road retracing the old packhorse route from Nagoya through Suwa to Tokyo. Local boosters had lobbied heavily to see that the project pass through the Ina valley, and its completion was widely heralded as the harbinger of a new age for the community.

While these innovations in transport have indisputably improved local residents' access to the goods and services of the metropolitan core, they have also re-created the region's peripheralization under new guises. With the coming of the railroad, local producers of consumer goods faced stiff new competition; by 1925, mass-produced textiles could be imported cheaply enough to destroy Iida's remaining weaving industry, which had survived until then behind transport barriers. Similar trade-offs accompanied the arrival of the expressway. Job creation has been slower than expected. While Suwa has parlayed its technical prowess in silk machinery into a leading role in electronics, and while the Tokai megalopolis has passed through a period of relentless expansion in automobile manufacturing and other heavy industries, the main development in Shimoina has been a further subordination to both of these former trading partners. Most new jobs in the Ina valley are in the vulnerable and relatively low-wage subcontracting sector. Meanwhile, the reduction of travel time to the Nōbi plain (from three days in the packhorse era to just two hours by car) has made Nagoya a favorite destination for Sunday shopping excursions, robbing the retail and entertainment districts in Iida—two of the last mainstays of the area's service economy—of some of their luster. Even outside gangs have taken advantage of the new highway to colonize the area; since the 1980s, members of a Nagoya crime syndicate have allegedly infiltrated the Iida city underworld.[42]

42. For a more general rumination on the contradictory effects of improved transport, which may extend the markets for a region's specialized exports but simultaneously subjects local producers to competition from other regions, see Pollard 1981:115.

As much as the local landscape has changed since the turn of the century, in other words, the terms of Iida's incorporation into the national economy have not been fundamentally altered. The region continues to develop as a periphery, oriented toward the more dynamic economic centers to its north and south. Recent years have witnessed a steady exodus of young adults, particularly from the outlying mountains, in search of better education and job opportunities; as a result, the county population has stagnated at roughly the level attained seven decades ago, when silk prices reached their zenith. Yet while they chafe against their backwater status within Japan, the people of Shimoina continue to enjoy innumerable advantages accruing from Japan's centrality in the expanding Pacific economy. From the national subsidies supporting local agriculture to the imported fuels that preserve the area's woodlands, Tokyo's ascendancy as a world capital is never far from view.

In the years ahead, the newfound wealth of an increasingly leisure-oriented society may provoke substantial changes in the local landscape. Already the area's more remote villages have built a number of showy tourist facilities, offering weekend visitors from Nagoya and Tokyo a place to fish in clear streams, sample the exotic flesh of bear and boar, witness a "traditional" village festival, or simply relax in the cool, clean air. Seeing in recreation the region's next major industry, village offices throughout the mountains are abuzz with plans to develop hot springs, vacation cottages, golf courses, even ski resorts. And the transport lobbies continue; as recently as 1991, a regional booster association was doing its best to have Iida designated as a stop on a proposed mag-lev (magnetic levitation) rail line between Tokyo and Osaka. But while any of these new developments would spell important changes in the local landscape, none portend a fundamental restructuring like the one Shimoina underwent in the late 1800s. At the start of the next century as at the close of the last, its identity as a peripheral region within a global core will continue to set the parameters of development in the southern Japanese Alps.

Bibliography

Abe Takeshi and Osamu Saitō.

    1988.   "From Putting-out to the Factory: A Cotton-weaving District in Late-Meiji Japan." *Textile History* 19 (2): 143–58.

Achi-mura.

    1984.   *Achi sonshi.* Nagano-ken Shimoina-gun Achi-mura.

Allinson, Gary.

    1975.   *Japanese Urbanism: Industry and Politics in Kariya, 1872–1972.* Berkeley and Los Angeles: University of California Press.

Amino Yoshihiko.

    1986.   *Chūsei saikō: Rettō no chiiki to shakai.* Tokyo: Nihon Editaa Sukūru.

Anan-machi.

    1987.   *Anan chō shi.* Nagano-ken Shimoina-gun Anan-machi.

Anderson, Perry.

    1974.   *Lineages of the Absolutist State.* London and New York: Verso.

Andō Keiichirō and Yamori Kazuhiko, eds.

    1972.   *Kunizakai no mura.* Tokyo: Gakuseisha.

Arasaratnam, S.

    1990.   "Weavers, Merchants and Company: The Handloom Industry in South-eastern India, 1750–1790." In *Merchants, Markets and the State in Early Modern India,* ed. Sanjay Subrahmanyam, 190–214. Delhi: Oxford University Press.

Baker, Alan R. H.

    1984.   "Reflections on the Relations of Historical Geography and the *Annales* School of History." In *Explorations in Historical Geography,* ed. Alan R. H. Baker and Derek Gregory, 1–27. Cambridge: Cambridge University Press.

    1992.   *Ideology and Landscape in Historical Perspective.* Cambridge: Cambridge University Press.

Baker, Alan R. H., and Derek Gregory, eds.
   1984.   *Explorations in Historical Geography.* Cambridge: Cambridge University Press.

Banno, Junji.
   1983.   "External and Internal Problems after the War." In *Japan Examined: Perspectives on Modern Japanese History,* ed. Harry Wray and Hilary Conroy, 163–69. Honolulu: University of Hawaii Press.

Beasley, W. G.
   1989.   "Meiji Political Institutions." In *The Cambridge History of Japan,* vol. 5, *The Nineteenth Century,* ed. Marius B. Jansen, 618–73. Cambridge: Cambridge University Press.

Berg, Maxine, Pat Hudson, and Michael Sonenscher, eds.
   1983.   *Manufacture in Town and Country before the Factory.* Cambridge: Cambridge University Press.

Bernstein, Gail Lee.
   1988.   "Women in the Silk-reeling Industry in Nineteenth-Century Japan." In *Japan and the World,* ed. Gail Lee Bernstein and Haruhiro Fukui, 54–77. New York: St. Martin's Press.

Berry, Mary Elizabeth.
   1986.   "Public Peace and Private Attachment: The Goals and Conduct of Power in Early Modern Japan." *Journal of Japanese Studies* 12 (2): 237–71.

Bix, Herbert P.
   1986.   *Peasant Protest in Japan, 1590–1884.* New Haven: Yale University Press.
   1987.   "Class Conflict in Rural Japan: On Historical Methodology." *Bulletin of Concerned Asian Scholars* 19 (3): 29–42.

Bowen, Roger W.
   1980.   *Rebellion and Democracy in Meiji Japan.* Berkeley and Los Angeles: University of California Press.

Braudel, Fernand.
   1982.   *The Wheels of Commerce.* Vol. 2 of *Civilization and Capitalism, Fifteenth–Eighteenth Century.* Translated by Sian Reynolds. New York: Harper & Row.
   1990.   *People and Production.* Vol. 2 of *The Identity of France.* Translated by Sian Reynolds. New York: Harper Collins.

Braun, Rudolph.
   1978.   "Early Industrialization and Demographic Change in the Canton of Zurich." In *Historical Studies of Changing Fertility,* ed. Charles Tilly, 289–334. Princeton, N.J.: Princeton University Press.

Brayshay, Mark.
   1991.   "Royal Post-Horse Routes in England and Wales: The Evolution of the Network in the Later-Sixteenth and Early-Seventeenth Century." *Journal of Historical Geography* 17 (4): 373–89.

Brennig, Joseph J.
1990.   "Textile Producers and Production in Late-Seventeenth-Century Coromandel." In *Merchants, Markets and the State in Early Modern India*, ed. Sanjay Subrahmanyam, 66–89. Delhi: Oxford University Press.

Brookfield, Harold.
1975.   *Interdependent Development*. London: Methuen.

Brown, Philip C.
1991.   "The Central-Peripheral Balance of Authority in Early Modern Japan." Paper presented at the 43d annual meeting of the Association for Asian Studies, New Orleans, April 14, 1991.
1993.   *Central Authority and Local Autonomy in the Formation of Early Modern Japan: The Case of Kaga Domain*. Stanford: Stanford University Press.

Chambliss, William Jones.
1965.   *Chiaraijima Village: Land Tenure, Taxation, and Local Trade, 1818–1884*. Tucson: Association for Asian Studies and University of Arizona Press.

Chiba Tokuji.
1957.   "Meiji jūgo-jūshichinen Tenryūgawa tsūsen kamotsuhyō." *Ina* 5 (4): 8–14.
1963.   "Meiji shichinen no Iida no bukka." *Ina* 11 (3): 1–2.

Chiyo-mura.
1965.   *Chiyo sonshi*. Nagano-ken Shimoina-gun Chiyo-mura.

Christaller, Walter.
1966 [1933].   *Central Places in Southern Germany*. Translated by Carlisle Baskin. Englewood Cliffs, N.J.: Prentice-Hall.

Chūbu Denryoku Kabushiki Kaisha Iida Shisha, ed.
1981.   *Inadani denki no yoake*. Iida: Chūbu Denryoku Kabushiki Kaisha Iida Shisha.

Chūō Tetsudō Inasen Iin, ed.
1894.   *Chūō tetsudō Ina, Kiso ryōsen no hikaku chōsa*. Tokyo: Chūō Tetsudō Inasen Iin.

Coleman, D. C.
1983.   "Proto-Industrialization: A Concept Too Many?" *Economic History Review*, 2d ser., 36: 435–48.

Corbridge, Stuart.
1989.   "Marxism, Post-Marxism, and the Geography of Development." In *New Models in Geography*, vol. 1, ed. Richard Peet and Nigel Thrift, 224–56. London: Unwin Hyman.

Cornell, L. L., and Akira Hayami.
1986.   "The *shūmon aratame chō*: Japan's Population Registers." *Journal of Family History* 11 (4): 311–28.

Cottrell, William F.
1955.   *Energy and Society: The Relation between Energy, Social Change, and Economic Development*. New York: McGraw-Hill.

Cox, Kevin R., and Andrew Mair.
    1988.   "Locality and Community in the Politics of Local Economic De-
        velopment." *Annals of the Association of American Geographers* 78
        (2): 307–25.
Craig, Albert M.
    1961.   *Choshu in the Meiji Restoration.* Cambridge, Mass.: Harvard Uni-
        versity Press.
Crawcour, E. Sydney.
    1988.   "Industrialization and Technological Change, 1885–1920." In *The
        Cambridge History of Japan,* vol. 6, *The Twentieth Century,* ed.
        Peter Duus, 385–450. Cambridge: Cambridge University Press.
Cronon, William.
    1986.   *Changes in the Land: Indians, Colonists, and the Ecology of New En-
        gland.* New York: Hill & Wang.
    1991.   *Nature's Metropolis: Chicago and the Great West.* New York: Norton.
Crosby, Albert.
    1986.   *Ecological Imperialism: The Biological Expansion of Europe,
        900–1900.* Cambridge: Cambridge University Press.
Crumley, Carole L., and William H. Marquardt.
    1987.   "Regional Dynamics in Burgundy." In *Regional Dynamics: Bur-
        gundian Landscapes in Historical Perspective,* ed. Carole L. Crum-
        ley and William H. Marquardt, 609–23. San Diego: Academic Press.
Culver, William W., and Cornel J. Reinhart.
    1989.   "Capitalist Dreams: Chile's Response to Nineteenth-Century
        World Copper Competition." *Comparative Studies in Society and
        History* 31 (4): 722–44.
De Vries, Jan.
    1984.   *European Urbanization, 1500–1800.* Cambridge, Mass.: Harvard
        University Press.
Dodd, Stephen.
    1993.   "An Embracing Vision: Representations of the Countryside in Early
        Twentieth-Century Japanese Literature." Ph.D. diss., Columbia
        University.
Dodgshon, Robert A.
    1987.   *The European Past: Social Evolution and Spatial Order.* London:
        Macmillan.
Duara, Prasenjit.
    1988.   *Culture, Power, and the State: Rural North China, 1900–1942.*
        Stanford: Stanford University Press.
Dublin, Thomas.
    1985.   "Women and Outwork in a Nineteenth-Century New England Town:
        Fitzwilliam, New Hampshire, 1830–1850." In *The Countryside in the
        Age of Capitalist Transformation,* ed. Steven Hahn and Jonathan
        Prude, 51–69. Chapel Hill: University of North Carolina Press.
Duus, Peter.
    1984.   "Economic Dimensions of Meiji Imperialism: The Case of Korea,
        1895–1910." In *The Japanese Colonial Empire, 1895–1945,* ed.

Ramon H. Myers and Mark R. Peattie, 128–71. Princeton, N.J.: Princeton University Press.

1988. "Introduction." In *The Cambridge History of Japan,* vol. 6, *The Twentieth Century,* ed. Peter Duus, 1–52. Cambridge: Cambridge University Press.

Duus, Peter, Ramon H. Myers, and Mark R. Peattie.

1989. *The Japanese Informal Empire in China, 1895–1937.* Princeton, N.J.: Princeton University Press.

Earle, Carville.

1987. "Regional Economic Development West of the Appalachians, 1815–1860." In *North America: The Historical Geography of a Changing Continent,* ed. Robert D. Mitchell and Paul A. Groves, 172–97. Totowa, N.J.: Rowman & Littlefield.

Ebato Akira.

1969. *Sanshigyō chiiki no keizai-chirigakuteki kenkyū.* Tokyo: Kokon Shoin.

Eley, Geoff.

1984. "The Social History of Industrialization: 'Proto-Industry' and the Origins of Capitalism." *Economy and Society* 13 (4): 519–39.

Eng, Robert Yeok-Yin.

1978. "Imperialism and the Chinese Economy: The Canton and Shanghai Silk Industry, 1861–1932." Ph.D. diss., University of California, Berkeley.

Eng, Robert Y., and Thomas C. Smith.

1976. "Peasant Families and Population Control in Eighteenth-Century Japan." *Journal of Interdisciplinary History* 6 (3): 417–45.

Epstein, S. R.

1991. "Cities, Regions and the Late Medieval Crisis: Sicily and Tuscany Compared." *Past and Present,* no. 130: 3–50.

Fox, Edward W.

1971. *History in Geographic Perspective: The Other France.* New York: Norton.

Francks, Penelope.

1984. *Technology and Agricultural Development in Pre-War Japan.* New Haven: Yale University Press.

Fraser, Andrew.

1986. "Local Administration: The Example of Awa-Tokushima." In *Japan in Transition: From Tokugawa to Meiji,* ed. Marius B. Jansen and Gilbert Rozman, 111–30. Princeton, N.J.: Princeton University Press.

Freeman, Michael.

1986. "Transport." In *Atlas of Industrializing Britain, 1780–1914,* ed. John Langton and R. J. Morris, 80–93. London and New York: Methuen.

Fujimori Ichirō.

1960. "Kinsei ni okeru Shinshū nōmin sōdō no bunseki." *Shinano* 12 (4): 235–51.

Fujita Yoshihisa.
    1981.   *Nihon no sanson*. Kyoto: Chijin Shobō.
Furusato no Imeeji Kankōkai, ed.
    1971.   *Furusato no imeeji hyakuwa*. Iida: Furusato no Imeeji Kankōkai.
Furushima Toshio.
    1944.   *Shinshū chūma no kenkyū*. Tokyo: Tokyo University Press.
    1951.   *Edo jidai no shōhin ryūtsū to kōtsū*. Tokyo: Ochanomizu Shobō.
    1956.   "Meiji shonen ni okeru nōminsō no bunka." In *Meiji shi kenkyū soshō*, vol. 2, *Chiso kaisei to chihō jichisei*, ed. Meiji Shiryō Kenkyū Renrakukai, 41–91. Tokyo: Ochanomizu Shobō.
    1960.   "Ina de shiritai Meiji no sangyō." *Ina* 8 (11): 19–23.
    1963.   "Meiji jū, nijū nendai Nagano-ken kikai seishi kōba kakuritsuki no ichi kōsatsu." *Shinano* 15 (9): 14–28.
    1974a [1938]."Ina kaidō ni okeru chūma no hattatsu." In *Furushima Toshio chosakushū*, vol. 4, *Shinshū chūma no kenkyū*, 364–416. Tokyo: Tokyo University Press. Originally published in *Shakai keizai shigaku* 9 (2): 41–72; 9 (3): 81–104; and 9 (4): 78–103 (three parts).
    1974b.   "Kaidai." In *Furushima Toshio chosakushū*, vol. 4, *Shinshū chūma no kenkyū*, 1–17. Tokyo: Tokyo University Press.
Garon, Sheldon.
    1987.   *The State and Labor in Modern Japan*. Berkeley and Los Angeles: University of California Press.
Genovese, Eugene D., and Leonard Hochberg, eds.
    1989.   *Geographic Perspectives in History*. Oxford and New York: Basil Blackwell.
Giddens, Anthony.
    1984.   *The Constitution of Society: Outline of the Theory of Structuration*. Berkeley and Los Angeles: University of California Press.
Gilbert, Anne.
    1988.   "The New Regional Geography in English and French-speaking Countries." *Progress in Human Geography* 12 (2): 208–28.
Gimpel, Jean.
    1976.   *The Medieval Machine: The Industrial Revolution of the Middle Ages*. New York: Penguin Books.
Gluck, Carol.
    1985.   *Japan's Modern Myths: Ideology in the Late Meiji Period*. Princeton, N.J.: Princeton University Press.
Gordon, Andrew.
    1991.   *Labor and Imperial Democracy in Prewar Japan*. Berkeley and Los Angeles: University of California Press.
Gregory, Derek.
    1982.   *Regional Transformation and Industrial Revolution: A Geography of the Yorkshire Woollen Industry*. Minneapolis: University of Minnesota Press.
    1988.   "The Production of Regions in England's Industrial Revolution." *Journal of Historical Geography* 14: 50–58.

Gullickson, Gay L.
1986. *Spinners and Weavers of Auffay: Rural Industry and the Sexual Division of Labor in a French Village, 1750–1850*. Cambridge: Cambridge University Press.

Gutmann, Myron P.
1988. *Toward the Modern Economy: Early Industry in Europe, 1500–1800*. New York: Knopf.

Gutmann, Myron P., and René Leboutte.
1984. "Rethinking Proto-Industrialization and the Family." *Journal of Interdisciplinary History* 14: 607–21.

Haga Noboru.
1977. *Tōsan no fudō to rekishi*. Tokyo: Yamakawa Shuppansha.

Hahn, Steven.
1985. "The 'Unmaking' of the Southern Yeomanry: The Transformation of the Georgia Upcountry, 1860–1890." In *The Countryside in the Age of Capitalist Transformation*, ed. Steven Hahn and Jonathan Prude, 179–204. Chapel Hill: University of North Carolina Press.

Hahn, Steven, and Jonathan Prude, eds.
1985. *The Countryside in the Age of Capitalist Transformation*. Chapel Hill: University of North Carolina Press.

Hall, John W.
1968. "The New Look of Tokugawa History." In *Studies in the Institutional History of Early Modern Japan*, ed. John W. Hall and Marius Jansen, 55–64. Princeton: Princeton University Press.

Hane, Mikiso.
1982. *Peasants, Rebels, and Outcastes: The Underside of Modern Japan*. New York: Pantheon Books.

Hanley, Susan B., and Kozo Yamamura.
1971. "A Quiet Transformation in Tokugawa Economic History." *Journal of Asian Studies* 30 (2): 373–84.
1977. *Economic and Demographic Change in Preindustrial Japan, 1600–1868*. Princeton, N.J.: Princeton University Press.

Harada Katsumasa.
1983. *Meiji tetsudō monogatari*. Tokyo: Chikuma Shobō.

Harootunian, Harry D.
1988. *Things Seen and Unseen: Discourse and Ideology in Tokugawa Nativism*. Chicago: University of Chicago Press.
1989. "Late Tokugawa Culture and Thought." In *The Cambridge History of Japan*, vol. 5, *The Nineteenth Century*, ed. Marius B. Jansen, 168–258. Cambridge: Cambridge University Press.

Harvey, David.
1982. *The Limits to Capital*. Oxford: Basil Blackwell.

Hauser, William B.
1974. *Economic Institutional Change in Tokugawa Japan: Osaka and the Kinai Cotton Trade*. London and New York: Cambridge University Press.

Havens, Thomas R. H.
   1974.   *Farm and Nation in Modern Japan: Agrarian Nationalism,
            1870–1940.* Princeton, N.J.: Princeton University Press.
Hayami, Akira.
   1979.   "Thank You, Francisco Xavier: An Essay in the Use of Micro-Data
            for Historical Demography of Tokugawa Japan." *Keio Economic
            Studies* 16 (1–2): 65–81.
   1989.   "Kinsei Nihon no keizai hatten to 'industrious revolution.' " In
            Tokugawa shakai kara no tembō: Hatten, kōzō, kokusai kankei, ed.
            Hayami Akira, Saitō Osamu, and Sugiyama Shin'ya, 19–32. Tokyo:
            Dōbunkan.
Hayami, Akira, and Nobuko Uchida.
   1972.   "Size of Household in a Japanese County throughout the
            Tokugawa Era." In *Household and Family in Past Time,* ed. Peter
            Laslett and Richard Wall, 473–515. Cambridge: Cambridge Uni-
            versity Press.
Hayashi Sakae.
   1962.   *Iida, Shimoina no some to ori.* Iida: Iida Shimoina no some to ori
            kankōkai.
Hayashi Tōmito.
   1987.   "Tenryūgawa no yusō: Mokuzai, kureki no unpan o chūshin ni."
            Ina 35 (2): 32–40.
Heibonsha, ed.
   1979.   *Nagano-ken no chimei.* Tokyo: Heibonsha.
   1985.   *Aichi-ken no chimei.* Tokyo: Heibonsha.
Hirano Yasushi.
   1980.   "Ina chihō kumiai seishi bokkōki ni okeru kosaku yōsan sengyō keiei
            no seiritsu to sono igi ni tsuite." *Nōgyō shi kenkyūkai kaihō* 8:
            12–29.
   1990.   *Kindai yōsangyō no hatten to kumiai seishi.* Tokyo: Tokyo Univer-
            sity Press.
Hirasawa Kiyoto.
   1950.   "Genroku jidai ni hitotsu no sen o hiku." *Shinano* 2 (8): 38–41.
   1952.   *Shimoina sanshigyō hattatsushi.* Nagano-ken Suwa-gun Shimo-
            suwa-machi: Kōyō Shobō.
   1953a. "Tenryū dankyū to Shimoina bunka." *Ina* 1 (2–8) (six parts).
   1953b. "Ton'ya." *Ina* 1 (12): 4–7.
   1953c. "Meiji jū, nijū nendai Nagano-ken kikai seishi kōba kakuritsuki no
            ichi kōsatsu." *Shinano* 5 (9): 14–28.
   1954.   "Hachijū nen mae no Ina gun no sonraku no hintō." *Ina* 2 (4):
            12–15.
   1955.   "Inadani no kinsei no okure." *Ina* 3 (10): 14–16.
   1958.   "Kazunomiya tsūkō to Shimoina." *Ina* 6 (6): 24–28 and 6 (7):
            12–17 (two parts).
   1959a. "Tenmasho, rikuun kaisha, chūgyūba-gaisha to chūma kasegi." *Shi-
            nano* 11 (4): 1–16 and 11 (5): 18–29 (two parts).

1959b. "Mura tenma: Bunka, Bunsei no Toraiwa-mura no baai." *Ina* 7 (3): 1–6.

1960a. "Shimoina chihō no yōsan." In *Nihon sangyō shi taikei*, vol. 5, ed. Chihōshi Kenkyū Kyōgikai, 170–84. Tokyo: Tokyo University Press.

1960b. "Shinshū Shimoina chihō no Meiji ishin zengō no ikki no shudō-ryoku to shite no motoyui koki." *Shinano* 12 (11/12): 654–59.

1963. "Meiji sanjū nen no kome sōdō." *Ina* 11 (10): 11–17.

1965a. "Iida chihō no orimono." *Ina* 13 (9): 13–20.

1965b. "Ajima-sama otenma: wakiōkan no tenma." *Ina* 13 (10): 1–8.

1967. "Shimoina kangyōkai to inasaku." *Ina* 15 (2): 10–16.

1968. "Edo jidai no Iida machi no jinkō, kosū." *Ina* 16 (11): 15–22.

1969. "Iida motoyui no koto nado." *Ina* 17 (6): 30–34, (7): 13–21, and (8): 3–8 (three parts).

1972a. *Hyakushō ikki no tenkai*. Tokyo: Azekura Shobō.

1972b. *Iida jō to kinsei no Iida machi*. Iida: Ina-shi Gakkai.

1978. *Kinsei Minami Shinano nōson no kenkyū*. Tokyo: Ochanomizu Shobō.

Hirschmeier, Johannes.

1965. "Shibusawa Eiichi: Industrial Pioneer." In *The State and Economic Enterprise in Japan*, ed. William W. Lockwood, 209–48. Princeton, N.J.: Princeton University Press.

Hirschmeier, Johannes, and Tsunehiko Yui.

1975. *The Development of Japanese Business, 1600–1973*. Cambridge, Mass.: Harvard University Press, 1975.

Hohenberg, Paul M., and Lynn Hollen Lees.

1985. *The Making of Urban Europe, 1000–1950*. Cambridge, Mass.: Harvard University Press.

Horie, Yasuzo.

1965. "Modern Entrepreneurship in Meiji Japan." In *The State and Economic Enterprise in Japan*, ed. William W. Lockwood, 183–208. Princeton, N.J.: Princeton University Press.

Horiguchi Sadayuki.

1972. *Ina sonraku shakai kenkyū shiwa*. Nagano-ken Kamiina-gun Minoze-machi: Santeisha.

Hoston, Germaine A.

1986. *Marxism and the Crisis of Development in Prewar Japan*. Princeton, N.J.: Princeton University Press.

1992. "The State, Modernity, and the Fate of Liberalism in Prewar Japan." *Journal of Asian Studies* 51 (2): 287–316.

Howell, David L.

1989. "The Capitalist Transformation of the Hokkaido Fishery, 1672–1935." Ph.D. diss., Princeton University.

1994. *Capitalism from Within: Economy, Society, and the State in a Japanese Fishery*. Berkeley and Los Angeles: University of California Press.

1992. "Proto-Industrial Origins of Japanese Capitalism." *Journal of Asian Studies* 51 (2): 269–86.

Howell, Philip.
  1991.   Review of *Regions and Industries: A Perspective on the Industrial Revolution in Britain,* edited by Pat Hudson (Cambridge: Cambridge University Press, 1989). *Economic Geography* 67 (2): 167–69.
Hudson, Pat, ed.
  1989.   *Regions and Industries: A Perspective on the Industrial Revolution in Britain.* Cambridge: Cambridge University Press.
  1989a. "The Regional Perspective." In *Regions and Industries: A Perspective on the Industrial Revolution in Britain,* ed. Pat Hudson, 5–40. Cambridge: Cambridge University Press.
  1989b. "Capital and Credit in the West Riding Wool Textile Industry, c. 1750–1850." In *Regions and Industries: A Perspective on the Industrial Revolution in Britain,* ed. Pat Hudson, 69–89. Cambridge: Cambridge University Press.
Huntington, Ellsworth.
  1907.   *The Pulse of Asia.* Boston: Houghton Mifflin.
Ichikawa, Masami, Shigemi Takayama, Seiji Horiuchi, and Isamu Kayane.
  1980.   "Inland Water and Water Resources in Japan." In *Geography of Japan,* ed. Association of Japanese Geographers, 73–98. Tokyo: Teikoku-Shoin.
Ichikawa Takeo.
  1955.   "Shimoina nōgyō no chiikisei." *Ina* 3 (12): 9–13.
  1961.   "Nagano-ken ni okeru chikusangyō no hattatsu: Basanchi no keizai-chirigakuteki kenkyū." *Shinano* 13 (8): 1–14.
Ichimura Minato.
  1966.   *Inadani no rekishi.* Iida: Inashi Gakkai.
Igara-mura.
  1973.   *Igara sonshi.* Iida: Nagano-ken Iida City Office.
Iida Bunkazai no Kai, ed.
  1969.   *Kyōdo no hyaku nen.* Iida: Iida Bunkazai no Kai.
Iida Shimoina Kashi Kumiai, ed.
  1982.   *Kashi no rekishi to Shimoina kashi kumiai no ayumi.* Iida: Iida Shimoina Kashi Kumiai.
Iida Shin'yō Kinkō, ed.
  1976.   *Inadani no rekishi no naka ni: Iida Shin'yō Kinkō gojū shūnen kinen shi.* Iida: Iida Shin'yō Kinkō.
Iioka Masatake.
  1977.   "Kinsei chūki no yōzai seisan shihō to saiun hi: Shinshū Kashio, Ōkawara yama de no Genbun, Kanpō ki shidashi o chūshin ni." *Tokugawa Rinseishi Kenkyūjo Kenkyū Kiyō,* March 1977, 107–24.
Ikeda Masataki.
  1965.   "Nagano-ken no ginkō to seishigyō." Part 1. *Shinano* 17 (5): 1–15.
Imai Genshirō.
  1953.   "Iida wan no ichiba ni deru made." Part 2. *Ina* 1 (12): 28–31.
Imamaki Hisashi.
  1959.   "Shinshū Shimoina chihō ni okeru kamisukigyō no hatten to kami ton'ya sōdō." *Shinano* 11 (4): 48–63 and 11 (5): 37–52 (two parts).

Ina Shiryō Kankōkai, ed.
>   1975.   *Shinpen Ina shiryō sōsho*. Vol. 5. Tokyo: Rekishi Toshosha.

Ishikawa Jun'ichirō.
>   1980.   *Tenryūgawa: sono fudō to bunka*. Shizuoka: Shizuoka Shinbunsha.

Ishikawa Takeo and Takeuchi Atsuhiko, eds.
>   1986.   *Nagano-ken no jiba sangyō*. Nagano: Nagano Kyōikukai.

Itō Gohei.
>   1955.   "Yōsan rōdō no jukyū chiiki." *Ina* 3 (11): 14–17 and 3 (12): 4–9 (two parts).

Itow, S.
>   1984.   "Secondary Forests and Coppices in Southwestern Japan." In *Man's Impact on Vegetation*, ed. W. Holzner, M. J. A. Werger, and I. Ikusima, 317–26. The Hague: Dr. W. Junk.

Itsubo Tatsurō.
>   1987.   "Kinsei zaimachi keisei no ichi sokumen." *Ina* 35 (4): 11–16 and 35 (5): 27–32 (two parts).
>   1988.   "Shōhin ryūtsū no shintō to Iida han no taiō." Part 1. *Ina* 36 (7): 24–30.

Iwashima Jun.
>   1967.   *Iida, Inadani hyaku nen no ayumi*. Iida: Gekkan Shinshūsha.

Jansen, Marius B.
>   1954.   *The Japanese and Sun Yat-sen*. Cambridge, Mass.: Harvard University Press.
>   1961.   *Sakamoto Ryōma and the Meiji Restoration*. Princeton, N.J.: Princeton University Press.
>   1986.   "The Ruling Class." In *Japan in Transition*, ed. Marius B. Jansen and Gilbert Rozman, 68–90. Princeton, N.J.: Princeton University Press.
>   1989.   *The Cambridge History of Japan*, vol. 5, *The Nineteenth Century*. Cambridge: Cambridge University Press.
>   1989a.   "Japan in the Early Nineteenth Century." In *The Cambridge History of Japan*, vol. 5, *The Nineteenth Century*, ed. Marius B. Jansen, 50–115. Cambridge: Cambridge University Press.
>   1992.   *China in the Tokugawa World*. Cambridge, Mass.: Harvard University Press.

Jansen, Marius B., and Gilbert Rozman, eds.
>   1986.   *Japan in Transition: From Tokugawa to Meiji*. Princeton, N.J.: Princeton University Press.

Johnson, Chalmers.
>   1982.   *MITI and the Japanese Miracle: The Growth of Industrial Policy, 1925–1975*. Stanford: Stanford University Press.

Johnston, R. J., J. Hauer, and G. A. Hoekveld, eds.
>   1990.   *Regional Geography: Current Developments and Future Prospects*. London and New York: Routledge.

Jones, Eric L.
>   1988.   *Growth Recurring: Economic Change in World History*. Oxford: Clarendon Press.

Kamijō Hiroyuki.
    1977.   *Chiiki minshū shi nōto: Shinshū no minken, fusen undō.* Nagano
            City: Ginka Shobō.
Kamisato-machi.
    1978.   *Kamisato shi.* Nagano-ken Shimoina-gun Kamisato-machi.
Kanae-machi.
    1969.   *Kanae chōshi.* Nagano-ken Shimoina-gun Kanae-machi.
Kanaya Hiroyuki.
    1987.   "Yagurataki to Tenryū tsūsen." *Ina* 35 (2): 3–6.
Kandatsu Haruki.
    1985.   "Sangyō kakumei to chiiki shakai." In *Kōza Nihon rekishi,* vol. 8,
            ed. Rekishigaku Kenkyūkai and Nihonshi Kenkyūkai, 121–65.
            Tokyo: Tokyo University Press.
Kawahara, Yukiko.
    N.d.   "Development of the Silk Industry in Meiji and Taishō Shimane."
            Unpublished paper.
Kawaji-mura.
    1988.   *Kawaji sonshi.* Nagano-ken Iida-shi Kawaji-shisho.
Kawaura Yasuji.
    1960.   "Kindai sangyō e no tenbō." In *Nihon sangyō shi taikei,* vol. 5, ed.
            Chihōshi Kenkyū Kyōgikai, 416–30. Tokyo: Tokyo University
            Press.
Kega-mura.
    1987.   *Kega shi.* Nagano-ken Iida.
Keirstead, Thomas.
    1992.   *The Geography of Power in Medieval Japan.* Princeton, N.J.: Prince-
            ton University Press.
Key, Bernard Merril.
    1971.   "The Role of Foreign Contributions in Japanese Capital Forma-
            tion, 1868–1936, with Special Reference to the Period 1904–
            1914." Ph.D. diss., University of California, Berkeley.
Kidd, Yasue Aoki.
    1978.   *Women Workers in the Japanese Cotton Mills.* Ithaca, N.Y: Cornell
            University East Asia Papers No. 20.
Kikuchi Kiyoto.
    1981.   *Saku no kōtsū.* Saku: Ichii Kunugi.
Kinsei Ina Shiryō Kankōkai, ed.
    1953.   *Chūma ikken kirokushū.* Vols. 1–3. Iida: Kinsei Ina Shiryō
            Kankōkai.
Kiyomizu Kiyoji.
    1973.   "Meiji no kikin to Han'i no washi." *Ina* 21 (3): 16–19.
Knox, Paul, and John Agnew.
    1989.   *The Geography of the World Economy.* London and New York: Ed-
            ward Arnold.
Kobayashi Kōji.
    1977.   *Manshū imin no mura: Shinshū Yasuoka-mura no Shōwa shi.*
            Tokyo: Chikuma Shobō.

Kobayashi Kōjin.
  1953.   "Iida no ginkō." *Ina* 1 (3): 20–21; 1 (4): 22–24; 1 (5): 14–17; 1
          (6): 13–16; and 1 (7): 20–24 (five parts).
  1959.   *Iida no shōsei sanbyaku nen.* Iida: Shinano Kyōdo Shuppansha.
  1967.   *Tenka no Itohei.* Iida: Shinano Kyōdo Shuppansha.
Kriedte, Peter.
  1983.   *Peasants, Landlords and Merchant Capitalists: Europe and the
          World Economy, 1500–1800.* Cambridge: Cambridge University
          Press.
Kriedte, Peter, Hans Medick, and Jürgen Schlumbohm.
  1981.   *Industrialization before Industrialization.* Translated by Beate
          Schempp. Cambridge: Cambridge University Press.
  1986.   "Proto-Industrialization on Test with the Guild of Historians: Re-
          sponse to Some Critics." Translated by Leena Tanner. *Economy and
          Society* 15 (2): 254–72.
Kusakabe Shin'ichi.
  1987.   "Tenryūgawa to tsūsen." *Ina* 35 (2): 24–32.
Kyōdo Shuppansha Henshūbu, ed.
  1983.   *Shashinshū, omoide no arubamu: Iida no Meiji, Taishō shi.* Mat-
          sumoto: Kyōdo Shuppansha Henshūbu.
Langton, John.
  1984.   "The Industrial Revolution and the Regional Geography of En-
          gland." *Transactions, Institute of British Geographers*, n.s., 9 (2):
          145–67.
Langton, John, and R. J. Morris, eds.
  1986.   *Atlas of Industrializing Britain, 1780–1914.* London and New
          York: Methuen.
Lees, Lynn.
  1989.   "Urban Decline and Regional Economies: Brabant, Castile, and
          Lombardy, 1550–1750." *Comparative Studies in Society and His-
          tory* 31 (3): 439–61.
Leupp, Gary.
  1992.   *Servants, Shophands, and Laborers in the Cities of Tokugawa Japan.*
          Princeton, N.J.: Princeton University Press.
Levine, David.
  1977.   *Family Formation in an Age of Nascent Capitalism.* New York: Ac-
          ademic Press.
Lewis, Michael.
  1990.   *Rioters and Citizens: Mass Protest in Imperial Japan.* Berkeley and
          Los Angeles: University of California Press.
Li, Lillian M.
  1981.   *China's Silk Trade: Traditional Industry in the Modern World,
          1842–1937.* Cambridge, Mass.: Harvard University Council on East
          Asian Studies.
Lipietz, Alain.
  1986.   "New Tendencies in the International Division of Labor: Regimes
          of Accumulation and Modes of Regulation." In *Production, Work,*

Territory: The Geographical Anatomy of Industrial Capitalism, ed. Allen J. Scott and Michael Storper, 16–40. Boston and London: Allen & Unwin.

Lockwood, William W., ed.
1965.   *The State and Economic Enterprise in Japan.* Princeton, N.J.: Princeton University Press.

Ludden, David.
1990.   "Agrarian Commercialism in Eighteenth-Century South India: Evidence from the 1823 Tirunelveli Census." In *Merchants, Markets and the State in Early Modern India,* ed. Sanjay Subrahmanyam, 215–41. Delhi: Oxford University Press.

McCallion, Stephen.
1989.   "Trial and Error: The Model Filature at Tomioka." In *Managing Industrial Enterprise: Cases from Japan's Prewar Experience,* ed. William D. Wray, 87–120. Cambridge, Mass.: Harvard University Council on East Asian Studies.

Mann, Michael.
1986.   *The Sources of Social Power,* Vol. 1: *A History of Power from the Beginning to A.D. 1750.* Cambridge: Cambridge University Press.

Mann, Susan.
1987.   *Local Merchants and the Chinese Bureaucracy, 1750–1950.* Stanford: Stanford University Press.

Mantoux, Paul.
1961 [1907].   *The Industrial Revolution in the Eighteenth Century.* New York: Harper & Row.

Marburg, Sandra Lin.
1984.   "Man's Role, Woman's Place: Images of Women in Human Geography." Ph.D. diss., University of California, Berkeley.

Marsland, Stephen E.
1989.   *The Birth of the Japanese Labor Movement: Takano Fusatarō and the Rōdō Kumiai Kiseikai.* Honolulu: University of Hawaii Press.

Masaki Keiji.
1973.   "Min'yū teppō to Iida jiken." *Ina* 21 (12): 18–23.
1974.   "Iida jiken no saiginmi." *Ina* 22 (5): 19–24, (6): 3–8, (7): 8–13, (9): 12–15; and (10): 9–14 (five parts).
1978.   *Tōkai to Ina.* Nagoya: Masaki Keiji.

Massey, Doreen.
1978.   "Regionalism: Some Current Issues." *Capital and Class* 6: 106–25.

Matsukawa-machi.
1965.   *Kamikatagiri sonshi.* Nagano-ken Shimoina-gun Matsukawa-machi.
1981.   *Matsukawa chōshi.* Nagano-ken Shimoina-gun Matsukawa-machi.

Matsuo-mura.
1982.   *Matsuo sonshi.* Nagano-ken Shimoina-gun Matsuo-mura.

Matsushima Nobuyuki.
1987.   "Inadani no uitachi." *Ina* 35 (9): 3–12, (10): 3–13, and (12): 3–13 (three parts).

Matsuzaki Shōji.
1977.   *Shinshū no tokusan to tetsudō*. Nagano: Ginka Shobō.
Meinig, Donald W.
1978.   "The Continuous Shaping of America: A Prospectus for Geographers and Historians." *American Historical Review* 83 (5): 1186–1205.
Mendels, Franklin F.
1972.   "Proto-Industrialization: The First Phase of the Industrialization Process." *Journal of Economic History* 32 (1): 241–61.
Meyer, David.
1989.   "Midwestern Industrialization and the American Manufacturing Belt in the Nineteenth Century." *Journal of Economic History* 49 (4): 921–37.
Miho-mura.
1988.   *Miho sonshi*. Nagano-ken Iida-shi Miho-shisho.
Minami, Ryoshin.
1976.   "The Introduction of Electric Power and Its Impact on the Manufacturing Industries, with Special Reference to Smaller-Scale Plants." In *Japanese Industrialization and Its Social Consequences*, ed. Hugh Patrick, 299–325. Berkeley and Los Angeles: University of California Press.
1987.   *Power Revolution in the Industrialization of Japan, 1885–1940*. Institute of Economic Research, Hitotsubashi University, Economic Research Series, No. 24. Tokyo: Kinokuniya.
Minamishinano-mura.
1976.   *Minamishinano sonshi: Tōyama*. Nagano-ken Shimoina-gun Minamishinano-mura.
Minamishinano-mura Rōjin Kurabu.
1979.   "Minwa." Mimeograph.
Misawa Yatarō.
1971.   "Suwako no hanran to Tenryūgawa tsūsen." *Shinano* 23 (2): 25–41.
Mitani Taiichirō.
1974.   *Taishō demokurashii ron*. Tokyo: Chūō Kōron.
Mitchell, Robert D., and Paul A. Groves, eds.
1987.   *North America: The Historical Geography of a Changing Continent*. Totowa, N.J.: Rowman & Littlefield.
Miura Hiroshi.
1964.   "Meiji chūki no Shinshū Komaba chihō ni okeru busshi no idō." *Ina* 12 (7): 19–22.
1988.   *Inadani sanson no henbō: sono rekishi chirigakuteki kenkyū*. Nagano: Shinano Kyōikukai.
Miyagawa Kiyoji.
1960.   "Chikuma, Azumi chihō no tabako." In *Nihon sangyō shi taikei*, ed. Chihōshi Kenkyū Kyōgikai, 158–63. Tokyo: Tokyo University Press.
Miyamoto Mataji.
1966.   *Kansai to Kantō*. Tokyo: Seiahō.

Miyamoto Tsuneichi.
     1967.   *Tenryūgawa ni sotte.* Tokyo: Dōyūkan.
Miyashita Ichirō.
     1972.   "Ina ken Shōsha no setsuritsu to Nankinmai konyū jiken." *Ina* 20
               (9): 7–12, (10): 3–7, (11): 7–13, and (12): 28–34 (four parts).
     1980.   *Shinshū Ina shakai shi.* Tokyo and Nagano: Reibunsha.
Mizoguchi Toshiyuki.
     1989.   "The Changing Pattern of Sino-Japanese Trade, 1884–1937." In
               The Japanese Informal Empire in China, 1895–1937, ed. Peter
               Duus, Ramon H. Myers, and Mark R. Peattie, 10–30. Princeton,
               N.J.: Princeton University Press.
Mizoguchi Tsunetoshi.
     1983.   "Kōshū ni okeru kinsei yakihata sonraku no seigyō." *Nagoya
               Daigaku Bungakubu Kenkyū Ronshū (Shigaku)* 29: 273–89.
Mokyr, Joel.
     1990.   *The Lever of Riches: Technological Creativity and Economic Progress.*
               Oxford and New York: Oxford University Press.
Molony, Barbara.
     1991.   "Activism among Women in the Taishō Cotton Textile Industry."
               In *Recreating Japanese Women, 1600–1945,* ed. Gail Lee Bernstein,
               217–38. Berkeley and Los Angeles: University of California Press.
Moriya, Katsuhisa.
     1990.   "Urban Networks and Information Networks." In *Tokugawa
               Japan: The Social and Economic Antecedents of Modern Japan,* ed.
               Chie Nakane and Shinzaburō Ōishi, 97–123. Tokyo: University of
               Tokyo Press.
Morris, Dana, and Thomas C. Smith.
     1985.   "Fertility and Mortality in an Outcaste Village in Japan,
               1750–1869." In *Family and Population in East Asian History,* ed.
               Susan B. Hanley and Arthur P. Wolf, 229–46. Stanford: Stanford
               University Press.
Morris, Dana, and Stephen Vlastos.
     1980.   Review of *Economic and Demographic Change in Preindustrial
               Japan, 1600–1868,* by Susan Hanley and Kozo Yamamura (Prince-
               ton, N.J.: Princeton University Press, 1977). *Journal of Asian Stud-
               ies* 39 (2): 361–68.
Mosk, Carl.
     1978.   "Fecundity, Infanticide, and Food Consumption in Japan." *Explo-
               rations in Economic History* 15: 269–89.
Mukaiyama Masashige.
     1959.   "Unsō o hiku." *Ina* 7 (4): 16–21.
     1969.   *Zoku Shinano minzoku ki.* Tokyo: Keiyūsha.
     1984.   *Ina nōson shi.* Tokyo: Keiyūsha.
Murasawa Takeo.
     1959.   "Inadani to tetsudō." *Ina* 7 (3): 13–16 and (4): 10–15 (two parts).
     1963.   "Inadani no tabako." Part 2. *Ina* 10 (4): 18–23.

1964.   "Iida jiken zengō." *Ina* 12 (12): 1–4; 13 (2): 21–23; 13 (3): 24–28; 13 (4): 6–10; and 13 (6): 11–19 (five parts).

1977–78.   "Shimoina seitō shi." *Ina* 25 (12): 15–20; 26 (2): 10–15; 26 (5): 34–41; 26 (6): 19–28; and 26 (7): 32–37 (five parts).

1983.   *Iida jōwa.* Iida: Minami Shinshū Shinbunsha.

Murase Masaaki.

1984.   "Mikawa to Shinshū o musubu shōhin ryūtsū no tenkai." *Shinano* 36 (7): 34–45.

Myers, Ramon H., and Mark R. Peattie, eds.

1984.   *The Japanese Colonial Empire, 1895–1945.* Princeton, N.J.: Princeton University Press.

Nagano-ken, ed.

1882.   *Meiji jūgonen kangyō nenpō.* Nagano: Nagano-ken.

1883.   *Meiji jūrokunen Nagano-ken tōkeisho.* Nagano: Nagano-ken.

1885.   *Meiji jūhachinen Nagano-ken kangyō nenpō.* Nagano: Nagano-ken.

1936.   *Nagano-ken chōson shi, nanshin hen.* Nagano: Nagano-ken. Cited in notes as NCS.

1971.   *Nagano kensei shi.* Vol. 1. Nagano: Nagano-ken.

1972.   *Nagano kensei shi.* Vol. 2. Nagano: Nagano-ken.

1978.   *Shinshū karamatsu zōrin hyaku nen no ayumi.* Nagano: Nagano-ken.

1983.   *Nagano ken shi: Kinsei shiryō hen.* Part 4, vol. 3, *Shimoina chihō.* Nagano: Nagano-ken.

1985.   *Nagano ken shi: Kindai shiryō hen, bekkan: Tōkei.* Vol. 2. Nagano: Nagano-ken.

1988.   *Nagano ken shi: Tsūshi hen.* Vol. 5 (*Kinsei* 2). Nagano: Nagano-ken.

Nagano-ken Achi-mura Rōjin Kurabu, ed.

1978.   *Korō wa kataru.* Tokyo: Sekibundō.

Nagano-ken Kyōiku Iinkai, ed.

1959.   *Chūmasei no kiroku.* Nagano: Nagano-ken Kyōiku Iinkai.

Najita, Tetsuo.

1967.   *Hara Kei in the Politics of Compromise, 1905–1915.* Cambridge, Mass.: Harvard University Press.

Najita, Tetsuo and J. Victor Koschmann, eds.

1982.   *Conflict in Modern Japanese History: The Neglected Tradition.* Princeton, N.J.: Princeton University Press.

Nakajima Katsumi.

1971.   "Taishō rokunen yori jūyonen made no Iida sakudō." *Ina* 19 (2): 22–27.

Nakamura, Masanori.

1988.   "The Japanese Landlord System and Tenancy Disputes: A Reply to Richard Smethurst's Criticisms." *Bulletin of Concerned Asian Scholars* 20 (1): 36–50.

Namiai-mura.

1984.   *Namiai sonshi.* Nagano-ken Shimoina-gun Namiai-mura.

Nishida, Yoshiaki.
    1989.   "Growth of the Meiji Landlord System and Tenancy Disputes af-
            ter World War I: A Critique of Richard Smethurst, *Agricultural
            Development and Tenancy Disputes in Japan, 1870–1940.*" *Journal
            of Japanese Studies* 15 (2): 389–415.
Nishioka Toranosuke and Hattori Shisō, eds.
    1956.   *Nihon rekishi chizu.* Tokyo: Zenkoku Kyōiku Tosho Kabushiki
            Kaisha.
Nolte, Sharon H., and Sally Ann Hastings.
    1991.   "The Meiji State's Policy toward Women, 1890–1910." In *Recre-
            ating Japanese Women, 1600–1945,* ed. Gail Lee Bernstein, 151–74.
            Berkeley and Los Angeles: University of California Press.
Norman, E. H.
    1975 [1940].   "Japan's Emergence as a Modern State." In *Origins of the
            Modern Japanese State: Selected Writings of E. H. Norman,* ed. John
            Dower, 109–316. New York: Pantheon Books.
Oguchi Yoshihiko.
    1960.   "Suwa chihō no sanshigyō." In *Nihon sangyō shi taikei,* vol. 5, ed.
            Chihōshi Kenkyū Kyōgikai, 185–96. Tokyo: Tokyo University Press.
Ohkawa, Kazushi, and Henry Rosovsky.
    1965.   "A Century of Japanese Economic Growth." In *The State and Eco-
            nomic Enterprise in Japan,* ed. William W. Lockwood, 47–92.
            Princeton, N.J.: Princeton University Press.
Ōkuma, Shigenobu.
    1910.   *Fifty Years of New Japan.* London: Smith, Eker. Reprint. New York:
            Kraus Reprint, 1970.
Onogi Kajō.
    1968.   "Shinano Matsumoto chihō ni okeru ryūtsū kikō no henshitsu katei:
            Chūma yusō no tenkai o chūshin to shite." *Shinano* 20 (7): 18–34.
Ōsawa Kazuo.
    1966.   "Gaisetsu Shimoina kōtsū shi." In *Kinsei Ina kōtsū shi kenkyū,* vol.
            4, ed. Shimoina Kyōikukai, 5–23. Iida: Shimoina Kyōikukai.
Ōshika-mura.
    1984.   *Ōshika sonshi.* Nagano-ken Shimoina-gun Ōshika-mura.
    1986.   *Kurashi no kataribe: Sakamichi.* Iida: Shimoina Nōgyō Kairyō
            Fukyūsho.
Ōshima Eiko.
    1980.   "Ryūsuisha no shin'yō, hanbai jigyō no tokushitsu: Kumiai seishi
            no ton'ya haijo undō to teirishikin to no kanren o chūshin ni."
            Kyōdo kumiai shorei kenkyū hōkoku 6: 135–58.
Perlin, Frank.
    1983.   "Proto-Industrialization and Pre-Colonial South Asia." *Past and
            Present,* no. 98: 30–95.
    1985.   "Scrutinizing Which Moment?" *Economy and Society* 14 (3):
            374–98.
Pollard, Sydney.
    1981.   *Peaceful Conquest: The Industrialisation of Europe, 1760–1970.* Ox-
            ford: Oxford University Press.

Pomeroy, Charles A.
    1967.   *Traditional Crafts of Japan.* New York and Tokyo: Walker/ Weatherhill.

Poni, Carlo.
    1985.   "Proto-Industrialization, Rural and Urban." *Review* 9 (2): 305–14.

Postan, Michael.
    1952.   "The Trade of Medieval Europe: The North." In *The Cambridge Economic History of Europe,* vol. 2: *Trade and Industry in the Middle Ages,* ed. M. Postan, 119–255. Cambridge: Cambridge University Press.

Pratt, Edward E.
    1991a.   "Village Elites in Tokugawa Japan: The Economic Foundations of the *Gōnō.*" Ph.D. diss., University of Virginia.
    1991b.   "Proto-Industry and Mechanized Production in Three Raw Silk Regions." Paper presented at the 43d annual meeting of the Association for Asian Studies, New Orleans, April 12, 1991.

Pred, Allan.
    1986.   *Place, Practice and Structure: Social and Spatial Transformation in Southern Sweden, 1750–1850.* Totowa, N.J.: Barnes & Noble.

Pudup, Mary Beth.
    1987.   "Land before Coal: Class and Regional Development in Southeast Kentucky." Ph.D. diss., Department of Geography, University of California, Berkeley.
    1988.   "Arguments within Regional Geography." *Progress in Human Geography* 12 (3): 369–90.
    1989.   "The Boundaries of Class in Preindustrial Appalachia." *Journal of Historical Geography* 15 (2): 139–62.

Pyle, Kenneth B.
    1973.   "The Technology of Japanese Nationalism: The Local Improvement Movement, 1900–1918." *Journal of Asian Studies* 33 (1): 51–65.
    1989.   "Meiji Conservatism." In *The Cambridge History of Japan,* vol. 5: *The Nineteenth Century,* ed. Marius B. Jansen, 674–720. Cambridge: Cambridge University Press.

Reddy, William M.
    1984.   *The Rise of Market Culture: The Textile Trade and French Society, 1750–1900.* Cambridge: Cambridge University Press.

Rein, J. J.
    1889.   *The Industries of Japan, Together with an Account of Its Agriculture, Forestry, Arts, and Commerce.* London: Hodder & Stoughton.

Reischauer, Haru Matsukata.
    1986.   *Samurai and Silk: A Japanese and American Heritage.* Tokyo: Tuttle.

Roberts, Luke.
    1991.   "The Merchant Origins of National Prosperity Thought in Eighteenth-Century Tosa." Ph.D. diss., Princeton University.

Rosenberg, Nathan, and L. E. Birdzell, Jr.
  1986.   *How the West Grew Rich: The Economic Transformation of the In-
           dustrial World*. New York: Basic Books.
Rozman, Gilbert.
  1973.   *Urban Networks in Ch'ing China and Tokugawa Japan*. Princeton,
           N.J.: Princeton University Press.
Sack, Robert David.
  1986.   *Human Territoriality: Its Theory and History*. Cambridge: Cam-
           bridge University Press.
Saga, Jun'ichi.
  1987.   *Memories of Silk and Straw: A Self-Portrait of Small-Town Japan*.
           Translated by Garry O. Evans. Tokyo and New York: Kodansha
           International.
Saitō, Osamu.
  1983.   "Population and the Peasant Family Economy in Proto-Industrial
           Japan." *Journal of Family History* 8 (1): 30–54.
  1986.   "The Rural Economy: Commercial Agriculture, By-Employment,
           and Wage Work." In *Japan in Transition*, ed. Marius B. Jansen
           and Gilbert Rozman, 400–420. Princeton, N.J.: Princeton Uni-
           versity Press.
  1992.   "Infanticide, Fertility, and 'Population Stagnation': The State of
           Tokugawa Historical Demography." *Japan Forum* 4 (2): 369–82.
Sakashita Hiroto.
  1964–65.   "Inadani kōtsū mondai no ima mukashi." *Ina* 12 (12): 18–22;
             13 (1): 21–27; 13 (2): 16–20; and 13 (3): 15–20 (four parts).
Sansom, George.
  1958.   *A History of Japan to 1334*. Stanford: Stanford University Press.
Sasaki Toshiji.
  1978.   *Nagano-ken Shimoina shakaishugi undō shi*. Iida: Shinshū Nippō.
Satō Soshin.
  1966.   "Sanshū kaidō to unsō basha." In *Kinsei Ina kōtsūshi kenkyū*, vol.
           4, ed. Shimoina Kyōikukai, 25–45. Iida: Shimoina Kyōikukai.
Satō Takayuki.
  1981.   "Kinsei chūki no bakufu zōrin seisaku to murakata no taiō." *Toku-
           gawa Rinseishi Kenkyūjo Kenkyū Kiyō*, March 1981, 167–200.
Sawada Takeshi.
  1981.   *Kunizakai no kiroku*. Tokyo: Dentō to Gendai Sha.
Sayer, Andrew.
  1989.   "The 'New' Regional Geography and Problems of Narrative." *En-
           vironment and Planning D: Society and Space* 7 (3): 253–76.
Schoenberger, Erica.
  1989.   "New Models of Regional Change." In *New Models in Geography*,
           vol. 1, ed. Richard Peet and Nigel Thrift, 115–41. London:
           Unwin Hyman.
Seidensticker, Edward.
  1983.   *Low City, High City*. New York: Knopf.

Seinaiji-mura.
1982. *Seinaiji sonshi*. Nagano-ken Shimoina-gun Seinaiji-mura.
Sekishima Hisao.
1960. "Sanshū, Enshū kaidō no kaishū." Part 2. *Ina* 8 (3): 7–12.
1972. "Chūō tetsudō Inasen no kōsō." *Ina* 20 (8): 32–35.
Semple, Ellen Churchill.
1911. *Influences of Geographic Environment*. New York: Henry Holt.
Shimada Kinzō.
1982. "Kinsei Tenryū ringyōchi ni okeru nenkiyama no genshō kcitai."
       Tokugawa Rinseishi Kenkyūjo Kenkyū Kiyō, March 1982, 36–68.
Shimazaki, Tōson.
1976. *The Family*. Translated by Cecelia Segawa Seigle. Tokyo: Tokyo
       University Press.
1987. *Before the Dawn*. Translated by William E. Naff. Honolulu: Uni-
       versity of Hawaii Press.
Shimohisakata-mura.
1973. *Shimohisakata sonshi*. Nagano-ken Shimoina-gun Shimohisakata-mura.
Shimoina Chiikishi Kenkyūkai, ed.
1982. *Shimoina no hyaku nen*. Nagano: Shinmai Shoseki Shuppan.
Shimoina-gun, ed.
1907. *Dai nikai Shimoina-gun tōkei ippan*. Iida: Shimoina-gun.
1911. *Shimoina-gun tokuyū bussan shi*. Iida: Shimoina-gun.
1923. *Nagano-ken Shimoina-gun tōkei ippan, dai jūshichikai* [Statistical
       Abstract of Shimoina County]. Iida: Shimoina-gun. Cited in notes
       as NSTI.
Shimoina-gun Rengō Fujinkai, ed.
1983. *Kataritsugu Ina no onna*. Iida: Shimoina-gun Rengō Fujinkai.
Shimoina Kyōikukai, ed.
1954. *Chūma ikken kirokushū*. Vols. 1–3. Iida: Shimoina Kyōikukai.
1962. *Shimoina no chishi: Namiai, Hiraya, Neba chihō*. Iida: Shimoina
       Kyōikukai.
1976. *Shimoina no chishitsu kaisetsu*. Iida: Shimoina Kyōikukai.
1987. *Shimoina Kyōikukai shi hyaku shūnen kinen*. Iida: Shimoina
       Kyōikukai.
Shimoina Nōchi Kaikaku Kyōgikai, ed.
1950. *Shimoina ni okeru nōchi kaikaku*. Iida: Shimoina Nōchi Kaikaku
       Kyōgikai.
Shimojō-mura.
1977. *Shimojō sonshi*. Nagano-ken Shimoina-gun Shimojō-mura.
Shinano Kyōikukai Shimoina Bukai, ed.
1934. *Shimoina no tokushu sangyō*. Iida: Shinano Kyōikukai Shimoina
       Bukai.
Shindatsu Haruki.
1985. "Sangyō kakumei to chiiki shakai." In *Kōza Nihon rekishi*, vol. 8,
       Kindai 2, ed. Rekishigaku Kenkyūkai and Nihonshi Kenkyūkai,
       121–65. Tokyo: Tokyo University Press.

Shinshū Minken Hyakunen Jitsugyō Iinkai, ed.
   1981.   *Shinshū minken undō shi.* Nagano City: Ginka Shobō.
Shin'yōsha, ed.
   1984.   *Iida no jiba sangyō.* Iida: Shin'yōsha.
Shiozawa Jinji.
   1982.   *Kajuen geishi: Nagano-ken Shimoina-gun Matsukawa-machi Ōshima.* Iida: Shin'yōsha.
   1987–88.   "Iida no jōkamachi to Yawata no monzenmachi." Parts 3, 5, and 6. *Ina* 35 (10): 13–21; 36 (3): 21–28, and (5): 21–25.
Shizuoka Tetsudō Kanri Kyoku, ed.
   1955.   *Iidasen.* Shizuoka: Shizuoka Tetsudō Kanri Kyoku.
Short, Brian.
   1989.   "The De-Industrialisation Process: A Case Study of the Weald, 1600–1850." In *Regions and Industries: A Perspective on the Industrial Revolution in Britain,* ed. Pat Hudson, 156–74. Cambridge: Cambridge University Press.
Sievers, Sharon L.
   1983.   *Flowers in Salt: The Beginnings of Feminist Consciousness in Modern Japan.* Stanford: Stanford University Press.
Skinner, G. William.
   1977.   "Cities and the Hierarchy of Local Systems." In *The City in Late Imperial China,* ed. G. William Skinner, 275–351. Stanford: Stanford University Press.
   1985.   "Presidential Address: The Structure of Chinese History." *Journal of Asian Studies* 44 (2): 271–92.
   1993.   "Conjugal Power in Tokugawa Japanese Families: A Matter of Life and Death." In *Sex and Gender Hierarchies,* ed. Barbara D. Miller, 236–70. Cambridge: Cambridge University Press.
Smethurst, Richard J.
   1986.   *Agricultural Development and Tenancy Disputes in Japan, 1870–1940.* Princeton, N.J.: Princeton University Press.
   1989.   "A Challenge to Orthodoxy and Its Orthodox Critics: A Reply to Nishida Yoshiaki." *Journal of Japanese Studies* 15 (2): 417–37.
Smith, Carol A.
   1976a.   "Regional Economic Systems: Linking Geographical Models with Socioeconomic Problems." In *Regional Analysis,* vol. 1, *Economic Systems,* ed. Carol A. Smith, 4–62. New York: Academic Press.
   1976b.   "Analyzing Regional Systems." In *Regional Analysis,* vol. 2, *Social Systems,* ed. Carol A. Smith, 3–20. New York: Academic Press.
   1976c.   "Exchange Systems and the Spatial Distribution of Elites." In *Regional Analysis,* vol. 2, *Social Systems,* ed. Carol A. Smith, 309–74. New York: Academic Press.
Smith, Neil.
   1984.   *Uneven Development: Nature, Capital and the Production of Space.* Oxford: Basil Blackwell.
   1989.   "Uneven Development and Location Theory: Toward a Synthesis." In *New Models of Geography,* vol. 1, ed. Richard Peet and Nigel Thrift, 142–63. London: Unwin Hyman.

Smith, Thomas C.
1955.   *Political Change and Industrial Development in Japan: Government Enterprise, 1868–1880.* Stanford: Stanford University Press.
1956.   "Landlords and Rural Capitalists in the Modernization of Japan." *Journal of Economic History* 16 (2): 165–81.
1959.   *The Agrarian Origins of Modern Japan.* Stanford: Stanford University Press.
1973.   "Pre-Modern Economic Growth: Japan and the West." *Past and Present,* no. 60: 127–60.
1977.   *Nakahara: Family Farming and Population in a Japanese Village, 1717–1830.* Stanford: Stanford University Press.
1986.   "Peasant Time and Factory Time in Japan." *Past and Present,* no. 111: 165–97.
1988 [1969].   "Farm Family By-Employments in Preindustrial Japan." In Thomas C. Smith, *Native Sources of Japanese Industrialization, 1750–1920,* 71–102. Berkeley and Los Angeles: University of California Press. Reprinted from *Journal of Economic History* 29 (4): 682–715.
Spagnoli, Paul G.
1983.   "Industrialization, Proletarianization and Marriage: A Reconsideration." *Journal of Family History* 8 (3): 230–47.
Steele, M. William.
1989.   "From Custom to Right: The Politicization of the Village in Early Meiji Japan." *Modern Asian Studies* 23 (4): 729–48.
Steiner, Kurt.
1965.   *Local Government in Japan.* Stanford: Stanford University Press.
Storper, Michael, and Allen J. Scott.
1986.   "Production, Work, Territory: Contemporary Realities and Theoretical Tasks." In *Production, Work, Territory: The Geographical Anatomy of Industrial Capitalism,* ed. Allen Scott and Michael Storper, 1–15. Boston: Allen & Unwin.
Taaffe, Edward J., and Howard L. Gauthier.
1973.   *The Geography of Transportation.* Englewood Cliffs, N.J.: Prentice-Hall.
Takagi-mura.
1979.   *Takagi sonshi.* Nagano-ken Shimoina-gun Takagi-mura.
Takahashi, Kohachiro.
1978.   "Contribution to the Discussion." In *The Transition from Feudalism to Capitalism,* by Paul Sweezy et al., 68–97. London: Verso.
Takahashi Tansui.
1917.   *Tenka no itohei.* Tokyo: Nittōdō.
Takamori-machi.
1972.   *Takamori chōshi.* Nagano-ken Shimoina-gun Takamori-machi.
Takatō-machi Kyōiku Iinkai, ed.
1966.   *Takatō chōshi.* Nagano-ken Ina-shi: Ina Mainichi Shinbunsha.
Takeuchi Toshimi, ed.
1976.   *Shinshū no sonraku seikatsu,* vol. 1, *Mura no keizai.* Tokyo: Meichō Shuppan.

Tashiro Kazui.
1981. *Kinsei Nitchō tsūkō bōeki shi no kenkyū.* Tokyo: Sōbunsha.
Tate Hideo.
1979. "Kiji monorui to tennen: Kiso-gun Nagiso." *Ina* 27 (2): 6–11.
1983. "Shinshū shikki no rekishi." Part 1. *Ina* 31 (8): 3–8.
Tatsue-mura.
1952. *Tatsue sonshi.* Nagano-ken Shimoina-gun Tatsue-mura.
Tatsuoka Kōminkan, ed.
1985. *Zoku, oka no kataribetachi.* Iida: Tatsuoka Kōminkan.
Tatsuoka-mura.
1968. *Tatsuoka sonshi.* Nagano-ken Iida-shi.
Tenryūsha, ed.
1984. *Kyōdo no ishizue Inadani no Tenryūsha: Mayu to kinu no rekishi.* Shimoina-gun Kanae-machi: Tenryūsha.
Thomas, William L., Jr., ed.
1965. *Man's Role in Changing the Face of the Earth.* 2 vols. Chicago: University of Chicago Press.
Thünen, Johann Heinrich von.
1966 [1826]. *Von Thünen's Isolated State.* Translated by Carla M. Wartenberg. Edited by Peter Hall. New York: Pergamon Press.
Toby, Ronald.
1991. "Both a Borrower and a Lender Be: From Village Moneylender to Rural Banker in the Tempō Era." *Monumenta Nipponica* 46 (4): 483–512.
Tokoro Mitsuo.
1973. "Inayama no unjō shidashi." *Shinano* 25 (9): 1–11 [693–703].
Tōkyō Nichinichi Shinbunsha Keizaibu, ed.
1930. *Keizai fudōki: Hokkaidō, Shin'etsu no maki.* Tokyo: Tōkō Shoin.
Tomioka Gihachi.
1978. *Nihon no shiomichi.* Tokyo: Kokon Shoin.
Topley, Marjorie.
1975. "Marriage Resistance in Rural Kwangtung." In *Women in Chinese Society,* ed. Margery Wolf and Roxanne Witke, 67–88. Stanford: Stanford University Press.
Totman, Conrad.
1967. *Politics in the Tokugawa Bakufu.* Berkeley and Los Angeles: University of California Press.
1980. *Collapse of the Tokugawa Bakufu.* Honolulu: University of Hawaii Press.
1983. "Logging the Unloggable: Timber Transport in Early Modern Japan." *Journal of Forest History* 27 (4): 180–91.
1986. "Tokugawa Peasants: Win, Lose, or Draw?" *Monumenta Nipponica* 41 (4): 457–76.
1989. *The Green Archipelago: Forestry in Preindustrial Japan.* Berkeley and Los Angeles: University of California Press.
Toyoda, Takeshi.
1969. *A History of Pre-Meiji Commerce in Japan.* Tokyo: Kokusai Bunka Shinkokai.

Toyoda Takeshi, Fujioka Kenjirō, and Ōfuji Tokihiko, eds.
    1978. *Ryūiki o tadoru rekishi*, vol. 4, *Chūbu-hen*. Tokyo: Gyōsei.
Toyooka-mura.
    1975. *Toyooka sonshi*. Nagano-ken Shimoina-gun Toyooka-mura.
Trewartha, Glenn T.
    1965. *Japan: A Geography*. Madison: University of Wisconsin Press.
Tsukada Masatomo.
    1974. *Nagano-ken no rekishi*. Tokyo: Yamakawa.
    1986. *Kinsei burakumin no kenkyū: Shinshū no gutai zō*. Kyoto: Buraku Mondai Kenkyūjo Shuppanbu.
Tsukahira, Toshio G.
    1966. *Feudal Control in Tokugawa Japan: The* Sankin Kōtai *System*. Harvard East Asian Monographs, No. 20. Cambridge, Mass.: Harvard University Press.
Tsukamoto Manabu.
    1979. "Kinsei Shinano no kakkyō to tōgō e no ugoki: Shinshū ron ni yosete." In *Nairiku chiiki sangyō, bunka no sōgōteki kenkyū: Chūkan hōkokusho*, ed. Shinshū Daigaku Jinbungakubu Keizaigakubu Tokutei Kenkyū kenkyūhan, 37–50. Matsumoto: Shinshū University.
Tsurumi, E. Patricia.
    1984. "Female Textile Workers and the Failure of Early Trade Unionism in Japan." *History Workshop Journal* 18: 3–27.
    1986. "Problem Consciousness and Modern Japanese History: Female Textile Workers of Meiji and Taishō." *Bulletin of Concerned Asian Scholars* 18 (4): 41–48.
    1990. *Factory Girls: Women in the Thread Mills of Meiji Japan*. Princeton, N.J.: Princeton University Press.
Tsutsui Taizō.
    1956–57. "Iida han no zaisei." *Ina* 3 (8)–4 (2) (five parts).
    1970. "Edo jidai kōki Iida hanryō ni okeru jinushisei tenkai to kōritsuna jinushi tokubun." *Shinano* 22 (8): 647–58.
    1972. "Ina gun no chiso kaisei." Part 3. *Ina* 20 (8): 50–56.
    1974a. "Edo jidai o minaosu." *Ina* 21 (7): 3–8 and 21 (8): 3–8 (two parts).
    1974b. "Shinshū Iida hanryō ni okeru shiyūrin no seiritsu katei." *Tokugawa Rinseishi Kenkyūjo Kenkyū Kiyo*, March 1974, 105–22.
Tucker, Richard P., and John F. Richards, eds.
    1983. *Global Deforestation in the Nineteenth-Century World Economy*. Durham: Duke University Press.
Umegaki, Michio.
    1986. "From Domain to Prefecture." In *Japan in Transition: From Tokugawa to Meiji*, ed. Marius B. Jansen and Gilbert Rozman, 91–110. Princeton, N.J.: Princeton University Press.
    1988. *After the Restoration: The Beginning of Japan's Modern State*. New York: New York University Press.

Unno Fukujū.
  1985. "Shokusan kōgyō to gōnōshō." In *Kōza Nihon rekishi*, vol. 7, *Kindai 1*, ed. Rekishigaku Kenkyūkai and Nihonshi Kenkyūkai, 171–210. Tokyo: Tokyo University Press.
Urry, John.
  1986. "Capitalist Production, Scientific Management and the Service Class." In *Production, Work, Territory: The Geographical Anatomy of Industrial Capitalism*, ed. Allen J. Scott and Michael Storper, 41–66. Boston and London: Allen & Unwin.
van Wolferen, Karel.
  1989. *The Enigma of Japanese Power: People and Politics in a Stateless Nation*. London: Macmillan.
Vance, James E.
  1970. *The Merchant's World*. Englewood Cliffs, N.J.: Prentice-Hall.
  1986. *Capturing the Horizon: The Historical Geography of Transportation*. New York: Harper & Row.
Vaporis, Constantine N.
  1986. "Post Station and Assisting Villages: Corvée Labor and Peasant Contention." *Monumenta Nipponica* 41 (4): 377–414.
  1994. *Breaking Barriers: Travel and the State in Early Modern Japan*. Cambridge, Mass.: Harvard University Council on East Asian Studies.
Vlastos, Stephen.
  1986. *Peasant Protests and Uprisings in Tokugawa Japan*. Berkeley and Los Angeles: University of California Press.
  1989. "Opposition Movements in Early Meiji, 1868–1884." In *The Cambridge History of Japan*, vol. 5: *The Nineteenth Century*, ed. Marius B. Jansen, 367–431. Cambridge: Cambridge University Press.
Vogel, Ezra.
  1980. *Japan as Number One: Lessons for America*. New York: Harper & Row.
Wakuda Yasuo.
  1981. *Nihon no shitetsu*. Tokyo: Iwanami Shoten.
Walthall, Anne.
  1989. "A Woman of the Restoration: Matsuo Taseko in Kyoto." Paper presented at the 41st Annual Meeting of the Association for Asian Studies, Washington, D.C., March 1989.
  1990. "The Family Ideology of the Rural Entrepreneurs in Nineteenth-Century Japan." *Journal of Social History* 23 (3): 493–514.
Walton, John K.
  1989. "Proto-Industrialisation and the First Industrial Revolution: The Case of Lancashire." In *Regions and Industries: A Perspective on the Industrial Revolution in Britain*, ed. Pat Hudson, 41–68. Cambridge: Cambridge University Press.
Warde, Alan.
  1985. "Spatial Change, Politics and the Division of Labour." In *Social Relations and Spatial Structures*, ed. Derek Gregory and John Urry, 190–212. New York: St. Martin's Press.

Waswo, Ann.
    1977.   *Japanese Landlords: The Decline of a Rural Elite.* Berkeley and Los Angeles: University of California Press.
    1988.   "The Transformation of Rural Society, 1900–1950." In *The Cambridge History of Japan,* vol. 6, *The Twentieth Century,* ed. Peter Duus, 541–605. Cambridge: Cambridge University Press.

Watanabe Kazutoshi.
    1979.   "Kinsei shukueki-zaichō seiritsuki ni okeru shōninsō no yakuwari." In *Kinsei no toshi to zaigō shōnin,* ed. Toyoda Takeshi, 249–82. Tokyo: Gannandō.

Waters, Neil.
    1983.   *Japan's Local Pragmatists: The Transition from Bakumatsu to Meiji in the Kawasaki Region.* Cambridge, Mass.: Council on East Asian Studies, Harvard University.

Westney, D. Eleanor.
    1987.   *Imitation and Innovation: The Transfer of Western Organizational Patterns to Meiji Japan.* Cambridge, Mass.: Harvard University Press.

White, James W.
    1988.   "State Growth and Popular Protest in Tokugawa Japan." *Journal of Japanese Studies* 14 (1): 1–25.

Whittlesey, Derwent.
    1954.   "The Regional Concept and the Regional Method." In *American Geography: Inventory and Prospect,* ed. Preston E. James and Clarence F. Jones, 19–69. Syracuse, N.Y.: Syracuse University Press for the Association of American Geographers.

Wickham, Chris.
    1988.   "Historical Materialism, Historical Sociology." *New Left Review,* no. 171: 63–78.

Wigen, Kären.
    1985.   "Common Losses: Transformations of Commonland and Peasant Livelihood in Tokugawa Japan, 1603–1868." M.A. thesis, University of California, Berkeley.
    1990.   "Regional Inversions: The Spatial Contours of Economic Change in the Southern Japanese Alps, 1750–1920." Ph.D. diss., University of California, Berkeley.
    1992.   "The Geographic Imagination in Early Modern Japanese History: Retrospect and Prospect." *Journal of Asian Studies* 51 (1): 3–29.
    N.d.   "Economic Integration in the Greater Nōbi Region: The Geography of Trade and Transport." Unpublished paper.

Worster, Donald, ed.
    1988.   *The Ends of the Earth: Perspectives on Modern Environmental History.* Cambridge: Cambridge University Press.

Wray, William D.
    1986.   "From Sail to Steam." In *Japan in Transition,* ed. Marius B. Jansen and Gilbert Rozman, 248–70. Princeton, N.J.: Princeton University Press.

Wright, Gavin.
    1986.   *Old South, New South.* New York: Basic Books.

Wrigley, E. A.
   1988.   *Continuity, Chance, and Change: The Character of the Industrial Revolution in England.* Cambridge: Cambridge University Press.
Yagi Haruo.
   1980.   *Okaya no seishigyō.* Tokyo: Nihon Keizai Hyōronsha.
Yamamoto Hirofumi.
   1972.   *Ishinki no kaidō to yusō.* Tokyo: Hōsei Daigaku.
Yamamoto Jishō.
   1958.   "Iida ga unda bōeki ton'ya no shiso Yamamoto Jirōsuke." *Ina* 6 (3): 11–14 and (4): 23–26 (two parts).
Yamamoto Shigemi.
   1977.   *Aa Nomugi Tōge.* Tokyo: Kadokawa Bunkō.
Yamamoto-mura.
   1957.   *Yamamoto sonshi.* Nagano-ken Iida-shi.
Yamamura, Kozo.
   1973.   "Toward a Reexamination of the Economic History of Tokugawa Japan, 1600–1867." *Journal of Economic History* 33 (3): 509–41.
   1974.   *A Study of Samurai Income and Entrepreneurship: Quantitative Analyses of Economic and Social Aspects of the Samurai in Tokugawa and Meiji Japan.* Cambridge, Mass.: Harvard University Press.
Yamauchi Naomi.
   1980.   "Ina Kaidō sankan mura ni okeru chūma kasegi no tenkai." *Shinano* 32 (7): 1–19.
Yasuba Yasukichi.
   1975.   "Anatomy of the Debate on Japanese Capitalism." *Journal of Japanese Studies* 2 (1): 63–82.
Yasuba Yasukichi, and Saitō Osamu, eds.
   1983.   *Puroto kōgyōkaki no keizai to shakai. Sūryō keizaishi ronshū* 3. Tokyo: Nihon Keizai Shinbunsha.
Yasuoka-mura.
   1984.   *Yasuoka sonshi.* Nagano-ken Shimoina-gun Yasuoka-mura.

Index

Compositor:	Bookmasters, Inc.
Text:	10/13 Galliard
Display:	Galliard
Printer:	Edwards Bros.
Binder:	Edward Bros.